Molecular Chemistry
of the
Transition Elements

Molecular Chemistry of the Transition Elements

An Introductory Course

François Mathey

Ecole Polytechnique, Palaiseau, France

Alain Sevin

Université Paris VI, France

JOHN WILEY & SONS

Chichester · New York · Brisbane · Toronto · Singapore

National   01243 779777
International   (+44) 1243 779777
e-mail (for orders and customer service enquires): cs-books@wiley.co.uk
Visit our Home Page on http://www.wiley.co.uk
                       or http://www.wiley.com

Other Wiley Editorial Offices

John Wiley & Sons, Inc., 605 Third Avenue,
New York, NY 10158-0012, USA

Jacaranda Wiley Ltd, 33 Park Road, Milton,
Queensland 4064, Australia

John Wiley & Sons (Canada) Ltd, 22 Worcester Road,
Rexdale, Ontario M9W 1L1, Canada

John Wiley & Sons (Asia) Pte Ltd, 2 Clementi Loop #02-01,
Jin Xing Distripark, Singapore 129809

Library of Congress Cataloging-in-Publication Data

Mathey, François.
    Molecular chemistry of the transition elements / François Mathey.
        p.   cm.   —   (Inorganic chemistry)
    Includes bibliographical references (p.   —   ) and index.
    ISBN 0-471-95919-7.   —   ISBN 0-471-95687-2 (pbk.)
    1. Organotransition metal compounds. I. Title. II. Series:
Inorganic chemistry (John Wiley & Sons)
QD411.8.T73M38 1996   95-48041
647'.056—dc20   CIP

British Library Cataloguing in Publication Data

A catalogue record for this book is available from the British Library

ISBN 0-471-95919-7 (cloth)
ISBN 0-471-95687-2 (paper)

Typeset in 10/12 pt Times by Techset Composition Ltd, Salisbury
Printed and bound in Great Britain by Bookcraft (Bath) Ltd, Midsomer Norton, Somerset

This book is printed on acid-free paper responsibly manufactured from sustainable forestation,
for which at least two trees are planted for each one used for paper production.

Contents

Preface

For several years now, the authors have given an introductory course in transition metal organometallic chemistry to the students of Ecole Polytechnique. This book is a considerably expanded translation of the text which accompanied the original course.

Students at Ecole Polytechnique have a strong background in mathematics and physics, so we were led to present transition metal chemistry using a qualitative theoretical description based upon frontier orbitals. This unconventional approach is probably the most characteristic feature of this book. It is counterbalanced by monographs on several representative compounds and illustrations of the uses of transition metal organometallic chemistry in organic synthesis and homogeneous catalysis, from laboratory through to industrial scale. It was not our intention to be comprehensive and subjects such as clusters and bioinorganic chemistry have been excluded from the text because of their inherent complexity. We do hope, nonetheless, to provide the students with a solid understanding of the field, and a feel for its *raison d'être*.

In preparing the book, we have been assisted by a number of people: Martine Rouyer typed the French coursebook: David Tune and Duncan Carmichael prepared the English translation; John Osborn and Igor Tkatchenko generously provided data on homogeneous catalysis; and Claire Walker and Vanessa Pomfret saw the final manuscript through to publication. The diagrams were made using the CSC ChemDraw Plus™ package by the authors, who would be very grateful for any suggestions for improvements to the text.

Finally, we dedicate this book to our wives, Dominique and Annette, for their great patience and unfailing support.

F.M., A.S.
Palaiseau, 5 April, 1996

Common Abbreviations

acac	acetylacetonate	$[CH_3C(O)CHC(O)CH_3]^-$
AO	atomic orbital	
bipy	2,2′-bipyridyl	
^nBu	normal butyl	$CH_3CH_2CH_2CH_2-$
^iBu	*iso*-butyl	$(CH_3)_2CHCH_2-$
^tBu	*tertio*-butyl	$(CH_3)_3C-$
cod	1,5-cyclooctadiene	
cot	cyclooctatetraene	
Cp	cyclopentadienyl (C_5H_5)	$\eta^1=$ $\eta^5=$
Cp*	pentamethylcyclopentadienyl (C_5Me_5)	
Cy	cyclohexyl	
δ	chemical shift (NMR)	
diphos	1,2-bis(diphenylphosphino)ethane	$(C_6H_5)_2P-CH_2CH_2-P(C_6H_5)_2$
DMF	dimethylformamide	$(CH_3)_2N-CH=O$
DMSO	dimethylsulfoxide	$(CH_3)_2S=O$
en	ethylenediamine	$H_2N-CH_2CH_2-NH_2$
Et	ethyl	CH_3CH_2-
HMPT	hexamethylphosphorotriamide	$[(CH_3)_2N]_3P=O$
HOMO	highest occupied molecular orbital	
L	neutral two-electron ligand	
LUMO	lowest unoccupied molecular orbital	
Me	methyl	CH_3-
MO	molecular orbital	
ν	frequency (IR)	
nbd	norbornadiene	
OAc	acetate	$CH_3C(O)O-$
Ph	phenyl	C_6H_5-
o-, m-, p-	*ortho-, meta-, para-*	W,X: *ortho*; W,Y: *meta*; W,Z: *para*
Pr	propyl	$CH_3CH_2CH_2-$
^iPr	*iso*-propyl	$(CH_3)_2CH-$
py	pyridine	

TBP	trigonal bipyramid	
THF	tetrahydrofuran	
TMEDA	tetramethylethylenediamine	$(CH_3)_2N-CH_2CH_2-N(CH_3)_2$
TMS	tetramethylsilane	
X	anionic one-electron ligand	

1 A Brief History of Organometallic Chemistry

In historical terms the carbon–transition metal bond is quite old, but most of its chemistry has been developed since the second world war. The first transition metal–ethylene complex K[Pt Cl$_3$(C$_2$H$_4$)] was discovered by the Danish pharmacist Zeise in 1827[1] but a whole century passed before the real significance of this compound was understood. From an industrial standpoint, the first applications of organometallic chemistry date from the discovery of nickel–tetracarbonyl by Langer and Mond in 1888.[2] The applications of this compound were understood much more rapidly and crude nickel has been refined by the Mond carbonylation–decarbonylation process for many decades:

$$Ni + 4\,CO \xrightarrow{50°C} [Ni(CO)_4] \xrightarrow[230°C]{\Delta} Ni + 4\,CO$$
$$\text{(impure)} \qquad\qquad\qquad\qquad \text{(pure)}$$

Paradoxically, this early discovery of Ni(CO)$_4$ inhibited the development of molecular transition metal chemistry. Ni(CO)$_4$ is a volatile, inflammable and extremely toxic liquid whose handling is both difficult and dangerous. For obvious reasons, many organic chemists concluded that the best way to deal with metal carbonyls and their chemistry was to ignore them completely.

The second metal carbonyl, Fe(CO)$_5$, followed in 1891[3] but then 30 years passed before further progress was made. In 1919, Hein's reaction of chromium trichloride and PhMgBr produced an important compound which he termed polyphenylchromium:[4]

$$CrCl_3 \xrightarrow{PhMgBr} [CrPh_n]^{0,+1} \quad (n = 2, 3, 4)$$

Once again, the nature and significance of the bond between chromium and the benzene ring was only recognized 35 years later.

The Fischer–Tropsch process[5] of 1925 was the next significant advance. It allows the conversion of synthesis gas (CO/H$_2$) into a mixture of hydrocarbons and was used in Germany for converting coal into petrol during the last world war. The iron- or cobalt-based heterogeneous catalyst produces species containing transient M–C or M=C bonds. Research into the mechanism of this transformation is still one of the most important areas of organometallic chemistry.

After alkene (1827) and arene (1919) complexes, the first complex of a diene with a transition metal, butadienetricarbonyliron, was discovered in 1930 by the German chemist Reihlen.[6] Those in Germany at this time also witnessed the birth of the 'oxo' process for transforming olefins into aldehydes through the addition of CO and H$_2$ to a double bond. This process has undergone continuous development since its introduction by Roelen in 1938,[7] and annually produces five million tons of aldehydes and their derivatives world wide.

The second world war saw a decline in research into transition metal organometallic chemistry. Nonetheless, Reppe developed a number of new catalytic reactions, including the $Ni(CN)_2$ catalysed tetramerization of acetylene to cyclooctatetraene.[8]

Transition metal molecular chemistry was dominated by German chemists up to the second world war. The British and then the Americans took the lead in the 1950s. The first great postwar breakthrough occurred in 1951, when bis(cyclopentadienyl)iron or 'ferrocene'[9] appeared in a famous *Nature* article published by Kealy and Pauson. They erroneously proposed a σ complex (structure A: iron bound to a single carbon of each ring); the correct formulation as a π complex (structure B: iron bound equally to all five carbons in each ring) was established a year later by Wilkinson and Woodward.[10]

It is amusing to note that industrial chemists were familiar with an orange substance deposited in iron tubing by cyclopentadiene vapour, but had largely ignored it. The discovery of ferrocene provided an enormous impetus to the development of organometallic chemistry, because it became clear for the first time that transition metals can employ bonding modes which are totally unknown to classical organic chemistry. The simple idea of taking a transition metal in a 'sandwich' between two 'slices' of an organic aromatic molecule was revolutionary.

The next surprise was due to Ziegler and Natta in 1955, who discovered that it is possible to polymerize olefins using soluble titanium- and aluminium[11]-based catalysts. It is hard to overstate the importance of this first trustworthy process for the preparation of polyethylene.

On a completely different note, the theoreticians Longuet-Higgins and Orgel predicted that it should be possible to stabilize an antiaromatic ring such as cyclobutadiene by forming a complex with a transition metal in 1956.[12] Their hypothesis was proven 2 years later by Criegee and Hübel[13] (formula C):

$M = Mo(CO)_4, Fe(CO)_3, Co(C_5H_5)$ etc...
(M is bonded to all 4 carbon atoms of the ring)

From then on, discoveries accelerated to an almost continuous flood. The following list is neither complete nor totally objective!

In 1964 Fischer discovered carbene complexes, which have a metal–carbon double bond.[14] Carbyne complexes, having a metal–carbon triple bond, followed from the same laboratory in 1973.[15] In 1964, Banks discovered the metathesis of olefins.[16] This kind of redistribution reaction is totally unknown in traditional organic chemistry. It obeys the following scheme:

$$2\ RCH{=}CH_2 \rightleftharpoons RCH{=}CHR + H_2C{=}CH_2$$

These discoveries appear totally unconnected, but it later became clear that complexes related to Fischer's carbenes are catalytic intermediates in the metathesis reaction.

In 1965, Wilkinson described his famous catalyst $RhCl(PPh_3)_3$, which permits hydrogenation at low temperatures and pressures.[17] Then in 1971, Monsanto presented a new industrial synthesis of acetic acid through carbonylation of methanol:[18]

$$CH_3OH \ + \ CO \quad \xrightarrow[\text{180°C, 30-40 atm}]{\text{catalyst: [Rh] + I}^-} \quad CH_3CO_2H \quad (99\%)$$

From 1982, groups led by Bergman,[19] Crabtree, Felkin, Graham, Watson, and others described the activation of C−H bonds in saturated hydrocarbons by transition metals:

$$R\text{-}CH_3 \ + \ [M] \quad \longrightarrow \quad R\text{-}CH_2\text{-}M\text{-}H$$

Finally, in 1984, Kubas showed for the first time that molecular hydrogen could coordinate to a transition metal without being cleaved:[20]

$$M \ + \ \overset{H}{\underset{H}{|}} \quad \longrightarrow \quad M \longleftarrow \overset{H}{\underset{H}{|}}$$

Although, in the strictest terms, this result does not concern the history of the metal–carbon bond, it shows how the formation of a complex between a metal and an intact σ bond is the first step in certain oxidative addition reactions:

$$M \ + \ \overset{A}{\underset{B}{|}} \quad \longrightarrow \quad M \longleftarrow \overset{A}{\underset{B}{|}} \quad \longrightarrow \quad M \overset{A}{\underset{B}{\diagdown}}$$

As one can see, the history of molecular transition metal chemistry is intimately bound up with the discovery of new bond types, new structures and reactions and new industrial processes. Its richness is clear even from this short summary, but the reader who is interested in a more detailed description of the birth of organometallic chemistry should consult reference 21.

1.1   References

1. W. C. Zeise, *Pogg. Ann. Phys. Chem.*, 1827, **9**, 632.
2. L. Mond, C. Langer and F. Quincke, *J. Chem. Soc.*, 1890, 749.
3. L. Mond and C. Langer, *J. Chem. Soc.*, 1891, 1090; M. Berthelot, *C. R. Acad. Sci. Paris*, 1891, 1343.
4. F. Hein, *Chem. Ber.*, 1919, **52**, 195.
5. F. Fischer and H. Tropsch, German patents 411416, 1922 and 484337, 1925.
6. H. Reihlen, A. Gruhl, G. von Hessling and O. Pfrengle, *Liebigs Ann. Chem.*, 1930, **482**, 161.
7. O. Roelen, German patent 849548, 1938.
8. W. Reppe, O. Schlichtung, K. Klager and T. Toepel, *Liebigs Ann. Chem.*, 1948, **560**, 104.
9. T. J. Kealy and P. L. Pauson, *Nature*, 1951, **168**, 1039.
10. G. Wilkinson, M. Rosenblum, M. C. Whiting and R. B. Woodward, *J. Am. Chem. Soc.*, 1952, **74**, 2125.
11. See: K. Ziegler, *Adv. Organomet. Chem.*, 1968, **6**, 1.
12. H. C. Longuet-Higgins and L. E. Orgel, *J. Chem. Soc.*, 1956, 1969.

13. W. Hübel and E. H. Braye, *J. Inorg. Nucl. Chem.*, 1958, **10**, 250; R. Criegee and G. Schroder, *Liebigs Ann. Chem.*, 1959, **623**, 1.
14. E. O. Fischer and A. Maasbol, *Angew. Chem., Int. Ed. Engl.*, 1964, **3**, 580.
15. E. O. Fischer, G. Kreis, C. G. Kreiter, J. Muller, G. Huttner and H. Lorenz, *Angew. Chem., Int. Ed. Engl.*, 1973, **12**, 564.
16. R. L. Banks and G. C. Bailey, *Ind. Eng. Chem. Res.*, 1964, **3**, 170.
17. J. F. Young, J. A. Osborn, F. H. Jardine and G. Wilkinson, *J. Chem. Soc., Chem. Commun.*, 1965, 131.
18. J. F. Roth, J. H. Craddock, A. Hershman and F. E. Paulik, *Chem. Tech.* 1971, **1**, 600.
19. A. H. Janowicz and R. G. Bergman, *J. Am. Chem. Soc.*, 1982, **104**, 352.
20. G. J. Kubas, R. R. Ryan, B. I. Swanson, P. J. Vergamini and H. J. Wasserman, *J. Am. Chem. Soc.*, 1984, **106**, 451.
21. G. B. Kauffman; in *Comprehensive Coordination Chemistry*, eds. G. Wilkinson, R. D. Gillard and J. A. McCleverty, Pergamon, Oxford, 1987, vol 1, pp 1–20. For an English translation of the earlier papers, see: G. B. Kauffman, *Classics in Coordination Chemistry, Part II: Selected papers (1798–1899)*, Dover, New York, 1976.

2 The Fundamental Concepts

2.1 The Different Ligands and Their Electron Counts in Transition Metal Complexes

The classical octet rule of main group chemistry is replaced by a so-called 18-electron rule for the transition metals, because they have incomplete, energetically accessible d subshells. The 18e configuration allows a transition metal to attain the favoured electronic configuration of the rare gas situated at the end of its row in the periodic table (Table 2.1). For example, titanium and the metals of the same period prefer a krypton configuration, which involves the filling of their 3d, 4s and 4p orbitals and requires the presence of $10 + 2 + 6 = 18$ electrons. This important point, which determines the structure of many complexes, will be treated in depth in Section 2.2.

To be able to define the electron count in a complex, it is necessary to know the electronic configuration of the metal atom and the number of electrons provided by its supporting ligands. An isolated transition metal atom generally has an electronic configuration $(n - 1)\, d^x\, ns^2$ (a few exceptions exist, such as $3d^5\, 4s^1$ for chromium) but, conventionally, all of the valence electrons are placed in the d subshell and the metal is termed d^{x+2}. Table 2.2 summarizes the resulting formal electron counts for each metal.

After definition of the number of electrons provided by the metal centre, two different formalisms may be used to count the total number of valence electrons in a metal complex. They are based upon covalent or ionic models. The covalent model is more useful for metals in low oxidation states whilst the ionic approach is more realistic for higher degrees of oxidation. If we imagine breaking a two-electron metal–ligand bond, the covalent model distributes the electrons to the metal and to the ligand so as to give two neutral fragments. In the same case, the ionic model gives both electrons to the ligand. If the covalent and ionic approaches lead to the same result, it means that the neutral ligand uses a lone pair of electrons to coordinate to the metal centre, which explains the frequent use of the representation $L \rightarrow M$. Where the ionic model furnishes a metallic cation and an anionic two-electron ligand (e.g. in the case of a metal–halogen bond), the anionic fragment is symbolized by X^- and the metal by M^+. The covalent model treats this case as the combination of an uncharged metal centre with a one electron donor, the $X^{\bullet}$ radical.

So far, we have only considered σ ligands. These are bound to the metal through a single atom. Other ligands, particularly those with delocalized π systems, can bind through several atoms simultaneously. Their hapticity, defined as the number of atoms (n) which are coordinated to the metal centre, is denoted using the prefix (η^n). From an electron counting standpoint, such ligands may be analysed as a combination of neutral (L) and anionic (X) components.

Table 2.3 summarizes some of the most common ligands.

Nitric oxide NO is unusual because it can contribute either one or three electrons to a metal centre. It adopts different geometries in each case. The linear form is as a three-

Table 2.1 Periodic table of the elements

The Modern Periodic Table of the Elements

IA																		NOBLE GASES

Atomic number
Atomic mass

VIII

* Lanthanide series
† Actinide series

PERIODS

Table 2.2 d^n configuration of the transition metals as a function of oxidation state (OS)

Metal	Ti	V	Cr	Mn	Fe	Co	Ni	Cu
	Zr	Nb	Mo	Tc	Ru	Rh	Pd	Ag
	Hf	Ta	W	Re	Os	Ir	Pt	Au
OS								
0	4	5	6	7	8	9	10	–
I	3	4	5	6	7	8	9	10
II	2	3	4	5	6	7	8	9
III	1	2	3	4	5	6	7	8
IV	0	1	2	3	4	5	6	7

Table 2.3 Number of electrons donated by common ligands

Bond	Fragmentation		Class	OS*
	Covalent model	Ionic model		
M–X X = H, R, Ar, F, Cl, Br, I, CN, OR, SR, NR$_2$, PR$_2$ ...	M + X (1e)	M$^+$ + X$^-$ (2e)	X	+1
M ← L L = CO, OR$_2$, SR$_2$, NR$_3$, PR$_3$ ...	M + L (2e)	M + L (2e)	L	0
M=O**	M + O (2e)	M^{2+} + O^{2-} (4e)	X$_2$	+2
M=CR$_2$ (electrophilic carbene)	M + CR$_2$ (2e)	M + CR$_2$ (2e)	L	0
M=CR$_2$ (nucleophilic carbene)	M + CR$_2$ (2e)	M^{2+} + CR$_2^{2-}$ (4e)	X$_2$	+2
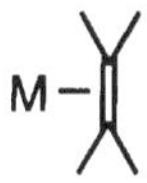 (η^2-alkene)	M +  (2e)	M +  (2e)	L	0
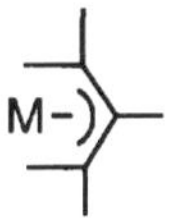 (η^3-allyl)	M + (allyl)°(3e)	M$^+$ + (allyl)$^-$(4e)	LX	+1
 (η^4-diene)	M  + (4e)	M  + (4e)	L$_2$	0
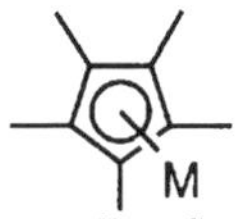 (η^5-cyclopentadienyl)	M + Cp°(5e)	M$^+$ + Cp$^-$ (6e)	L$_2$X	+1
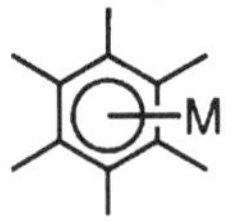 (η^6-arene)	M  + (6e)	M  + (6e)	L$_3$	0

* OS, contribution of the ligand to the oxidation state of the metal.
** Other electronegative atoms or groups such as S, NR, are treated as oxygen.

electron ligand in covalent formalism; the bent complexes have a one-electron formulation because a lone pair remains localized on the nitrogen atom. For linear NO ligands, ionic terminology reflects the electrophilic properties of the nitrogen atom and the bond is therefore considered to be polarized $M^- + NO^+$. The NO^+ moiety is isoelectronic with CO and behaves as a two-electron ligand. Complexes where the ligand is bent are defined $M^+ + NO^-$ (2e) in the ionic model.

$$M-N \overset{\curvearrowright}{\underset{\diagdown O}{}} \quad \text{(1e, OS +1)} \qquad\qquad M=N=O \quad \text{(3e, OS -1)}$$

Many ligands can act as bridges between two or several (n) metal atoms. To denote a ligand in bridging mode, the prefix μ^n is used. In these cases, electron counting is a little more difficult. For example, a chlorine atom $Cl^•$ may use its unpaired electron and a lone pair (3e in total) to bridge two metals: $M-Cl-M'$. Generally it is easiest to evaluate the overall electron count in the bridged complex by calculating the valence electrons of M and M' individually using the covalent model and subsequently adding the three chlorine electrons to complete the total. This technique is particularly useful for bridging hydrogens or alkyl groups, which can only contribute a single electron to the bridge (to give bonds with three electrons and three centres analogous to those found in boron hydrides). The most common μ^2 ligands are shown below:

$$\mu^2 \text{ (1e) H, } CH_3, \dots \text{[ionic model: } H^-, CH_3^- \text{ (2e)]}$$
$$\mu^2 \text{ (3e) F, Cl, Br, I, OR, SR, } NR_2, PR_2 \text{ [ionic model: } LX^- \text{ (4e)].}$$

The covalent model is also preferable when the complex contains metal–metal bonds:

$$M-M \quad \longrightarrow \quad \underset{\text{(1e)}}{M^\circ} + \underset{\text{(1e)}}{M^\circ}$$

The main advantage of the ionic model is that it defines a metal oxidation state by placing formal charges upon the metal centre. These formal oxidation states may be calculated by summing the hypothetical positive charges resulting from the presence of anionic ligands and then adding the global charge of the complex. However, it should be noted that formal oxidation states do not always allow a prediction of the chemical behaviour of a metal centre.

The following examples give a comparison of the two models.

Complex	Covalent model	Ionic model	OS
$[Fe(CN)_6]^{4-}$	$Fe^{4-} + 6$ CN	$Fe^{2+} + 6$ CN^-	+2
	$12 + 6 = 18$	$6 + (6 \times 2) = 18$	
$(OC)_5Mn-Mn(CO)_5$	$Mn + 5$ CO $+ Mn(CO)_5$	$Mn + 5$ CO $+ Mn(CO)_5$	0
	$7 + (5 \times 2) + 1 = 18$	$7 + (5 \times 2) + 1 = 18$	
$[Fe(\eta^5-C_5H_5)_2]$	$Fe + 2$ Cp°	$Fe^{2+} + 2$ Cp^-	+2
	$8 + (2 \times 5) = 18$	$6 + (2 \times 6) = 18$	
$[Co(\eta^5-C_5H_5)_2]*$	$Co + 2$ Cp°	$Co^{2+} + 2$ Cp^-	+2
	$9 + (2 \times 5) = 19$	$7 + (2 \times 6) = 19$	
$[TiCl_2(\eta^5-C_5H_5)_2]$	$Ti + 2$ $Cp^\circ + 2$ $Cl^•$	$Ti^{4+} + 2$ $Cp^- + 2$ Cl^-	+4
	$4 + (2 \times 5) + 2 = 16$	$0 + (2 \times 6) + (2 \times 2) = 16$	
$[RhCl(PPh_3)_3]$	$Rh + 3$ $PPh_3 + Cl^\circ$	$Rh^+ + 3$ $PPh_3 + Cl^-$	+1
	$9 + (3 \times 2) + 1 = 16$	$8 + (3 \times 2) + 2 = 16$	

*Cobaltocene does not obey the 18-electron rule; it is paramagnetic because it has an unpaired electron.

2.2 Molecular Orbitals in Metal Complexes: Theoretical Aspects

Atomic Orbitals (AOs) of the Transition Metals

The valence atomic orbitals (AOs) of the transition metals are more complicated than in their main-group counterparts. Consequently, they are usually approximated using a useful mathematical simplification known as the Slater-type orbital. The wavefunction of a Slater AO takes the form:

$$\Psi = N \, r^{(n-1)} \, e^{-\rho r} Y(\theta, \Phi)$$

where N is a normalization factor, n is the principal quantum number, ρ is an exponential which defines the atom and energy level under consideration and $Y(\theta, \Phi)$ describes the form and orientation of the orbital in space. This equation for the AO therefore comprises a *radial component*, which is a function of ρ and the distance r and an *angular component*. In the case of a hydrogen-like orbital having a single electron of charge $-e$ orbiting a nucleus of charge Ze, the exponential ρ is proportional to Z/n and the Schrödinger equation can be solved directly to give an exact description of the electron density in space. For polyelectronic atoms, it is necessary to use an effective nuclear charge, where $Z_{\text{eff}} < Z$, because the electrons in the inner subshells screen the nuclear charge from the valence electrons.

To represent an AO schematically, we draw a boundary C of the function $Y(\theta, \Phi)$ which contains 95% (or a similar percentage) of $\int_0^c \Psi \Psi dv$. This representation of the AO carries two different pieces of information: the phase properties of Ψ are given along with a picture of the electron density which is available from the integral of the squared function. (Note that only the square of Ψ has any observable meaning). The d orbital functions which we obtain from $Y(\theta, \Phi)$ are more complicated than the familiar s and p AOs; they are centrosymmetric and have *two nodal planes*: (note that Ψ has a different sign in the white and grey areas).

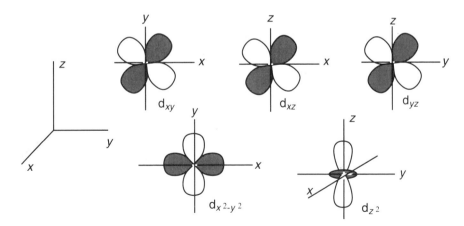

The shapes of these d atomic orbitals are the same for any ion or atom, so changing the metal only affects the radial extension of the orbitals involved. The valence set comprises the nd, $(n+1)$s and $(n+1)$p orbitals, $(3d + 4s + 4p$ in the iron period, for example). This means that $5 + 1 + 3 = 9$ valence AOs are available and *the valence shell is saturated by 18 electrons*. Thus, whilst the p-block elements have only four valence orbitals available for

bonding to other atoms, the transition metals can employ nine valence AOs to form bonds
to their ligands.

5 AO (*n*)d	**1** AO (*n*+1)s	**3** AO (*n*+1)p

The d-type AOs may be combined and hybridized in the same way as the s and p orbitals
of the lighter elements, which allows metals to bind their ligands using MOs which are
built from hybrid combinations of d, s and p AOs.

As with the lighter elements, the radial exponent and energy of each AO in a transition
metal varies as a function of Z_{eff}/n; n is constant for a given period but Z_{eff} increases with
increasing atomic number, so that the further one goes to the right of a row, the more
compact any given AO becomes. Consequently, the metal d, s and p orbital ionization
energies increase from left to right of any given row of the periodic table because the
electrons become more firmly bound, as a result of the increase in the effective nuclear
charge (see the following table).

Modified Slater orbitals having the form

$$\Psi = N \, r^{(n-1)} \, (c_1 e^{-\rho 1 r} + c_2 e^{-\rho 2 r}) Y(\theta, \Phi)$$

give a better radial representation of the d orbitals than the simple formula given above.
The coefficients c_1 and c_2 are very similar, but the two exponents are different, with the
larger one being tailored to give a good fit at short range and the smaller being more
accurate at longer distances. The following tables compare the valence orbitals and
electronegativities of the lighter elements with three first-row transition metals.

	H	C	N	O	F
$\rho(s,p)$	1.3	1.6	1.9	2.2	2.4
$E(s)$	−13.6	−21.4	−26.0	−32.3	−40.0
$E(p)$		−11.4	−13.4	−14.8	−18.1
Electronegativity	2.2	2.5	3.0	3.5	4.1

	ρ_s	ρ_p	ρ_{1d}	ρ_{2d}	$E(s)$	$E(p)$	$E(d)$
Mn	1.45	0.90	5.15	1.7	−8.6	−5.0	−11.6
Fe	1.57	0.98	5.35	1.8	−8.9	−5.1	−12.2
Co	1.7	1.05	5.55	1.9	−9.2	−5.2	−13.2

(Energies are in electronvolts (eV), where 1 eV = 23.06 kcal/
mol = 96.49 kJ/mol; electronegativity is given in arbitrary units.)

Electronegativity is an experimentally determined parameter, somewhat linked to orbital
energies, which measures the aptitude of an atom to attract electrons. The charge $+Ze$ on
the metal nucleus increases linearly from the left to the right of the period, but the overall
effect upon the electronegativity of the valence electrons is non-linear, because of the

attenuating effect of the inner electron shells upon the charge transmitted to the outermost orbitals. The following electronegativities are observed experimentally for the transition metal block:

Ti	V	Cr	Mn	Fe	Co	Ni
1.32	1.45	1.56	1.60	1.64	1.70	1.75
Zr	Nb	Mo	Tc	Ru	Rh	Pd
1.22	1.23	1.30	1.36	1.42	1.45	1.55
Hf	Ta	W	Re	Os	Ir	Pt
1.23	1.33	1.40	1.46	1.52	1.55	1.44

These data clearly show that all of the *transition metals are less electronegative than the lighter elements* and that the *differences between the metals are relatively small*. Consequently, it is possible to construct a generalized molecular orbital scheme which is valid for all transition metals; the precise identity of the metal within a complex is only considered when we fill the available molecular orbitals as a function of the number of valence electrons, n, which it possesses. These valence electrons are conventionally considered as being in 'd-type' orbitals (we invariably refer to a d^n metal centre, where n is the *total* number of valence electrons) because the changes in the electronic structure of the metal during complexation render precise descriptions of their initial localization unnecessary. Therefore, by emphasizing electron count and distribution, this perturbation approach provides a powerful tool for simplifying and unifying the apparently disparate chemistry of these elements, even if each metal has subtly different characteristics which are only distinguishable using more sophisticated approaches.

The Transition Metal in a Complex

We use the term 'complexes' to define the compounds formed by the coordination of ligands to transition metals: it indicates that the M–L interaction is less strong than the covalent bonds between the lighter elements which give rise to more stable compounds which we term 'molecules'. This relative weakness results from a substantial difference between the ligand HOMO and the metal valence orbital energies. Nonetheless, these M–L interactions are strong enough to cause a redistribution of the metal electron density, which gives rise to new MOs which do not necessarily confer the same magnetic or chemical properties as those in the parent metal fragment. Thus, the metal changes its properties significantly upon coordination. The most important characteristic of ML bonds is their marked electronic dissymmetry: the electronegativity tables given above clearly demonstrate that M−X bonds have the strongly polarized structure $M^{\delta+}-X^{\delta-}$, whenever X is a light element, a ligand having lone pairs, or a ligand employing π-type MOs.

Electronic Aspects of the Formation of Metal–Ligand Bonds in Complexes

Metal–ligand bonds may be usefully described by second-order perturbation theory because the metal and the ligand generally possess frontier orbitals which have distinctly different energies. Therefore, the shapes and energies of the participating orbitals are only moderately changed by the M–L interaction. This allows us to use molecular orbital representations for the product orbitals which are recognizably related to their parent components in the isolated metal and ligand fragments. When two orbitals interact, M for the metal (energy E_M, orbital Φ_M), and L for the ligand (energy E_L, orbital Φ_L), we have the following scheme:

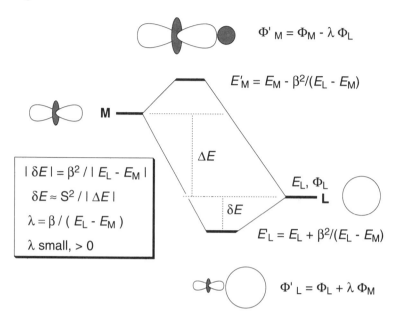

$$\Phi'_M = \Phi_M - \lambda \Phi_L$$

$$E'_M = E_M - \beta^2/(E_L - E_M)$$

$$|\delta E| = \beta^2 / |E_L - E_M|$$

$$\delta E \approx S^2 / |\Delta E|$$

$$\lambda = \beta / (E_L - E_M)$$

$$\lambda \text{ small, } > 0$$

$$E_L, \Phi_L$$

$$E'_L = E_L + \beta^2/(E_L - E_M)$$

$$\Phi'_L = \Phi_L + \lambda \Phi_M$$

A d_z^2 orbital is used in this illustration, but any other valence orbital would give a similar energy diagram. The most important general conclusions are the following:

Two atomic orbitals (AOs) interact to generate two new molecular orbitals (MOs). The lower is stabilized (at $E'_L = E_L + \beta^2/\{E_L - E_M\}$) and the higher is destabilized (at $E'_M = E_M - \beta^2/\{E_L - E_M\}$) with respect to the isolated fragments. In these equations the resonance integral, β, is negative and proportional to the overlap of the two atomic orbitals, S. Consequently, E'_L is more negative (more stable) than E_L whilst E'_M is less negative than E_M.

The stabilized combination is a bonding molecular orbital which results from in-phase mixing of the two atomic orbitals; it is similar to the ligand atomic orbital in form and energy, but contains a small contribution from the metal-centred orbital. The upper level is an antibonding MO combination resulting from an out-of-phase mixing of the M and L AOs; it strongly resembles the original metal orbital, but contains a small contribution from the ligand.

This bonding model can be used universally, provided that the energy levels of the metal and the ligand are realistic. If the metal orbital is at lower energy than the ligand, then their roles are reversed: the bonding orbital resides on the metal and the antibonding combination is mainly ligand-centred. The rules which govern the nature of the interaction do not change.

It is obvious that the optimum electronic configuration for stabilizing a complex has two electrons in each metal–ligand bond. There are two ways in which this can be achieved and two types of M–L bond can then be defined:

- Case 1: the ligand has two electrons available in Φ_L. Here, a small portion of the ligand electron density is transferred to the metal orbital, which acts purely as an acceptor. This case is very common, particularly amongst ligands possessing lone pairs: $:NR_3$, $:PR_3$ and $:CO$ are well known examples.

- Case 2: the ligand has a single electron available, as for $^{\cdot}Cl$. Here, at least formally, the metal must donate one electron to form the bond.

This classification gives the two limiting schemes below:

$$M+ : L \longrightarrow M \underline{\quad\quad} : L$$
$$d^n \quad\quad\quad\quad\quad\quad d^n$$
$$M^\circ + {}^\circ Cl \longrightarrow M^+ \underline{\quad\quad} : Cl^-$$
$$d^n \quad\quad\quad\quad\quad\quad d^{n-1}$$

This molecular orbital approach to bonding confirms the results which were deduced qualitatively above: the metal does not change oxidation state upon coordination by a two-electron ligand, but becomes formally oxidized by one unit for every one-electron ligand which is present. There is, therefore, a certain parallel between the d^n notation *measuring the number of electrons remaining on the metal once the metal–ligand bonds are formed*, and the traditional oxidation state which *measures the formal charge of the metal in a purely ionic complex*. This relationship is only a formal one because, although the M—X bonds are polarized, they are rarely ionic in the sense of compounds such as NaCl. The concept of the oxidation state is practical and applied universally; the equally useful notion of d^n is subtle and has different consequences. The transformation $d^n \rightarrow d^{n-1}$ is associated with a change in the occupation of the frontier orbitals and, therefore, a change in the character of the metal atom.

Electron Counting in a Transition Metal Complex

Electron counting is a fundamental procedure. As with oxidation state (which will be discussed later), the valence electron count follows a formal procedure which takes into account the molecular orbital scheme described previously. Take a d^n metal (nine valence orbitals) and combine it with p identical ligands, each of which has an MO of σ symmetry capable of bonding to the metal. Assume that each ligand uses only one orbital to form an M—L bond and that the metal employs only one valence orbital to interact with each ligand. Then, we find the simple interaction of a ligand with a transition metal orbital (analysed above) repeated p times over. The sum of these p interactions gives the structure of the complex. For every bond which is formed, irrespective of the precise shape of the resulting MOs, there will be a bonding orbital at low energy and a corresponding antibonding level at high energy.

The p ligand orbitals give rise to p ligand-centred bonding MOs and the metal orbitals give p metal-centred antibonding MOs. The $9 - p$ unperturbed metal levels accommodate the metal electrons which are not used in the formation of the M—L bonds.

It is simple to deduce that $9 - p$ largely unperturbed metal orbitals and p metal–ligand hybrid orbitals, making *9 orbitals in all*, are available to accept the electrons which were originally localized upon the metal and ligand fragments. Therefore the complex is saturated when it accommodates a total of 18 metal and ligand electrons. This 18e configuration is synonymous with the octet rule for lighter atoms in as much as it implies 'all valences satisfied', or 'all electrons localized in bonding, non-bonding or very weakly antibonding orbitals'.

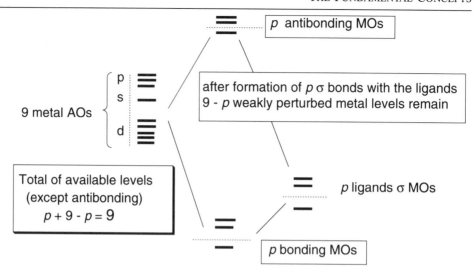

Before studying the shape and form of the MOs in a complex, it is important to describe the different types of bond which may occur between a transition metal and a ligand. The two principal bond types differ in their local orbital symmetry, which may be of either the σ or π type which we meet routinely in organic chemistry.

The Metal–Ligand σ Bond

σ (axial) symmetry implies facile rotation about the bond in question, a situation found in the C–X or C–C linkages in saturated hydrocarbons, for example. It does not guarantee *free* rotation, because steric interactions through space may result in blocked configurations, but it does imply that no important electronic barrier to rotation is associated with the bonding orbital itself. The rotational symmetry of the interacting orbitals about the bond axis maximizes both their overlap and the stability of the resulting bonding orbital.

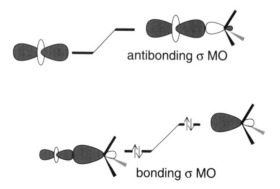

σ bonds are the most important binding forces in the majority of complexes (pure π complexes, where they are absent, will be examined later). The precise geometry of a complex imposes symmetry constraints upon the ensemble of bonding orbitals available;

this symmetry-controlled global description is termed the *overall ligand field* of the complex. We will examine its effect in detail later.

In the example above, the σ bonding MO can be conveniently described by:

$$\Phi_\sigma = \cos\theta(\mathrm{L}) + \sin\theta(\mathrm{M})$$

where L and M represent the MOs of the ligand and metal respectively. The normalization factor θ is small because the large L to M orbital energy gap ensures that the electronic perturbation associated with the bonding interaction is also small. The electron density takes the form $\Phi_\sigma \cdot \Phi_\sigma = \{\cos^2\theta\,(\mathrm{L})(\mathrm{L}) + \sin^2\theta\,(\mathrm{M})(\mathrm{M}) + 2\sin\theta\,\cos\theta\,(\mathrm{L})(\mathrm{M})\}$. Before interaction we have two electrons in L and none in M; after interaction, there is a transfer of electron density through the bonding MOs from L to M, proportional in magnitude to the term in $\sin^2\theta$. This bonding interaction will be cancelled out if the antibonding orbitals are filled. Should this be the case, no bond will be formed because the repulsion engendered by filled antibonding orbitals is slightly greater than the corresponding attraction of their filled bonding combinations (for an example of this phenomenon, see MO diagram of an ML_6 complex below).

π-type Metal–Ligand Bonds

So far we have only considered σ bonds, their spatial orientation and their electronic fields. Ligands having π-type MOs also interact with metal orbitals, but the resulting M$-$L bonds are weaker. Two broad classes of π bond may be discerned; these have two (first case below) or three centres (second case).

Imagine as a first example a σ-bound metal–ligand ensemble whose ligand has additional π MOs which can be orientated to overlap with metal-centred AOs of the same local symmetry (see figure below). The σ-bonding framework may then develop secondary π interactions. Their strength is defined by the relative energies of the orbitals involved. The possibilities are:

1. *The ligand π level lies at high energy and is empty,* as is usually found for an antibonding MO in a ligand such as PR_3 (interaction 1 in the figure at the bottom of the next page). Here, the metal-to-ligand π orbital energy gap is large; furthermore, the ligand π MO is directed mainly towards the substituents at phosphorus. Thus, the overall interaction proportional to $S^2/\Delta E$ is very small and the π interaction can be neglected in a first approximation.

2. *The ligand orbital is at high energy and is occupied.* This is the case of a lone pair on a metal-bound halide; interaction 2 shows the influence of chloride π orbitals. If a metal orbital of appropriate symmetry is empty, then the system will be stabilized. Should the metal d MO be filled, it will be repelled to higher energy, resulting in an overall destabilization of the system. This case will be studied in more detail in our discussion of the spectrochemical series in ML_6 complexes.

3. *The ligand orbital is energetically low-lying and empty.* This is the situation depicted in interaction 3, which shows the π^* MO of a metal-bound carbonyl ligand. Here, the corresponding metal d orbital will be stabilized should it be occupied. An electron transfer from the ligand to the metal occurs during the formation of the metal ligand σ bond, but the π *electron transfer is in the reverse direction, from the metal to the ligand.* Thus we have complementary processes where the σ bond involves a transfer of

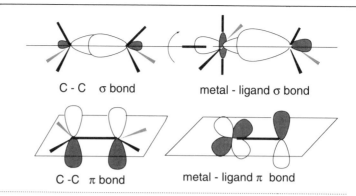

C - C   σ bond   metal - ligand σ bond

C -C   π bond   metal - ligand π bond

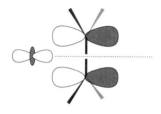
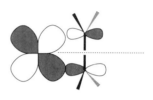

1st bonding MO (symmetrical)   2nd bonding MO (antisymmetrical)

Φ1 = π (alkene) + λ (M)   Φ2 = M' + λ' π* (alkene)

electron donation : π → M   electron donation : M → π*

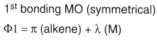

three-centre metal – alkene bonding

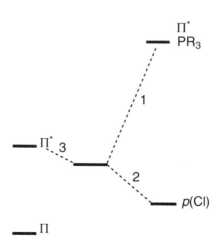

electron density from ligand to metal, whilst the π bond involves metal-to-ligand backdonation.

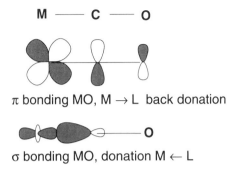

π bonding MO, M $\rightarrow$ L back donation

σ bonding MO, donation M $\leftarrow$ L

The bond between a metal and an alkene can be viewed in a similar fashion to this third case: it involves a three-centre interaction, which has one σ and one π analogue of the bonds described above. The isosceles triangle formed by the metal and alkene carbon atoms is bisected perpendicularly by a molecular symmetry plane which passes through the metal (see following scheme). The MOs involved in the M–alkene bond may then be classified as symmetric (S) or antisymmetric (A) with respect to this plane, so that the π bonding MO combination of the alkene is S type and the π^* alkene MO is of A type. These symmetrical and antisymmetrical combinations are orthogonal to each other and do not mix.

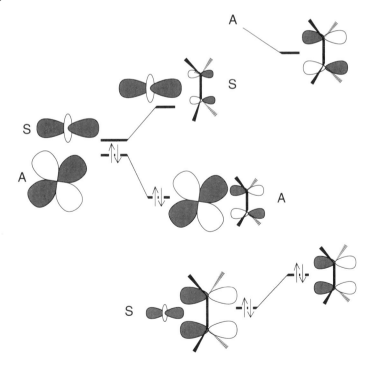

As in the case above, the S combination is responsible for an electronic transfer from the ligand towards the metal, while the A combination involves metal-to-ligand backdonation.

The bond types which we have just studied provide a framework for understanding the vast majority of metal–ligand bonds involving the transition metals. The fuller treatments of more complicated π systems presented below only add details to these fundamental results.

Metal Complexes: Stability and Geometry

The following study is designed to provide a simple and qualitative understanding of the structures adopted by transition metal complexes, the shape and form of the MOs involved, and the geometry and stability of a complex as a function of the number of its valence electrons 'd^n'. As far as possible, we will employ a method originally developed by Hoffmann and collaborators.[1,2] It begins with a thorough analysis of the orbital configurations in an octahedral ML_6 complex and then moves on to smaller complexes by sequentially removing ligands from their orbitals in the ML_6 configuration. This approach reveals families of MOs which are common to various different ML_n geometries. This simplifies the understanding of the reverse process, where a ligand and an empty metal orbital interact to give a bond in a new complex, and also underlines the relationships between MOs in complexes having different coordination spheres.

A precise methodology is required for such an analysis. Strictly speaking, we must employ *group theory*, a rigorous mathematical description of the symmetry of any kind of array in space. The data available from group theory can only be used with precision if the reference axes of the complexes under investigation are clearly defined. We will not derive the specific point groups of the complexes here, but those interested in this procedure may consult the appendix, which also gives orbital diagrams for typical transition metal complexes and very approximate energy scales (the exact values are quite ligand dependent). The form of each of these diagrams is invariant to a first approximation, which allows them to be used as a starting point for the discussion of orbital levels in any transition metal complex having the appropriate geometry. When applied to the model complexes $M(H)_n{}^{n-}$ (where $L = H^-$) they provide a respectable ligand field energy diagram for pure σ ligands; use of $L = CO$ gives a more realistic system where the influence of π-type interactions is also included. In these model calculations iron has been used, but a change in the metal alters mainly the absolute orbital energies; in general, *the relative orbital ordering does not change*. Thus, to obtain the electronic description of any particular complex, it is only necessary to fill the appropriate orbital diagram with the number of electrons available, starting from the bottom.

Below, we will analyse the variation in molecular orbital energy as a function of changing geometry for a number of typical complexes. This treatment provides a useful framework for the interpretation of reaction mechanisms, in processes such as oxidative addition, which we will encounter later.

The ML_6 Fragment

The octahedral geometry is favoured for ML_6 because it minimizes the electrostatic interactions between the six anionic ligands arranged about the positively charged centre. It is a very frequently observed geometry, found in ionized structures and true 'complexes' in both the solid and liquid phases. We will describe its bonding interactions in detail, because the conclusions which we draw can almost invariably be applied to complexes where the six ligands are not truly identical. The σ MOs of an octahedral complex are:

1. six metal–ligand bonding MOs;

2. five d orbitals which are divided into two degenerate groups: three non-bonding low-energy MOs which are unperturbed by the σ field, and two higher-energy MOs which are slightly antibonding;

3. four high-energy, strongly antibonding MOs.

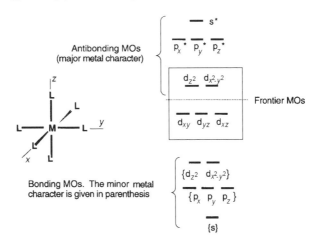

Any population of the antibonding d_{z^2} and $d_{x^2-y^2}$ MOs will destabilize the complex, so they are generally empty in the ground state. Thus the metal valence set comprises the $6 \times 2 = 12$ electrons of the metal–ligand bonding orbitals, together with three non-bonding levels which are capable of accepting up to six more electrons. When these non-bonding orbitals are also completely filled, the resulting eighteen electron complex is electronically saturated and cannot coordinate further ligands, for both steric and electronic reasons. (Very few exceptions to this general rule are known.)

Pictorial representations of the most important frontier orbitals are given in the following figure (pp. 19 and 20). The metal–ligand σ bonds are dominated by powerful contributions from the metal s and p orbitals. Amongst the d set, only d_{z^2} and $d_{x^2-y^2}$ form bonds to the ligands; as a consequence, they must have corresponding ligand-antibonding combinations. These appear as the doubly degenerate LUMO.

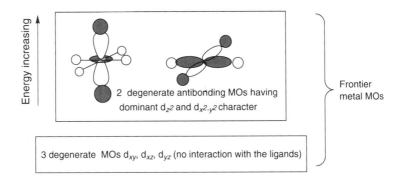

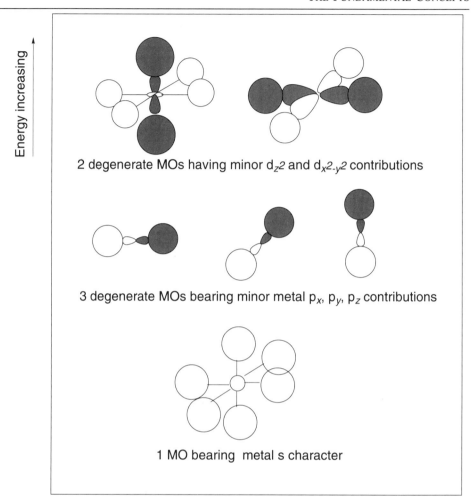

2 degenerate MOs having minor d_{z^2} and $d_{x^2-y^2}$ contributions

3 degenerate MOs bearing minor metal p_x, p_y, p_z contributions

1 MO bearing metal s character

(the high energy antibonding MOs are not shown)

Pure σ ligands do not possess low lying MOs of π symmetry and are therefore incapable of perturbing orbitals which are antisymmetrical about the M–L axis, such as d_{xy}, d_{yz}, and d_{xz}. Thus, the energy of these orbitals is completely unaffected by complexation of the metal to σ ligands. (Their interactions with π-bonding fields will be covered below.)

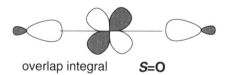

overlap integral   **S=0**

The energy gap between the filled d_{xy}, d_{yz}, d_{xz} MOs and the empty d_{z^2} and $d_{x^2-y^2}$ constitutes the fundamental parameter in *crystal field theory*. Consider the ferrocyanide ion $Fe(CN)_6^{4-}$, an ionic 18-electron d^6 complex of iron (II) surrounded by six cyanide anions.

The interactions between Fe and CN are strong; this means that the bonding combinations of d_{z^2} and $d_{x^2-y^2}$ with the cyanides are very stable (low energy) and, therefore, that their antibonding combinations are displaced to very high energy. As a consequence, the gap between the d_{xy}, d_{yz}, d_{xz} non-bonding MOs and the empty d_{z^2} and $d_{x^2-y^2}$ antibonding combination is large. In this 'strong field' case, the six iron d electrons strongly favour a low-spin configuration, because the energy required for promotion of electron density into d_{z^2} and $d_{x^2-y^2}$ is prohibitive. For ligands which form weaker bonds to the metal, the gap between the non-bonding and antibonding MO groups is small (weak field), and higher spin electronic conformations involving the population of the d_{z^2} and $d_{x^2-y^2}$ orbitals become possible if the electron pairing energy is large with respect to the non-bonding to antibonding promotion energy:

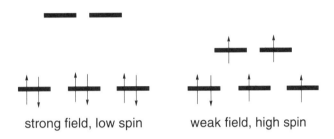

strong field, low spin   weak field, high spin

The '3 + 2' arrangement of the 'metal' d MOs persists to a large extent when the octahedral shape is deformed, for example by replacement of some of the ligands by closely related species. This causes a partial lifting of the degeneracy (i.e. equivalence) of the orbitals, but this effect is often small and we will ignore it for convenience at present. Although we have taken an 'ionized' model here, the general sensitivity of the HOMO–LUMO gap to the nature of the metal and the ligands is also observed for non-ionic compounds. After including π bonding effects, the overall ligand influence upon the HOMO to LUMO gap in a complex may be classed within a 'spectrochemical series'.

The Spectrochemical (or Spectroscopic) Series of ML$_6$ Complexes

For practical purposes, it is often useful to be able to compare the effects of different ligands upon a given metal centre having an octahedral geometry. This is possible through comparison of the spectroscopic absorption frequencies which provoke the transition of an electron from the triply degenerate non-bonding $d_{xy,\ xz,\ yz}$ combinations to the doubly degenerate MO $d_{x^2-y^2}$, or d_{z^2} levels. (Emission spectra, which measure the reverse relaxation of an electron from the $d_{x^2-y^2}$, or d_{z^2} orbitals to the ground state, are equally useful but less easy to measure.)

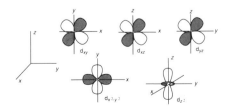

This approach provides an experimental comparison of the field strength of different ligands: the shorter the absorbed wavelength, the higher the value of hv and the greater the energy gap between the two sets of MOs. Ligands are generally classified, by hv value, in units of cm^{-1}. (This unit, the wavenumber, cm^{-1}, is defined as the reciprocal of the wavelength $1/\lambda$ measured in centimetres; it is related to the frequency, measured in s^{-1}, by $v = c/\lambda$.) Therefore, a high wavenumber corresponds to a strong field. The spectrochemical series in order of increasing field strength is as follows:

$$I^- < Br^- < Cl^- < F^- < OH^- < H_2O < NH_3 \ll CN^- < PR_3 < CO$$

This scale is approximate, because it incorporates values obtained in both aqueous and non-aqueous solution. It is influenced by two different effects:

1. $-\sigma$ *bonding effects.* Here, strong σ field ligands cause a large displacement of the antibonding set to high energy, as was discussed above. This effect is most clearly seen for phosphines PR_3;
2. $-\pi$ *bonding effects.* These may be further subdivided into two general classes:
   (a) occupied π MOs on the ligand (as for the halides): these raise the energy of the triply degenerate non-bonding orbitals; predictably, the most electronegative cases (e.g. F^-) have the smallest ligand field effects.
   (b) unoccupied π MOs on the ligand (as in CO): the filled non-bonding metal orbitals are stabilized by interaction with the vacant ligand MOs which leads to an increase in the non-bonding $\rightarrow$ antibonding orbital energy gap.

Except in the case of the hydride ligand, the concept of a pure σ donor is an oversimplification because all of the other common ligands have π-type MOs. Thus, these π effects are almost always present to some degree.

The following figure summarizes the principal effects which influence the spectro-chemical series:

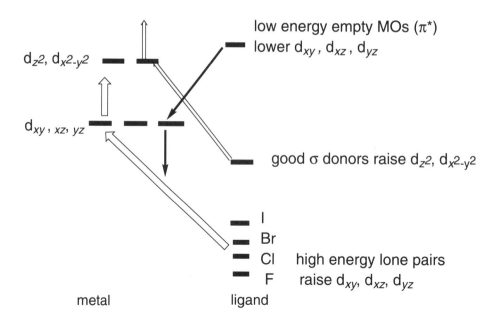

The ML$_5$ Fragment

Conceptually, this fragment may be obtained from an ML$_6$ complex by lengthening one of the L $\rightarrow$ M bonds: as the metal–ligand distance becomes infinite, the ML$_6$ configuration becomes ML$_5$. If the bond which we remove lies on the z-axis, the d$_{z^2}$ M–L antibonding orbital will be stabilized because it now has only one antibonding interaction with a ligand lone pair, rather than the two which are present in ML$_6$. In terms of an orbital diagram, we have the following:

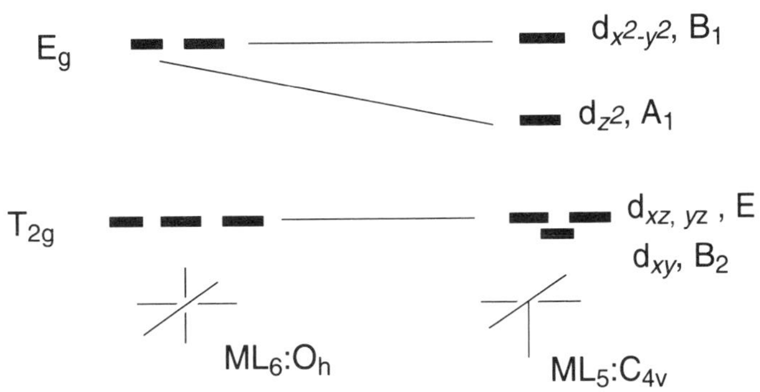

A stable, saturated d^6 octahedral structure is thereby transformed into an ML$_5$ d^6 fragment having only 16 electrons. The complex is therefore able to accept two more electrons into the vacant d$_{z^2}$ orbital, which produces a saturated ML$_5$ d^8 metal centre possessing a high energy *d$_{z^2}$* HOMO. This 'lone pair' means that the metal is a good nucleophile, particularly along the z-axis where there is no ligand to shield the electron density from an incoming electrophile.

The conformation of the ML$_5$ system is very flexible and several readily interconverting geometries are known; they are related by the phenomenon of 'pseudo-rotation'. Rather than looking intimately at the electronic mechanism of this process, we confine ourselves to a graphical representation of one of the known interconversion mechanisms:

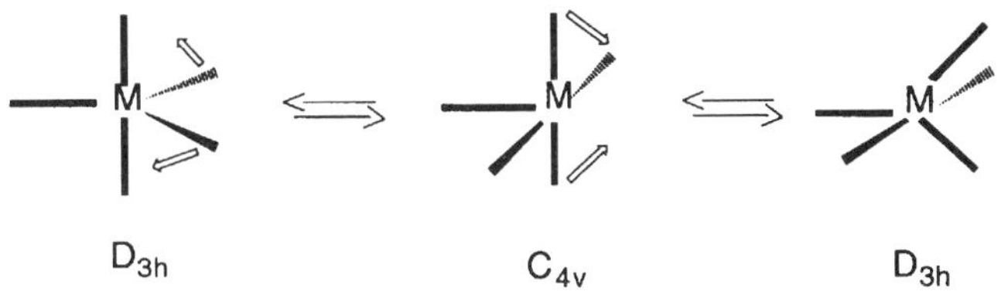

The MOs of the trigonal bipyramidal D$_{3h}$ structure, which is another stable configuration for d^8 metals, are given in the appendix. The other, less common, structures are treated elsewhere by Hoffmann and collaborators.

The ML$_4$ Fragment

We include this fragment because it is particularly important in a wide variety of catalytic processes. Our treatment outlines its principal electronic properties.

The orbitals of an ML_4 complex can be derived from the systems above by removal of the second ligand found along the z-axis. The effect here is predictable: the suppression of two antibonding $L \to M$ interactions means that the d_{z^2} MO appears at much lower energy. It still lies slightly above the degenerate non-bonding set because of antibonding interactions of its central crown with the four ligands in the xy plane, but the separation is small because this region constitutes only a minor part of the orbital.

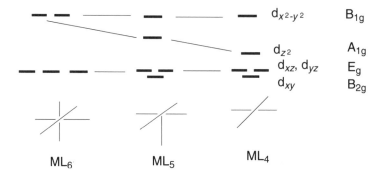

As always, the frontier MOs determine the properties of the complex and merit a detailed examination. Starting at high energy, the $d_{x^2-y^2}$ antibonding MO is strongly influenced by its out-of-phase overlap with the four ligands situated in the xy plane. In principle, it remains every bit as destabilized as in the octahedral case, because exactly the same orbital interactions are operating. The next orbital, d_{z^2}, appears at relatively low energy, as noted above. The positions of d_{xy}, d_{xz}, and d_{yz} depend upon the nature of the ligands. For pure σ ligands, these three MOs are equivalent and energetically degenerate. However, in a complex having π acceptor ligands such as $M(CO)_4$, d_{xy} is stabilized by four interactions whilst d_{xz} and d_{yz} only benefit from two:

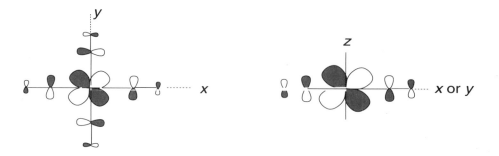

In such a case, the three lowest MOs are very close in energy but not completely degenerate (as confirmed by their irreducible representation: see Appendix).

From this analysis, we can see that a square-planar ML_4 complex can be stable for d^8 metal centres because the d_{xy}, d_{xz}, d_{yz} and d_{z^2} orbitals are all at energetically reasonable levels. However, this still corresponds to only *16 electrons* in the coordination sphere. The 18e configuration is not favoured because it would require 10 metal electrons, which would necessitate the filling of an orbital which is strongly antibonding. The LUMO in these cases is not always the $d_{x^2-y^2}$ MO, but often a metal-based p_z orbital situated just below the $d_{x^2-y^2}$. The d_{z^2} HOMO is not particularly hindered by the ligands lying in the xy plane and frequently behaves as a classical two-electron nucleophile. Thus, although the system is unsaturated, its 16 electrons provide a reasonable stability and a significant ability to react with electrophiles.

Compounds of the ML_4 type are found frequently. Classical examples include the Vaska and Wilkinson type d^8 complexes of general formula MXL_3 (where $L = CO$, PR_3 and $M = Ir$, Rh). If we are to understand the catalytic importance of the ML_4 fragment, it is essential to establish how its frontier orbitals change in energy as the square planar arrangement of the ligands is distorted. As an illustration, we will look at the case where the two ligands in the yz plane fold towards each other, whilst the metal and the ligands in the x-axis remain stationary. (Note that the xz and yz planes are identical but independent of each other; should such a folding occur simultaneously in both planes, then the effects will be additive.)

The three frontier orbitals which are affected by our ligand folding process are the d_{z^2}, $d_{x^2-y^2}$ and d_{yz}. As the ligands fold away from the xy plane, their antibonding σ overlap with the central band of d_{z^2} is reduced, causing a major stabilization of d_{z^2}. The same effect is seen to a greater degree for $d_{x^2-y^2}$, because of its greater localization in the xy plane. The d_{yz} level, which is non-bonding with respect to the ligands in the planar configuration, becomes increasingly destabilized by an out-of-phase overlap with the ligands as they fold. This provokes a strong repulsion towards high energy as the distortion increases and results in a crossing of the energies of the d_{z^2} and d_{yz} orbitals:

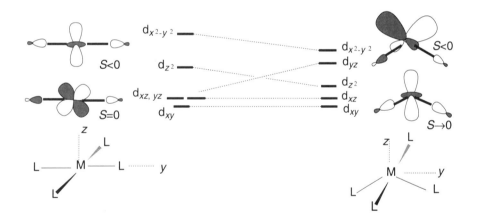

Where eight metal electrons are present, as in many important d^8 Rh(I) or Ir(I) catalysts, this folding imposes an important change in the frontier MOs: through inverting d_{z^2} and d_{yz} levels, it causes a transformation of the HOMO from being symmetrical to antisymmetrical. This has important consequences for the mechanism of the oxidative addition reactions which we will consider later.

Finally, we will look at ligand substitution reactions of square planar ML_4 complexes, which are generally governed by the '*trans effect*'.[3] If we take a complex such as $PtCl_4^{2-}$ (or other Pt(II), Pd(II), or Rh(I) ... compounds) we find that two successive replacements of chloride ions by ammonia lead to a *cis*-configured diammine platinum complex. However, the interaction of chloride with $Pt(NH_3)_4^{2+}$ cations gives a *trans*-diammine product. In both cases, the displaced ligand lies *trans* to a chloride, indicating that the chloride exerts a higher '*trans effect*' than the ammonia. In general, such *trans* effects are high for ligands which have strong π interactions with the metal.

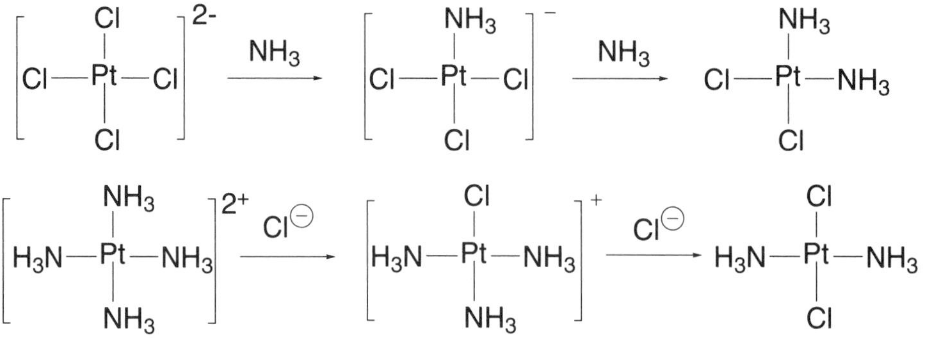

$$\left[\begin{array}{c} Cl \\ | \\ Cl-Pt-Cl \\ | \\ Cl \end{array}\right]^{2-} \xrightarrow{NH_3} \left[\begin{array}{c} NH_3 \\ | \\ Cl-Pt-Cl \\ | \\ Cl \end{array}\right]^{-} \xrightarrow{NH_3} \begin{array}{c} NH_3 \\ | \\ Cl-Pt-NH_3 \\ | \\ Cl \end{array}$$

$$\left[\begin{array}{c} NH_3 \\ | \\ H_3N-Pt-NH_3 \\ | \\ NH_3 \end{array}\right]^{2+} \xrightarrow{Cl^{\ominus}} \left[\begin{array}{c} Cl \\ | \\ H_3N-Pt-NH_3 \\ | \\ NH_3 \end{array}\right]^{+} \xrightarrow{Cl^{\ominus}} \begin{array}{c} Cl \\ | \\ H_3N-Pt-NH_3 \\ | \\ Cl \end{array}$$

The explanation of this behaviour can be found in the intimate substitution mechanism, which, for these 16e metal centres, is associative (see Section 2.3). The principle of microscopic reversibility requires that if the ligand which enters occupies an equatorial position in the pentacoordinated trigonal bipyramidal intermediate, then the ligand which leaves must also be equatorial. If we admit that the ligand L* with the highest *trans* effect tends to occupy an equatorial site, then we rationalize the preferential departure of Y.

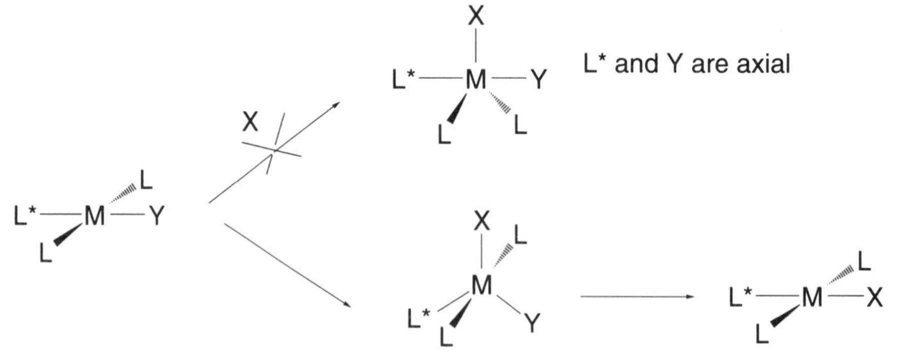

The ML₃ Fragment

It is interesting to look at the transformation planar $ML_3 \rightarrow$ pyramidal ML_3, because the two structures are quite common. The following figure qualitatively summarizes the orbital energy changes as a function of the pyramidal angle θ.

The pyramidalization changes neither the threefold rotation axis of the molecule nor the degeneracies of the orbital groups involved. However, the relative orbital energies are very sensitive to θ. From consideration of the energy diagram as a function of θ, it is clear that

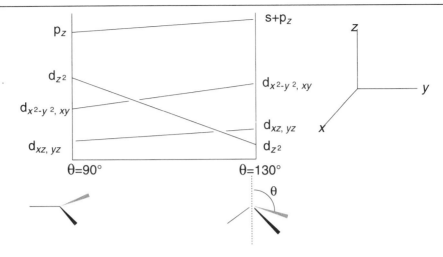

d^6 fragments, such as $Cr(CO)_3$ or $Mn(CO)_3^+$ will be most stable in a C_{3v} geometry (θ around 120–130°). Their three empty MOs, which typically have the following structures,

2 degenerate MOs $d_{xz, yz}$ $s+p_z$

are capable of accepting three more two-electron donors to make up the 18e valence set. Similarly $Fe(CO)_3$ (d^8) can accept two ligands, offering a total of four electrons. The following compounds, for example, are stable:

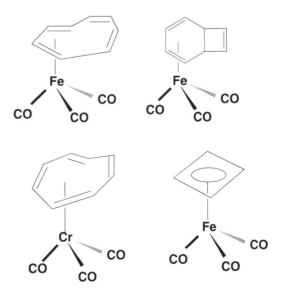

The ML_2 Fragment

These fragments are highly unsaturated because the ligands only supply four electrons; their orbital energies are as follows. It is instructive to note the similarity between this orbital diagram and the MO levels which we derived from the folded ML_4 structure.

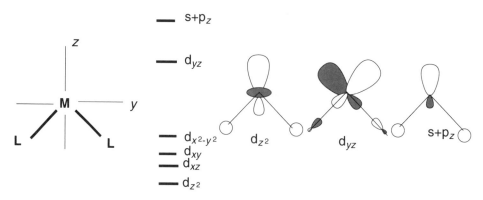

2.3   The Principal Types of Reaction in Transition Metal Chemistry

We have described the composition of a transition metal complex in the preceding chapters. We will now study how it reacts with its environment (intermolecular reactions) and how it reorganizes (intramolecular reactions). The reader interested in a more detailed description of this subject should consult reference 4.

The great majority of reactions, including those that are apparently the most complex, can be reduced to a limited number of elementary steps. These involve ligand substitution, oxidative addition (and its inverse, reductive elimination), insertion (and its reverse, elimination) and finally nucleophilic and electrophilic attacks upon ligands within a metal coordination sphere.

Ligand Substitution

This obeys the following scheme:

$$L_nM-L \; + \; L' \longrightarrow L_nM-L' \; + \; L$$

and has been widely studied. For example, with the metal carbonyls we see:

$$Ni(CO)_4 \; + \; PR_3 \longrightarrow R_3P \rightarrow Ni(CO)_3 \; + \; CO$$

In such a process, the metal coordination number, oxidation level and number of valence electrons remain unchanged. The two limiting cases which can be distinguished may be defined as *dissociative* or *associative* mechanisms. The dissociative mechanism resembles the SN_1 substitution in organic chemistry and proceeds as follows:

Slow step: $L_nM - L$ $\xrightarrow{k}$ $[L_nM]$ + L
  (18e)   (16e)

Fast step: $[L_nM]$ + L' $\longrightarrow$ $L_nM - L'$
  (16e)

Accordingly, its kinetics take the form: $v = k\,[L_nM-L]$. The entropy term ΔS is positive because the transition state is less ordered than the reagents themselves. This dissociative mechanism is favoured by the presence of labile ligands L (THF for example). Alternatively, it may be provoked by electrochemical reduction of the starting complex; this generates an unstable 19-electron complex which dissociates more easily. In general, neutral ligands L tend to dissociate more readily than anionic ligands X. For halides, the dissociation may be promoted by using thallium Tl^+ or silver Ag^+ salts:

$L_nM - Cl$ + Ag^+ $\longrightarrow$ $AgCl$ + $[L_nM]^+$
  (18e)   (16e)

Reactions involving dissociation may proceed either with retention of stereochemistry or with racemization, depending upon the rate of trapping of the intermediate by the incoming ligand. Take the case of an octahedral complex, for example:

Square Pyramid (SP)

Trigonal Bipyramid (TBP)

If the second step is very rapid, the 16-electron intermediate has no time to evolve and L_nM-L' has the same stereochemistry as L_nM-L^*. However, if the second step is slow, the first-formed square pyramidal structure may rearrange to a trigonal bipyramid, which permits racemization.

The associative mechanism obeys the following scheme:

Slow step: L_nM + L' $\xrightarrow{k}$ $[L_nM - L']$

Fast step: $[L_nM - L']$ $\longrightarrow$ $L_{n-1}M - L'$ + L

Here, the kinetics take a different form: $v = k[L_nM][L']$, resembling the SN_2 substitution in organic chemistry. The negative entropy change ΔS reflects a transition state which is more ordered than the starting materials. This mechanism is favourable for electron-deficient complexes, (16 or 17 valence electron compounds, for example) but it is not totally excluded for 18 electron compounds. (Stable complexes such as the 20-electron $Ni(C_5H_5)_2$ demonstrate that the existence of a 20e intermediate is not energetically impossible.) The following examples illustrate the two different substitution mechanisms:

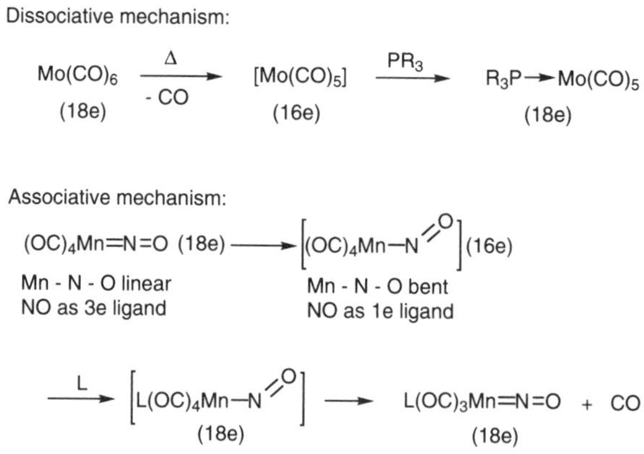

Dissociative mechanism:

$$Mo(CO)_6 \xrightarrow[-CO]{\Delta} [Mo(CO)_5] \xrightarrow{PR_3} R_3P{\rightarrow}Mo(CO)_5$$
$$\text{(18e)} \qquad\qquad \text{(16e)} \qquad\qquad \text{(18e)}$$

Associative mechanism:

$$(OC)_4Mn{=}N{=}O \;\text{(18e)} \longrightarrow \left[(OC)_4Mn{-}N{\nearrow}^{O}\right]\text{(16e)}$$

Mn - N - O linear   Mn - N - O bent
NO as 3e ligand   NO as 1e ligand

$$\xrightarrow{\;L\;} \left[L(OC)_4Mn{-}N{\nearrow}^{O}\right] \longrightarrow L(OC)_3Mn{=}N{=}O \;+\; CO$$
$$\text{(18e)} \qquad\qquad\qquad \text{(18e)}$$

Oxidative Addition

This follows the scheme below:

$$L_nM \;+\; A{-}B \longrightarrow L_nM{\overset{A}{\underset{B}{<}}}$$

A and B are one-electron ligands and L_nM is a complex having no more than 16 valence electrons. During oxidative addition, the metal valence electron count, oxidation level and degree of coordination all increase by two units. This reaction, which usually involves complexes in low oxidation states (0, 1, 2), is applicable to a wide range of σ bonds: H$-$H, H$-$Si, H$-$X, H$-$C, C$-$X, etc. The traditional Grignard synthesis is, at least formally, a well known example:

$$R{-}X \;+\; Mg(0) \longrightarrow R{-}Mg{-}X$$
$$[Mg(+2)]$$

so, organic chemists have been carrying out oxidative addition for the last 90 years without recognizing it!

Four main mechanisms govern this type of reaction. They are:

1. the three-centre concerted mechanism
2. the SN$_2$ mechanism (analogous to that in organic chemistry)
3. the radical mechanism
4. the ionic mechanism.

The concerted three-centre mechanism obeys the following scheme:

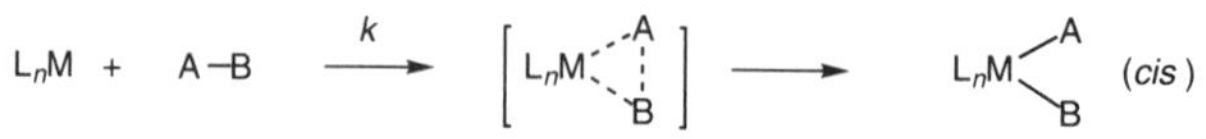

$$L_nM \;+\; A{-}B \xrightarrow{\;k\;} \left[L_nM{\overset{A}{\underset{B}{\cdots}}}\right] \longrightarrow L_nM{\overset{A}{\underset{B}{<}}} \;\;(cis)$$

It operates for weakly polarized bonds such as H$-$H, O$-$O, H$-$Si, H$-$C and gives an initial product which has a *cis* geometry. As would be expected for a concerted process, the second-order reaction kinetics, $v = k[L_nM][A{-}B]$, are largely unaffected by solvent polarity. The η^2–H_2 complexes described by Kubas (see Section 3.1) are good models for the transition state in this type of oxidative addition. Examples of this mechanism include the oxidative addition of oxygen or hydrogen to Vaska's complex:

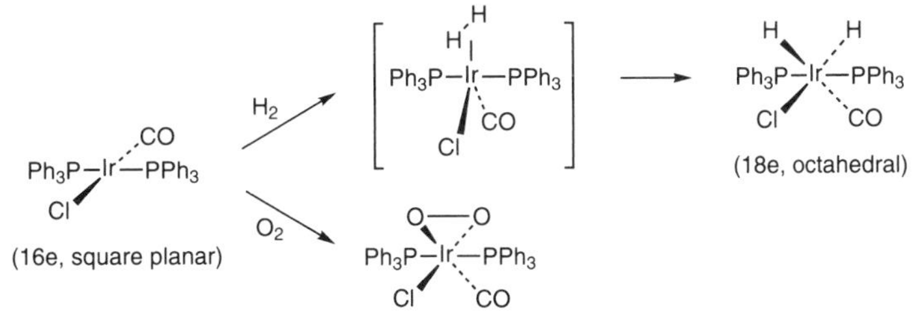

In the SN_2 mechanism the complex plays the role of a nucleophile:

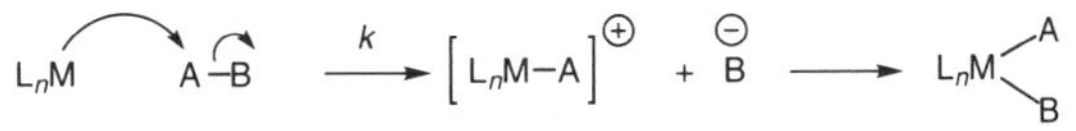

It requires a highly polar A–B bond, such as an alkyl halide. As in a traditional nucleophilic substitution, the rate varies enormously according to the potency of the metal nucleophile and the quality of the leaving group B:

$$OTs > I > Br > Cl$$

As in the three-centre mechanism, the kinetics are second order and ΔS is negative. However, unlike the three-centre case, the final product can have either *cis* or *trans* geometry and the reaction is accelerated by polar solvents. It should be noted that this mechanism is available to 18-electron complexes, because the number of metal valence electrons does not increase as the reaction progresses. Vaska's compound again provides a good example of this mechanism:

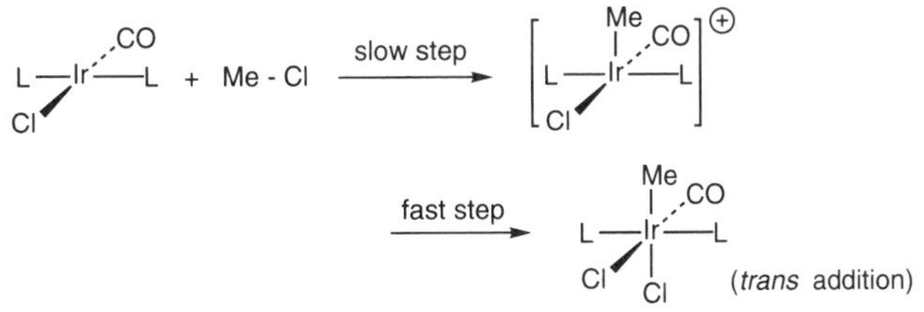

Radical mechanisms are generally observed when the participating complex is a strong reducing agent:

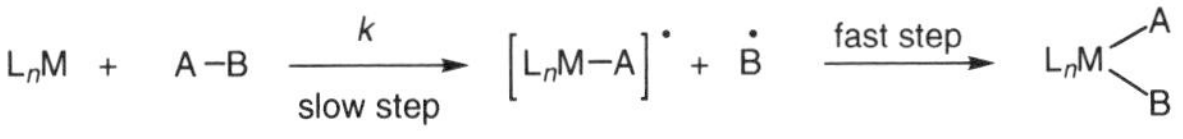

and are favoured by increasing stability of the radical $B^•$. In certain cases, radical-chain processes are observed; they are often initiated by O_2 or peroxides and can be blocked by inhibitors such as bulky phenols. The following example is characteristic of the binuclear oxidative additions frequently encountered with 17-electron complexes:

$$[Co(CN)_5]^{3-} + R\text{-}X \xrightarrow{\text{slow}} [X\text{-}Co(CN)_5]^{3-} + R^{\bullet}$$
$$\text{(17e)} \qquad\qquad\qquad\qquad \text{(18e)}$$

$$[Co(CN)_5]^{3-} + R^{\bullet} \xrightarrow{\text{fast}} [R\text{-}Co(CN)_5]^{3-}$$
$$\qquad\qquad\qquad\qquad\qquad \text{(18e)}$$

Ionic mechanisms generally require the presence of a strongly dissociated protic acid H−X as a reagent; this means that a polar solvent is necessary. Two cases are possible:

- *First case:* attack of H^+ upon a metal complex which acts as a Lewis base

$$L_nM + H^{\oplus} \xrightarrow{k} \left[L_nM\text{-}H\right]^{\oplus} \xrightarrow{X^{\ominus}} L_nM\overset{H}{\underset{X}{<}}$$

If X is not nucleophilic (BF_4^-, PF_6^-, ...), the reaction stops after the first step. The kinetics are second order: $v = k\,[L_nM][H^+]$, and the reaction is favoured by high electron density at the metal centre. Thus, it usually involves metals in low oxidation states and is promoted by the presence of strong donor ligands such as trialkyl phosphines.

- *Second case:* attack on the complex by X^-

$$L_nM + X^{\ominus} \xrightarrow{k} \left[L_nM\text{-}X\right]^{\ominus} \xrightarrow{H^{\oplus}} L_nM\overset{H}{\underset{X}{<}}$$

The rate is expressed differently: $v = k\,[L_nM][X^-]$. This less common mechanism is favoured by metals in high oxidation states and complexes having strong acceptor ligands, which confer Lewis acid character. The following examples illustrate the two different versions nicely:

$$[Pt(PPh_3)_3] + HCl \longrightarrow [H\text{-}Pt(PPh_3)_3]^{\oplus} \xrightarrow[\text{- PPh}_3]{\text{+ Cl}^{\ominus}} [PtH(Cl)(PPh_3)_2]$$
$$\text{(16e)} \qquad\qquad\qquad \text{(16e)}$$
$$v = k\,[Pt]\,[H^+]$$

$$\left[IrL_2\!\left(\bigcirc\right)\right]^{\oplus} + HCl \longrightarrow \left[IrClL_2\!\left(\bigcirc\right)\right] \xrightarrow{H^{\oplus}} \left[IrH(Cl)L_2\!\left(\bigcirc\right)\right]^{\oplus}$$
$$\text{(16e)} \qquad\qquad\qquad\qquad \text{(18e)}$$
(high Lewis acidity due
to the positive charge)
$$v = k\,[Ir]\,[Cl^-]$$

Reductive Elimination

This conforms to the general scheme below:

$$L_nM\overset{A}{\underset{B}{<}} \longrightarrow L_nM + A\text{-}B$$

and is the opposite of oxidative addition. Consequently, the metal valence electron count, oxidation level and coordination number are all reduced by two units during the process.

As expected, reductive elimination is promoted by high metal oxidation states and coordination numbers and facilitated if the A−B bond is strong. The most favourable cases are: R−H, R−R, RC(O)−H, RC(O)−R, R−X, ... This type of reaction is often the last step in catalytic cycles because it allows the final product to be expelled from the metal coordination sphere.

In general, the mechanism proceeds through a three-centre transition state:

$$L_nM\!\!\begin{array}{c}\diagup A\\\diagdown B\end{array} \longrightarrow \left[L_nM\cdots\begin{array}{c}A\\\vdots\\B\end{array}\right] \longrightarrow L_nM \; + \; A\text{-}B$$

A and B must be direct neighbours within the metal coordination sphere, which imposes a *cis* geometry. This mechanism also implies a retention of stereochemistry at A and B, which is important should either of these atoms be chiral. Thus, in the equation above the optical configuration at A will be the same when linked to the metal and later to B. The decarbonylation of acid chlorides by Wilkinson's catalyst provides a good example of reductive elimination:

$$[RhClL_3] \; + \; RC(O)Cl \xrightarrow{\text{oxidative add.}} [RC(O)\text{-}RhCl_2L_3] \xrightarrow{\text{- L}} [R\text{-}RhCl_2(CO)L_2]$$

(16e)   (18e)   (18e)

$$\xrightarrow{\text{reductive elim.}} R\text{-}Cl \; + \; [RhCl(CO)L_2]$$

(16e)

Bimolecular reductive eliminations are observed in a few cases:

$$[ArC(O)\text{-}Mn(CO)_5] \; + \; [MnH(CO)_5] \longrightarrow ArCHO \; + \; [Mn_2(CO)_{10}]$$

Oxidative Coupling and Reductive Decoupling

These reactions can be considered as special cases of reductive elimination and oxidative addition. The scheme for the mechanism of oxidative coupling is the following:

In such a process, the two two-electron π ligands are transformed into a new chelating ligand giving 2 × 1 electrons. This causes a decrease of two units in the metal electron count, an increase of two units in the oxidation state and no change in the degree of coordination. When using the metallacyclopropane representation for the π complexes, the analogy between this process and reductive elimination becomes clear.

Generally, this mechanism works better for the coupling of alkynes than alkenes. It provides a straightforward access to metalloles:

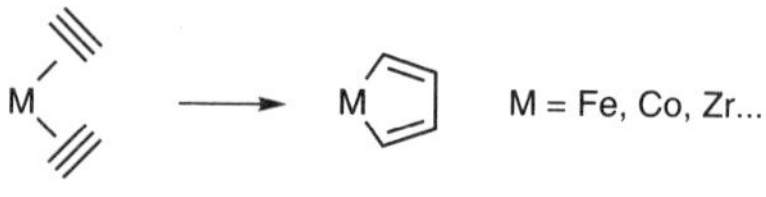

$$M = Fe, Co, Zr...$$

For alkenes, the presence of electron-attracting substituents or other factors favouring the *decoupling* of the π electrons aids the coupling process. In general, the more electron rich the metal, the easier the oxidative coupling becomes. It is often necessary to treat the mixture with a two-electron ligand L; this stabilizes the product by occupying the vacant site generated during the reaction. The following examples are typical:

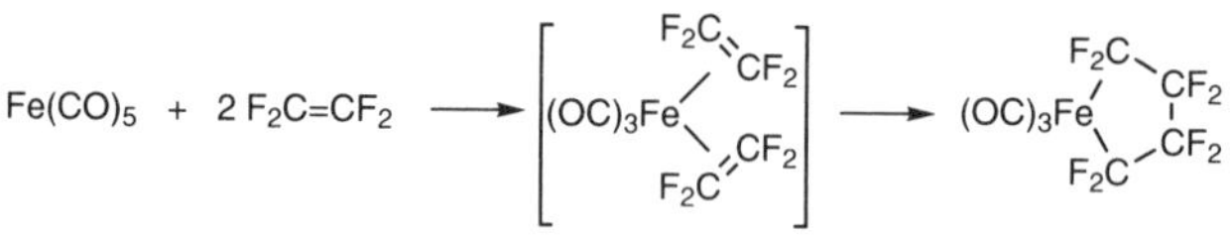

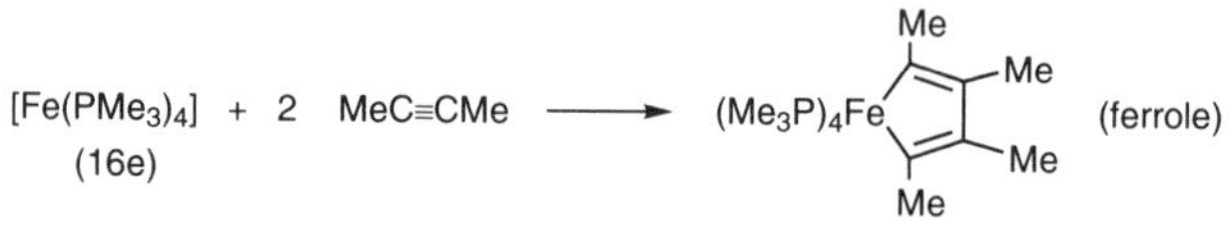

Finally, note that it is possible to couple two carbenes or a carbene and an olefin in a similar fashion:

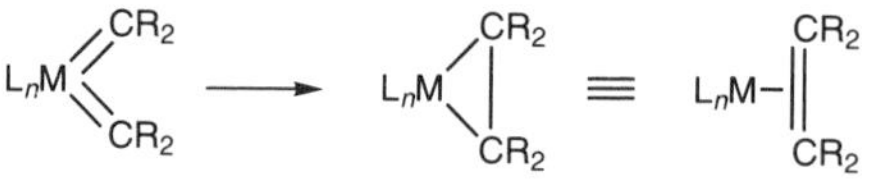

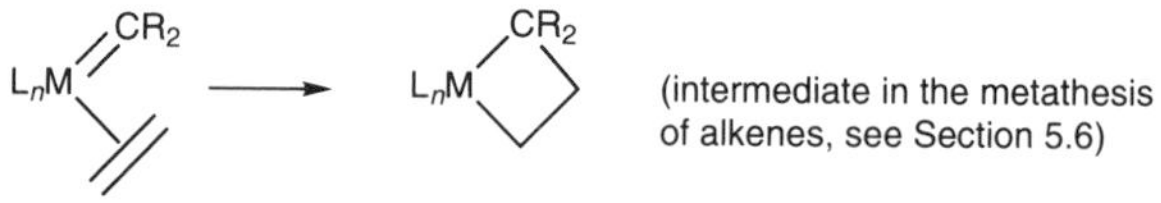

(intermediate in the metathesis of alkenes, see Section 5.6)

Insertion and Deinsertion

Overall, an insertion can be defined as the incorporation of an unsaturated two-electron ligand into a M–X σ bond (X being a one-electron ligand). Two insertion types (1,1) and (1,2) are possible:

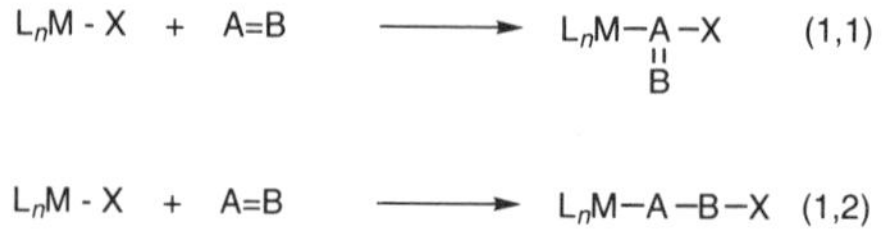

$$L_nM\text{-}X \;+\; A{=}B \;\longrightarrow\; L_nM{-}\underset{\underset{B}{\|}}{A}{-}X \qquad (1,1)$$

$$L_nM\text{-}X \;+\; A{=}B \;\longrightarrow\; L_nM{-}A{-}B{-}X \quad (1,2)$$

Each proceeds through two distinct steps. The first involves incorporating A=B as a ligand within the metal coordination sphere:

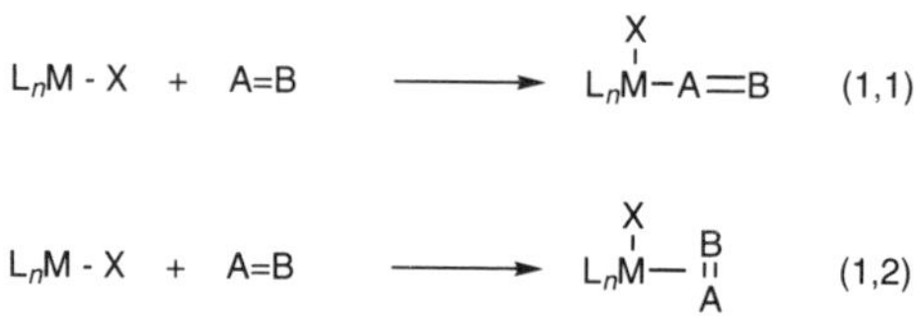

$$L_nM\text{-}X \;+\; A{=}B \;\longrightarrow\; L_nM{-}\overset{\overset{X}{|}}{A}{=}B \qquad (1,1)$$

$$L_nM\text{-}X \;+\; A{=}B \;\longrightarrow\; L_nM{-}\overset{\overset{X}{|}}{\underset{\underset{A}{\|}}{B}} \qquad (1,2)$$

Consequently, $L_nM{-}X$ must have no more than 16 electrons because the reaction requires a vacant coordination site. The second step, often called migration-insertion, incorporates $A{=}B$ into the $M{-}X$ bond:

$$L_nM\underset{|}{\overset{X}{A}}{-}A{=\!=}B \quad \longrightarrow \quad L_nM{-}\underset{|}{\overset{X}{A}}{=\!=}B \qquad (1,1)$$

$$L_nM{-}\underset{A}{\overset{X}{\underset{\|}{|}}}B \quad \longrightarrow \quad L_nM{-}A{-}B{-}X \qquad (1,2)$$

In both cases, the metal oxidation state remains unchanged. However, because the oxidation state of A changes by two units in the (1,1) case, this atom must be capable of modifying its valency. As usual, the migration of X to $A{=}B$ requires that these ligands adopt a *cis* geometry within the metal coordination sphere; obviously, in a case where many unsaturated ligands are present, X can migrate to sites other than $A{=}B$. Finally, during the migration-insertion, the number of electrons and the metal coordination number decrease by two and one units respectively. This means that for the process to operate, the addition of a two-electron ligand is often required to stabilize the complex which is produced.

Carbon monoxide CO is the ligand which most commonly undergoes (1,1) insertions. SO_2 can sometimes behave in the same way:

$$L_nM\text{-}R \ + \ CO \quad \longrightarrow \quad L_nM{-}\underset{O}{\overset{\|}{C}}{-}R$$

$$L_nM\text{-}R \ + \ SO_2 \quad \longrightarrow \quad L_nM{-}\overset{O}{\underset{O}{\overset{\|}{\underset{\|}{S}}}}{-}R$$

(SO_2 also gives 1,2-insertion products)

Alkenes and alkynes invariably give (1,2) insertions; the ligand X in these cases is frequently a hydride or hydrocarbon group. These inherently reversible reactions will be discussed in more detail in the chapters dedicated to the $M{-}H$ and $M{-}C$ bonds.

Nucleophilic and Electrophilic Attack on Coordinated Ligands

Let's look at a ligand L' coordinated to a metal centre L_nM in a complex $L_nM{-}L'$. If L_nM is an electron-deficient centre having Lewis acid character (positive charge, π acceptor ligands L, high metal oxidation state), then the sensitivity of L' to nucleophilic attack will be increased by its incorporation into the complex. To take an example, arenes, which are normally inert to nucleophiles, easily undergo nucleophilic attack when π-coordinated to $Cr(CO)_3$ (see Section 4.5). On the other hand, if L_nM is an electron-rich Lewis base (recognizable through characteristics such as a negative charge, σ donor ligands and a low metal oxidation state), a corresponding increase in the susceptibility of a ligand to electrophilic attack is seen upon complexation.

The attack of an external nucleophile Nu^- or electrophile E^+ upon L' can proceed either via *addition*, giving new ligands $L'Nu$ or $L'E$ which are retained within the metal coordination sphere, or alternatively by *abstraction*, which involves partial or complete departure of L' from the metal. The following examples illustrate these processes:

- Nucleophilic addition

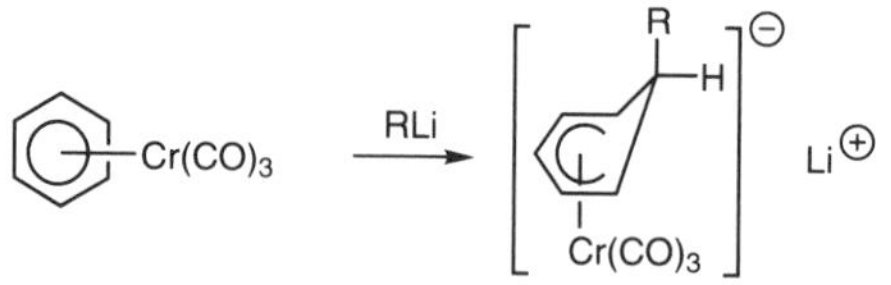

For obvious steric reasons, R^- approaches the face of the arene which is remote from the metal; this imposes an *exo* stereochemistry upon R in the anionic complex. This *exo* attack is a general rule, which is valid for all the reactions considered in this section.

- Nucleophilic abstraction

A good example of nucleophilic abstraction appears in a key step of the Pd(0)-catalysed carbonylation of Ar−X:

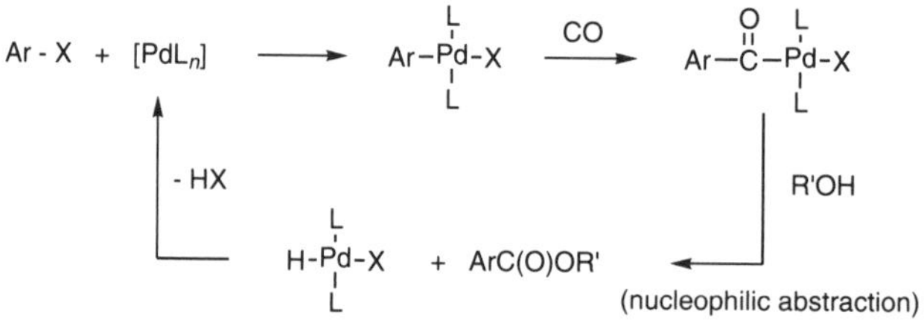

We will review a few important special cases.

1. Nucleophilic attack on coordinated CO

   Due to the large electronegativity difference between carbon and oxygen, the carbon of CO is significantly electrophilic even when complexed to metals. Consequently, it is sensitive to nucleophiles, as below:

Naturally, the ease of attack depends on the nature of L_nM. Fischer's preparation of the first carbene complexes employed this approach, with $Nu^- = RLi$ (see Section 3.4). Amine oxides $R_3N^+-O^-$, which are extremely useful reagents for liberating coordination sites on 18-electron metal carbonyl complexes, react by a similar mechanism.

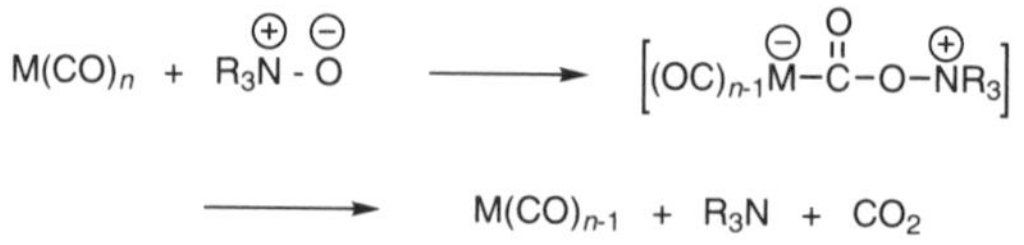

The analogous attack of OH^- on a metal carbonyl sometimes leads to a hydride, with loss of CO_2:

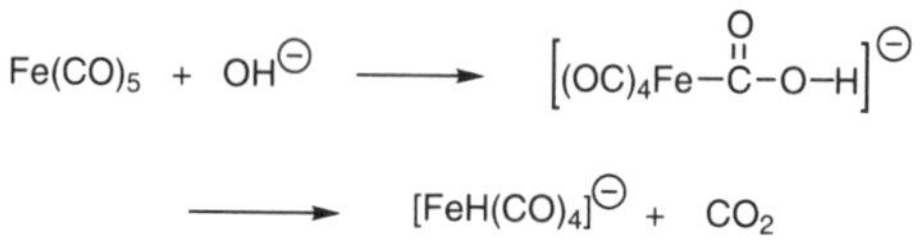

2. Nucleophilic attack on complexed polyenes (see reference 5)

   These attacks obey a few straightforward empirical rules:

(a) The polyenes ($2n$ π electrons) are more reactive than the polyenyls ($2n + 1$ π electrons). This is because complexed polyenyls carry a substantial negative charge (see Section 3.6) which tends to repel incoming nucleophiles.

(b) Open polyenyl ligands are more reactive than cyclic polyenyls.

(c) Open ligands react at their termini. This is invariably true for the polyenes ($2n$ electrons) but only applies to polyenyls if the metallic centre is very electrophilic. It is clear that steric influences play a major role in rules (b) and (c) because the end groups are more accessible than the central carbons. However, the termini of complexed polyene ligands carry a substantial positive charge, which also favours the approach of the incoming nucleophile.[5]

(d) Attack always takes place at the face of the polyene which is oriented away from the metal. We have already described the steric origins of this phenomenon.

These rules are illustrated nicely by the following examples:

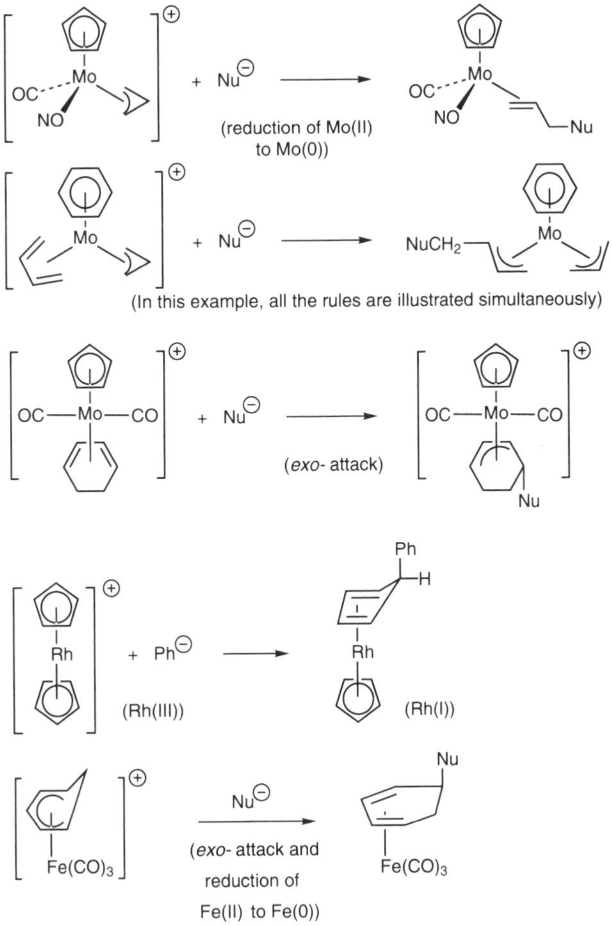

(reduction of Mo(II) to Mo(0))

(In this example, all the rules are illustrated simultaneously)

(exo- attack)

(Rh(III))   (Rh(I))

(exo- attack and reduction of Fe(II) to Fe(0))

Few exceptions to these rules are known. However:

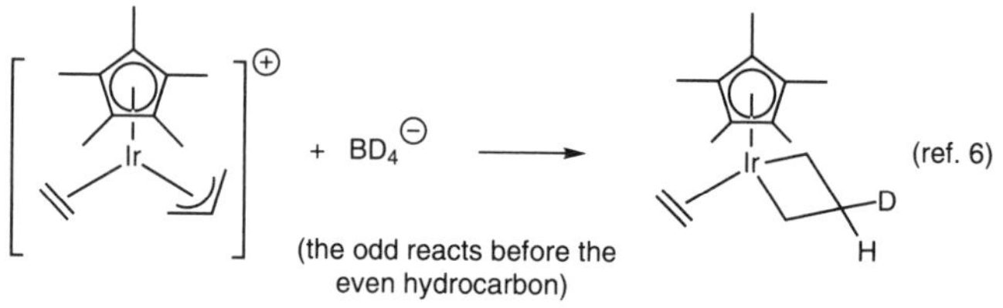

+ BD$_4^{\ominus}$ ⟶

(the odd reacts before the
even hydrocarbon)

(ref. 6)

3. Electrophilic abstraction of alkyl groups

Electrophilic metal centres such as Hg^{2+} can cleave M–C bonds:

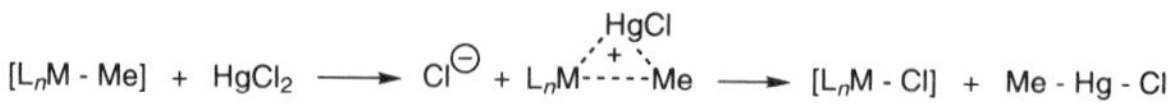

$$[L_nM\text{-}Me] + HgCl_2 \longrightarrow Cl^{\ominus} + L_nM\text{-----}Me \longrightarrow [L_nM\text{-}Cl] + Me\text{-}Hg\text{-}Cl$$

as can other electrophiles such as H^+, Ph_3C^+ and Br_2. For example:

$$Cp_2TaMe_3 + [Ph_3C]^{\oplus} \longrightarrow [Cp_2TaMe_2]^{\oplus} + Ph_3C\text{-}Me$$

$$[CpFe(CO)_2\text{-}R] + 0.5\,Br_2 \longrightarrow [CpFe(CO)_2\text{-}R]^{+\bullet} + Br^{\ominus}$$
(1e oxidation)

$$[CpFe(CO)_2\text{-}Br] \xleftarrow{0.5\,Br_2} [CpFe(CO)_2]^{\bullet} + R\text{-}Br$$

With vinyl groups, an inversion of stereochemistry is sometimes observed, because of a reversible electrophilic attack at the carbon of the double bond.

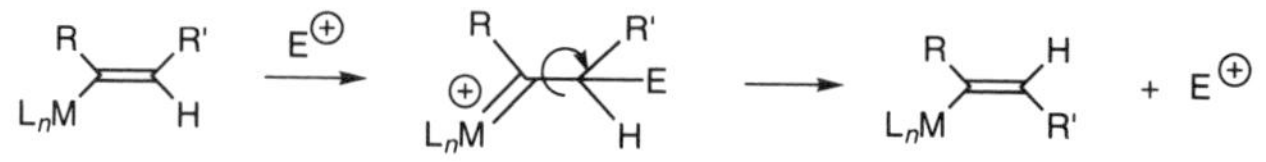

When the hydrogens on the α carbons are labile, an electrophile can also abstract H^- to generate a carbene:

$$[Cp(NO)(Ph_3P)Re\text{-}CH_2Ph] \xrightarrow{[Ph_3C]^{\oplus}} [Cp(NO)(Ph_3P)Re=CHPh]^{\oplus}$$

2.4 Theoretical Approach to Oxidative Addition and Reductive Elimination

In the following pages we will present the qualitative theoretical approach required to understand one of the more puzzling phenomena in organometallic chemistry. To take a striking example, how can a σ bond as strong as in H_2 (103 kcal/mol), have such a low activation energy towards cleavage that it can be broken by a transition metal complex at room temperature? (The same question, not treated here, also applies to activation at a metallic surface.) The answers have been hard to obtain and satisfactory explanations only began to appear in the 1980s. A number of important question marks still remain in this area.

Initially we will examine the most fundamental criteria: the MOs and the number of electrons involved in the process.

General Considerations

As our model system, we take the general case of a d^n, ML_r complex interacting with the σ bond in H_2. Later, we will look at what happens should we replace H_2 by a less symmetrical ligand $A-B$. Employing the classical formalism which attributes the electrons to the ligands, we can write the process, formally, as:

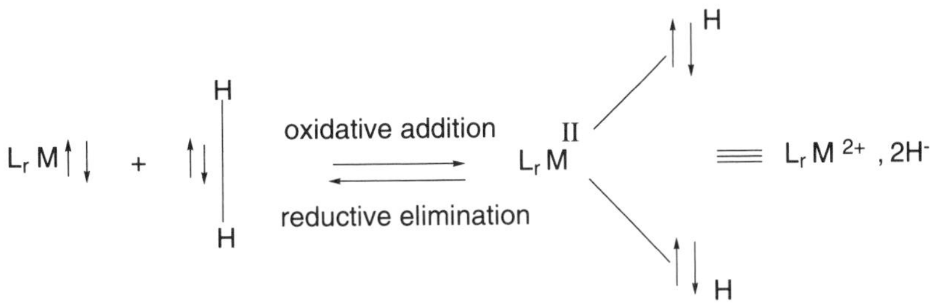

In the most widely applied conceptual model, the metal transfers two electrons to H_2, thus generating two hydride ions (H^-). Thus, the metal acts as a reducing agent which is oxidized from d^n to d^{n-2} during the course of the reaction. The reverse process occurs in reductive elimination. Further examples of this type of pathway are given below: the forward reactions are oxidative additions whilst the reverse are reductive eliminations. We will not treat one-electron processes here.

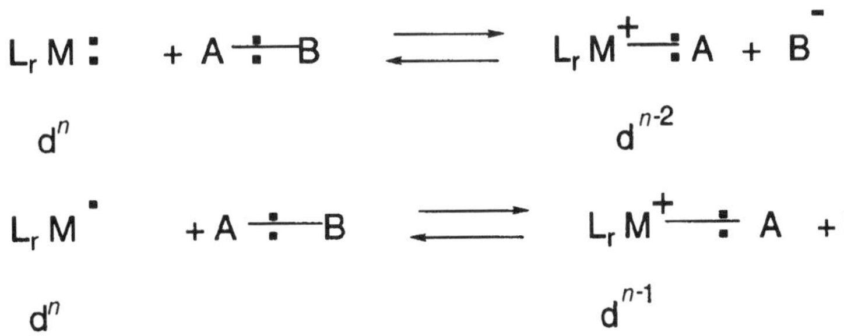

We have assumed that the ligands are totally passive spectators in these schemes. As we will see later, this assumption is not always strictly true.

A priori, a number of mechanistic pathways could explain the oxidative addition process. For a molecule such as H_2, it is very difficult to imagine a charge separated (H^-/H^+) mechanism passing through ionic intermediates in a two-stage addition as, for example, in the case of the addition of Br_2 to alkenes. As *with all apolar σ bonds, a concerted addition* of the two hydrogens *is the most plausible mechanism*, by far. Thus, the reaction must give a product having *cis*-configured H ligands. With molecules of the A−B type, e.g. CH_3−Cl, where A and B have different electronegativities, this is not necessarily the case. Here, if a polar mechanism operates, a *trans*-configured product becomes possible.

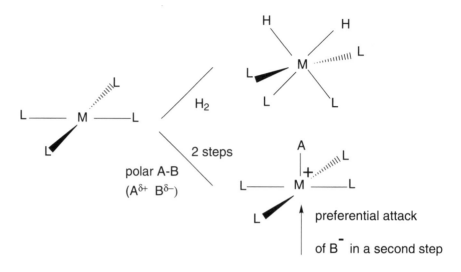

The square-planar starting material in the previous figure is appropriate because the most efficient substrates for oxidative addition generally have this geometry. The limiting polar and apolar cases will be explored in detail below, through analysis of the MOs in the products and reagents of the reaction.

Orbital Aspects of the Non-polar Case: Oxidative Addition of H_2

Consider the archetypal square-planar complex ML_4, whose orbitals were treated above. We can combine the orbitals in the reagents and products as in the approximate correlation diagram on the next page.

During the reaction, *the number of bonding MOs increases by one*. The reaction is governed by the shape and form of the MOs, their match in the reagents and products, and their evolution during the reaction. We will employ *correlation diagrams* to model the evolution of the orbitals graphically throughout the reaction trajectory. This approach, devised by Woodward and Hoffmann to study carbon chemistry, may be applied easily to other elements. The correlation diagrams are governed by a fundamental *non-crossing rule* for levels of the same symmetry; this rule simply states that the potential energy surfaces of orbitals having the same symmetry may not cross.

Correlation Diagrams: the Non-crossing Rule for Levels of the Same Symmetry

Imagine a process where two orbitals, MOs Φ_1 and Φ_2, of energies E_1 and E_2 respectively, see their energy vary as a function of the reaction coordinate q. Suppose, as is frequently the case, that E_1 starts by decreasing whilst E_2 rises. The question which we address is: what happens when E_1 tends towards E_2: how do the two curves behave?

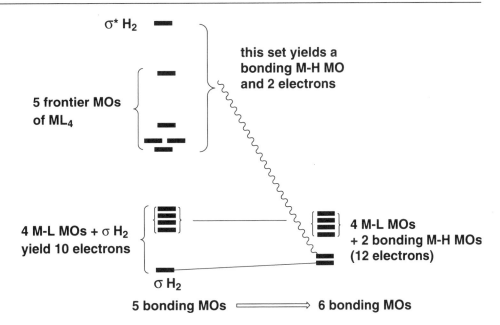

5 frontier MOs of ML$_4$

this set yields a bonding M-H MO and 2 electrons

4 M-L MOs + σ H$_2$ yield 10 electrons

σ H$_2$

4 M-L MOs + 2 bonding M-H MOs (12 electrons)

5 bonding MOs ⟹ **6 bonding MOs**

The following diagram outlines the problem:

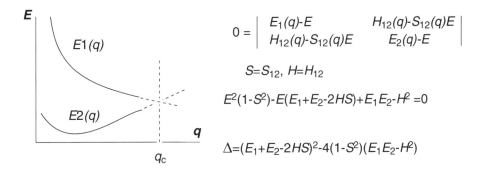

$$0 = \begin{vmatrix} E_1(q)\text{-}E & H_{12}(q)\text{-}S_{12}(q)E \\ H_{12}(q)\text{-}S_{12}(q)E & E_2(q)\text{-}E \end{vmatrix}$$

$$S = S_{12}, \ H = H_{12}$$

$$E^2(1\text{-}S^2)\text{-}E(E_1+E_2\text{-}2HS)+E_1E_2\text{-}H^2 = 0$$

$$\Delta = (E_1+E_2\text{-}2HS)^2 - 4(1\text{-}S^2)(E_1E_2\text{-}H^2)$$

If we define q_c as the point on the reaction coordinate where $E_1 = E_2$, then $\Delta(q_c)$ is zero by definition at this point. After defining the common value where $E_1 = E_2$ as E_c, an approximated mathematical treatment gives the solution: $E_c^2 S_{12}^2 - 2\,E_c H_{12} S_{12} + H_{12}^2 = 0$ (this is reflected in simple Hückel theory, where if $S_{12} = 0$, H_{12} must be 0). Since we know that H_{12} is *always proportional to* S_{12}, then *two energy values can never be intermixed unless S_{12} is precisely zero*, no matter what degree of calculational complexity is employed. Conversely, if the two orbitals have the same symmetry (by having the same irreducible representation in the molecular point group) then Δ can never be zero and the system always has two distinct roots. Since the only exception to this rule is the trivial case where the interfragment distance is infinite and the orbital overlap S_{12} is zero, we can state that *energy levels can only cross for MOs which have different symmetry*. For two orbitals of the same symmetry, simple Hückel theory gives an energy diagram as a function of q which is consistent with the above conclusions:

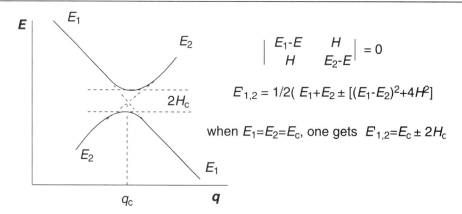

$$\begin{vmatrix} E_1\text{-}E & H \\ H & E_2\text{-}E \end{vmatrix} = 0$$

$$E_{1,2} = 1/2(\ E_1 + E_2 \pm [(E_1\text{-}E_2)^2 + 4H^2]$$

when $E_1 = E_2 = E_c$, one gets $\ E_{1,2} = E_c \pm 2H_c$

It is clear that the two curves converge, but as they approach q_c, they repel and exchange. In quantum mechanical terminology, *their crossing is avoided.*

This digression clearly underlines the constraints imposed by symmetry upon the behaviour of the evolving orbitals and their reaction pathways.

MO Correlations during Oxidative Addition

The clarification of the previous paragraph allows us to return to the correlation diagram for oxidative addition. We will look at this process in terms of orbitals which are symmetrical (S) or antisymmetrical (A) with respect to a plane lying perpendicular to the

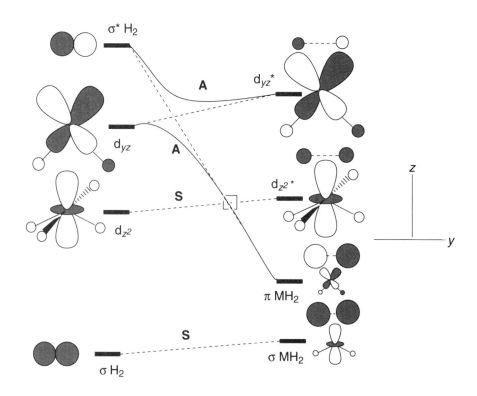

xz axis.[7-12] Initially, we will correlate the orbitals according to a criterion of 'strongest similarity' without worrying too much about the possibility of energy crossing. As it is clear that there is pronounced hydride character in the two bonding orbitals formed by the oxidative addition of dihydrogen to a metal centre, it is logical to connect the σ (H_2) to the bonding MO σ (MH_2) and σ^* (H_2) to π (MH_2) by dotted lines. The MOs which have high d character remain essentially unchanged. The previous figure illustrates these correlations.

Having established these relationships, we see that the two antisymmetric levels would have to cross during the course of the reaction; this is avoided through the exchange of the d_{yz} orbital with the σ^* of H_2. Thus, as the reaction progresses, the d_{yz} electron density flows into σ^* H_2, weakening the H—H bond and thereby causing the hydrogens to separate. In turn, this results in an enhanced geometry for further d_{yz} to σ^* H_2 electron donation. This self-reinforcing cycle eventually results in complete cleavage of the H—H linkage and the formation of two M—H bonds.

We have considered this fundamental reaction in detail because it plays a crucial role in many catalytic processes. It has also allowed us to show how MO correlation diagrams may be used to describe the intimate mechanism involved in the oxidation of the metal: *through populating the σ^* antibonding orbital of H_2, the metal progressively loses two electrons which are made available to the incipient ligands.* This mechanism may be generalized to other common σ bonds (C—H, C—C, etc.) with only minor modifications to take account of the exact shape of the MOs that are involved.

Case of a Polar Molecule $A^{\delta-}B^{\delta+}$

The limiting case of a polar substrate is quite different from the concerted non-polarized process which we have just described. Conceptually, this polar case can be easily understood in terms of an SN_2 attack of a metal on A—B (see overleaf).

Using our standard set of coordinate axes, the 'substrate' A—B approaches M from the z direction. The square-planar complex behaves as a nucleophile, so it is important to focus upon the site of its highest available electron density, the d_{z^2} HOMO. Equally, the characteristics of the LUMO of the A—B electrophile are of critical importance. If we arbitrarily assign the charges $A^- - B^+$, the two frontier MOs in this species will be distinctly non-symmetrical: the σ-bonding HOMO will be mainly localized on A with the antibonding LUMO residing on B. Therefore, the metal orbital which transfers the two electrons into σ^* A—B must be symmetrical, because it must have a form which can overlap with the symmetrical B-centred LUMO. Thus we immediately see an important difference between the polar and non-polar limiting cases in oxidative addition: the antisymmetrical d_{yz} level responsible for the electron transfer in non-polar concerted oxidative addition of compounds such as H_2 is not employed in the polar case; it is orthogonal to the LUMO giving rise to bond formation, and therefore incapable of reacting with it.

The correlations are established as above. The apparent simplicity of the reaction coordinates is deceptive: the two-electron transfer from the metal to the ligand is complex, because all of the three orbitals involved in the scheme have the same symmetry and all intermix. For simplicity, our diagram shows in dotted lines the correlation which would be established if the overlaps between the MOs were restricted to only the strongest interactions. Furthermore, two avoided cross-overs are found instead of the single one in the apolar case. This complexity is reflected in the evolution of the lowest energy MO,

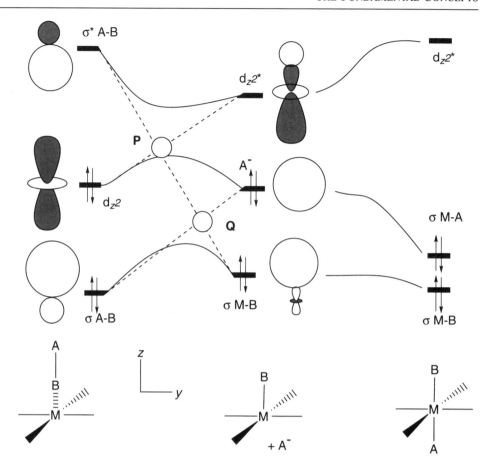

where electron density is transferred from A to B as the reaction proceeds. This requires a significant activation energy because the natural polarity of the A–B unit is reversed as the reaction advances.

The second step of the reaction, the capture of A^- by MB^+, is controlled by straightforward electrostatic forces because the occupied MO of the anion and the empty LUMO of the cation have the correct symmetry to form a σ bond. Overall *trans* addition is favoured by the orientation of the metal-centred LUMO which governs the second step. This spontaneous process occurs rapidly by comparison with the first step because it does not require any electronic redistribution. Thus, the first, rate-limiting step resembles an organic SN_2 substitution, whilst the subsequent reaction is readily understood in terms of the capture of a charged species by an SN_1 intermediate.

2.5 Addition of a Coordinated Ligand (R^-) to a CO or an Alkene: Theoretical Aspects

To understand this important reaction, it will be necessary to revise the electronic structure of CO which we looked at in our discussion of backdonation.

The CO Molecule

Schematically, the 10 valence electrons in carbon monoxide are shared according to the scheme below:

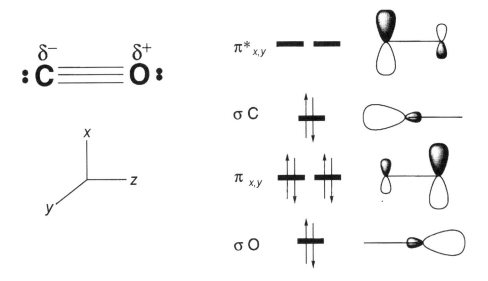

We see two lone pairs oriented along the CO axis. One of these is at low energy because it is localized on the highly electronegative oxygen atom; the other, localized at carbon, is the HOMO. Four electrons are found in the degenerate π_x and π_y orbitals and two are in σ_c. If we assume that the electrons in each bond are equally shared between the C and O atoms, then we have five electrons at C and five at O. This gives the counterintuitive polarity $C^- \ldots O^+$:

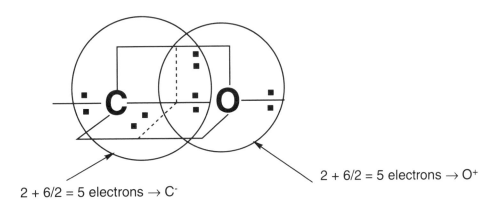

formal Lewis electron counts

This formal charge imbalance is largely offset by the electronegativity difference between the two atoms, which polarizes the bonding electron clouds and causes a shift of electron density towards the oxygen. Thus the overall polarity of CO is reduced but the carbon retains a small negative charge.

In our discussion of metal carbonyl backbonding, we saw that the empty π^* MOs of CO are good electrophiles which can accept electron density from the metal; equally, they can interact with any other part of the coordination sphere that is electron rich. Thus, they are well adapted for interaction with neighbouring H^- or R^- groups in the metal coordination sphere.

A Study of the Insertion Reaction

Let's look at a plausible reaction:

$$cis - M(R)(CO) \longrightarrow M-CO-R$$

This process is more complicated than it appears and it is difficult to follow the evolution of the electrons involved, or even to count them with certainty, at a sophisticated level. Consequently, we will adopt a simplified approach, wherein we consider only the changes occurring within the frontier orbital set as the reaction proceeds. The crucial point in our analysis is the understanding of how the π-type frontier orbitals develop as the $M-C-O$ axis bends; this is summarized in the following figure:

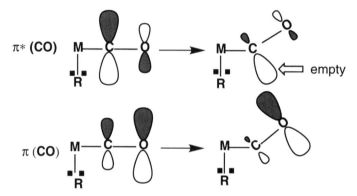

The bonding and antibonding combinations evolve differently. The rehybridization of the bonding π MO involves a shift of electron density towards the oxygen atom, which causes the formerly π-bonding electrons to become localized as an oxygen-centred lone pair of the type normally associated with aldehydes and ketones. Simultaneously, the antibonding π^* orbital moves to the carbon centre, where it confers strong carbocationic character. Thus, the bent product strongly resembles a ketone which has lost an alkyl substituent, as underlined by the comparison with the frontier MOs of a typical ketone shown below:

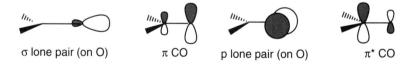

σ lone pair (on O)   π CO   p lone pair (on O) π^* CO

Thus, by conceptually localizing the participating electron pairs, we can describe the migration of an alkyl group onto a metal-bound carbonyl ligand quite easily, as is shown in the figure below. We can then draw a correlation between the original sites of electron density and the final configuration:

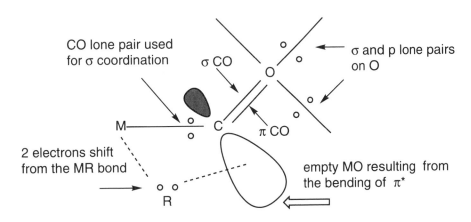

This very approximate approach can obviously be refined in more thorough and rigorous molecular orbital treatments, but these only add details to the points which we have deduced using our simple model. We already have the essentials: during the insertion *the metal does not change oxidation state*, because the R^- to $RC{=}O^-$ transformation only converts one anionic ligand into another, and *the number of electrons in the metal coordination sphere is decreased by two*.

Other Electronic Aspects

A detailed study of MOs in any given complex is necessary to permit a precise prediction of its reactivity, but several useful guidelines for its behaviour can be deduced from first principles. In the previous reaction, we studied the attack of R^- upon an adjacent CO electrophile, which was more or less activated by bending away from its equilibrium linear configuration. The energies of the CO orbitals have a complex relationship with the other ligands in the metal coordination sphere and are difficult to estimate without recourse to calculations. However, the factors governing the MOs associated with R^- are simpler to understand, which permits us to predict two effects which should favour insertion:

1. If the energy difference between the M—R bonding HOMO and the CO π^* LUMO is minimized, the reaction will be kinetically favoured because the energy barrier will be lowered.

2. If the metal can easily release electron density to its ligands, the reactivity of the R^- group will be enhanced, thus thermodynamically promoting its reaction with the electrophilic CO.

Consequently, to provoke the migration of R and favour the insertion reaction, the metal should be in a reasonably low oxidation state and surrounded by electron donor ligands. In molecular orbital terms, we can summarize these considerations as follows:

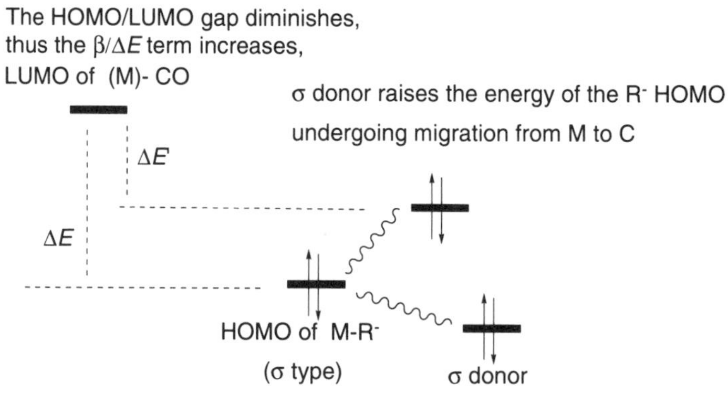

The HOMO/LUMO gap diminishes,
thus the $\beta/\Delta E$ term increases,
LUMO of (M)- CO

σ donor raises the energy of the R⁻ HOMO

undergoing migration from M to C

$\Delta E'$

ΔE

HOMO of M-R⁻

(σ type)

σ donor

Or, put simply, ligands raising the energy of the R⁻ HOMO assist the migration.

The Nucleophilic Attack of R⁻ on a Metal-bound Alkene

The description of the insertion of alkenes into the MR bond has some similarities with the insertion of CO; nonetheless, it should be noted that the orbitals which are employed are quite different and the reverse β-elimination process is often favoured when R = H. The frontier orbitals are given for the general case below:

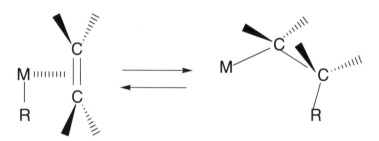

Here, R may be equal to H.
  The principal orbitals are:

(a) the M—R σ-bonding MO, which serves as the nucleophilic centre;
(b) The C=C π-bonding MO, which provides the electron density required to constitute the metal–carbon bond in the product;
(c) the other metal-centred MOs. If occupied, these orbitals tend to inhibit the insertion reaction, because their ability to backdonate electron density into the antibonding orbitals on the olefin lowers its susceptibility to attack by the nucleophilic R⁻

Consequently, it is clear that the metal must be capable of balancing two opposing effects: the provision of sufficient electron density to provoke the migration of the R group and the withdrawal of sufficient electron density from the alkene to render it susceptible to nucleophilic attack.

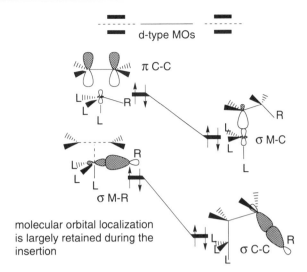

The diagram above shows schematically how the localization of the four highest energy electrons evolves as the reaction progresses. For the sake of clarity, the precise orbital correlations are not shown, but they resemble those which we saw in the study of oxidative addition. Overall, we can see that *the metal remains in the same oxidation state throughout the olefin insertion reaction*, because the alkyl group consumed in the reaction (σ M$-$R) is replaced by an alkyl group formed as the product (σ M$-$C).

The insertion reaction of olefins is inherently reversible and the position of the equilibrium depends upon the nature of the metal complex involved, a characteristic which it shares with the oxidative addition–reductive elimination couple. Studies have shown that when R $=$ H, *β-elimination is favoured if the metal is electron-rich, but insertion dominates if it is electron-poor*. The implications of this simple result are discussed in detail in Section 5.5, but are well worth underlining. In d^0 Ti(IV) compounds, where the alkene is coordinated to a very electron-poor centre, many thousands of insertions may occur into the metal alkyl bond before it is ejected from the metal coordination sphere by a β-elimination reaction. Thus, this system is an olefin polymerization catalyst. On the other hand, electron-rich complexes of the late transition metals such as L_nNi strongly favour the β-elimination process, which means that they suppress polymerization and catalyse the production of dimers, trimers and low molecular weight oligomers.

Finally, the so-called 'agostic' methyl groups are interesting because they show how two schemes can be combined to understand an unusual bonding condition. These complexes have deformed bond lengths and valence angles because of 'complexation' of a C−H bond by the metal.[13] Conceptually, these compounds may be rationalized as a combination of the insertion of a hydride into an olefin and the reductive elimination of a hydrocarbon from a metal centre. They give an approximate model for a possible transition state during hydrogen migration:

2.6 References

1. (a) M. Elian and R. Hoffmann, *Inorg. Chem.*, 1975, **14**, 1058. (b) For an extensive study, see T. A. Albright, J. K. Burdett and M. H. Whangbo, *Orbital Interactions in Chemistry*, Wiley, New York, 1985, and references cited therein.
2. C. J. Ballhausen, *Ligand Field Theory*, McGraw-Hill, New York, 1962.
3. R. H. Crabtree, *The Organometallic Chemistry of The Transition Metals*, 2nd ed., Wiley, New York, 1994.
4. J. P. Collman, L. S. Hegedus, J. R. Norton and R. G. Finke, *Principles and Applications of Organotransition Metal Chemistry*, University Science Books, Mill Valley, California, 1987.
5. S. G. Davies, M. L. H. Green and D. M. P. Mingos, *Tetrahedron*, 1978, **34**, 3047.
6. J. B. Wakefield and J. M. Stryker, *J. Am. Chem. Soc.*, 1991, **113**, 7057 (this work has been brought to the attention of the authors by M. L. H. Green).
7. R. Hoffmann, M. M. L. Chen and D. L. Thorn, *Inorg. Chem.* 1977, **16**, 503.
8. A. Dedieu and A. Strich, *Inorg. Chem.*, 1979, **18**, 2940.
9. A. Sevin, *New J. Chem.*, 1981, **5**, 233.
10. J. Y. Saillard and R. Hoffmann, *J. Am. Chem. Soc.*, 1984, **106**, 2006.
11. H. Rabaa, J. Y. Saillard and R. Hoffmann, *J. Am. Chem. Soc.*, 1986, **108**, 4327.
12. Various examples of H_2 complexes of transition metals, where the oxidative addition seems 'frozen' half-way are also well known. See: G. J. Kubas, *Acc. Chem. Res.*, 1988, **21**, 120, and references cited therein.
13. (a) M. Brookhart and M. L. H. Green, *J. Organomet. Chem.*, 1983, **250**, 395. (b) A theoretical treatment and a complete review of this area are given in: O. Eisenstein and Y. Jean, *J. Am. Chem. Soc.*, 1985, **107**, 1177. The influence of this effect upon alkene polymerization catalysts has been discussed in a recent review: R. H. Grubbs and G. W. Coates, *Acc. Chem. Res.*, 1996, **29**, 85.

3   The Main Functional Groups in Organometallic Chemistry

3.1   The Hydrides

$H_2Fe(CO)_4$,[1] and $HCo(CO)_4$,[2] the first known compounds having a metal–hydrogen bond, were discovered at the beginning of the 1930s by Hieber. Their structures were the subject of a prolonged controversy. For many years it was thought, from inaccurate electron diffraction data, that the metal occupied the centre of a tetrahedron composed of four carbon atoms, with the hydrogen atoms bound to oxygens. The correct structures, having metal–hydrogen bonds, were only established[3] after the second world war.

The hydrides play an essential role in transition metal molecular chemistry, being involved in the mechanism of many catalytic (hydrogenation and hydroformylation of alkenes, ...) and stoichiometric processes (reduction, activation of C–H bonds, etc.). It is thus logical to study them in detail.

Hydrogen as a Coordinating Group

The following coordination modes are known:

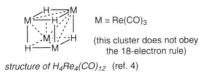

The great majority of cases fall within the first category, with the hydride playing the role of a terminal one-electron ligand. The other commonly encountered mode, the μ^2-H bridge, may be considered as a protonated metal–metal bond (two electrons delocalized over three centres). Its M–H–M angle is very variable (between 78° and 108°) and the IR stretching frequency $v(M–H)$ of between 800 and 1600 cm^{-1} is much lower than in terminal hydrides. μ^3 bridging hydrogens are usually confined to cluster chemistry:

M ≡ Re(CO)$_3$

(this cluster does not obey the 18-electron rule)

structure of $H_4Re_4(CO)_{12}$ (ref. 4)

Recently, a μ^4-H bridge has been unambiguously characterized.[5]

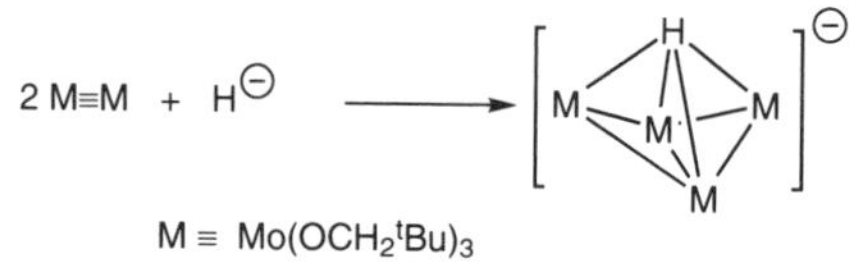

$$2\,M\equiv M \;+\; H^{\ominus} \longrightarrow \left[\text{(cluster)} \right]^{\ominus}$$

$$M \equiv Mo(OCH_2{}^tBu)_3$$

In certain cases a hydrogen has been 'caged' within a large (10–20 atom) metallic cluster,[6] as is the case for $[Rh_{13}(CO)_{24}H_3]^{2-}$. Proton NMR indicates that the hydrogens are bound to each of the rhodium atoms; this means that they move sufficiently quickly through the interior of the cage that NMR, having a relatively long timescale, cannot locate them with precision. They have obvious analogies with the interstitial hydrides (Raney nickel, Pd_2H, . . .) used as heterogeneous hydrogenation catalysts.

Last but not least, η^2 dihydrogen complexes[7] must be considered. Their recent discovery, by Kubas in 1984, had major mechanistic and theoretical impact. In the mechanistic area, it is very likely that this type of complex constitutes a stage in the oxidative addition of hydrogen to an unsaturated metallic centre:

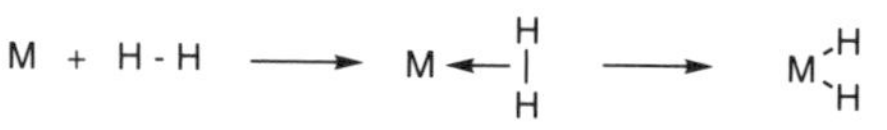

$$M + H\text{-}H \longrightarrow M \leftarrow \begin{array}{c} H \\ | \\ H \end{array} \longrightarrow M \begin{array}{c} {}^{,H} \\ {}^{\backslash H} \end{array}$$

From the theoretical standpoint, these were the first compounds to show clearly that a σ bond can play a coordinating role in the same way as a π bond. The orbital description of these dihydrogen complexes also resembles the more traditional π complexes. It brings into play bonding and backbonding interactions which populate the σ^* level of H_2 and stretch the H–H bond.

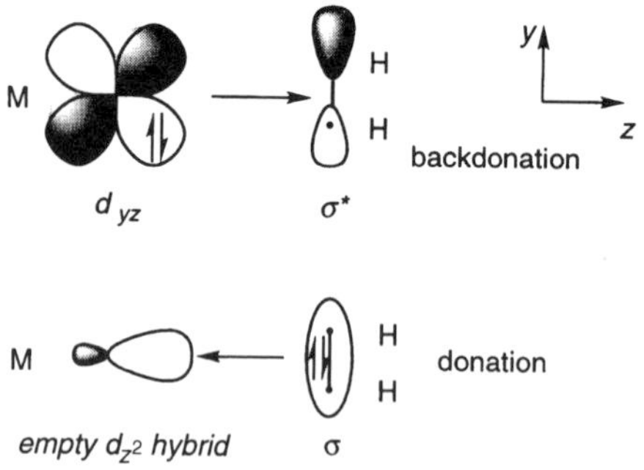

If this backbonding is too strong, H–H bond fission will occur. We describe how the first complex of this type, $[WH_2L_2(CO)_3]$, $L = (Me_2CH)_3P$, was characterized, as an illustrative example of the class as a whole.

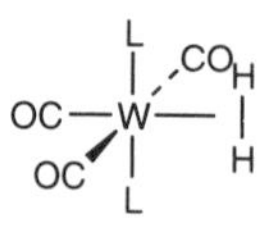

A neutron diffraction study of the complex indicates an H−H distance of 0.84 Å, which may be compared with 0.74 Å in H_2 gas. The IR spectrum of the complex contains an H−H stretching band at 2695 cm^{-1}. The replacement of H_2 by an H−D ligand allows the measurement of an NMR coupling constant $^1J(H-D)$ of 33.5 Hz versus 43.2 Hz in H−D gas. As one would imagine, the hydrogen is easily displaced from its complex by any two-electron coordinating group, even N≡N. The simple dissociation reaction is only very weakly endothermic:

$$[WH_2L_2(CO)_3] \longrightarrow [WL_2(CO)_3] + H_2$$

(18e)   (16e) ΔH = +9.9 kcal/mol

The Synthesis of Metallic Hydrides

Undoubtedly the simplest way to prepare a hydride is through the addition of hydrogen to an unsaturated complex. We give an example below from iridium chemistry which we discussed in Section 2.3.

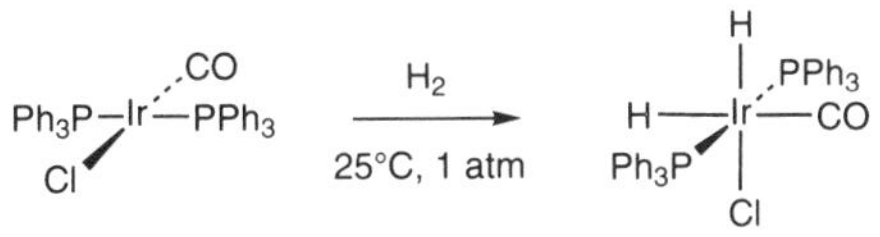

The *cis* geometry of the dihydride suggests that the reaction passes through a η^2–H_2 intermediate, similar to those above. We should also note that even though this reaction is a hydrogenation, it is classed as an oxidative addition. For a saturated, 18-electron complex to react with hydrogen, the displacement of one of the coordinating groups is necessary to liberate a site for attack: the reaction then becomes much more difficult:

$$[Os(CO)_4(PPh_3)] + H_2 \xrightarrow[\text{80 atm}]{100°C} [OsH_2(CO)_3(PPh_3)] + CO$$

Hydrogen gas can also be used to hydrogenolyse a metal–alkyl or metal–metal bond. The following example has special industrial importance (see Section 5.3):

$$Co_2(CO)_8 + H_2 \longrightarrow 2 HCo(CO)_4$$

The hydrogenolysis of thorium–methyl bonds has also been used to prepare a dinuclear hydride having Th−H−Th bridges.[8]

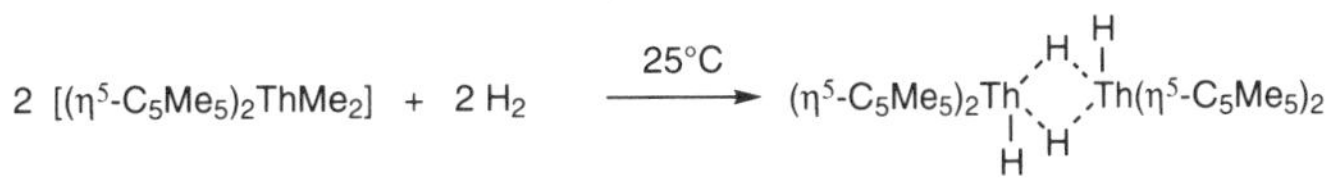

The hydride source can also be a protic acid H−X. Acids may either undergo oxidative addition at the metal centre, or protonate metal complexes which show Lewis base character.

$[Pt(PPh_3)_3] + H\text{-}CN \longrightarrow [PtH(CN)(PPh_3)_2] + PPh_3$

$[Mn(CO)_5]^{\ominus} \xrightarrow[\text{THF}]{CH_3SO_3H} HMn(CO)_5$

$[Fe(CO)_4]^{2-} + H^+ \longrightarrow [HFe(CO_4)]^{\ominus} + H^+ \longrightarrow H_2Fe(CO)_4$
$\quad\quad\quad\quad\quad\quad\quad\quad pK_a = 12.7 \quad\quad\quad\quad\quad\quad pK_a = 4.0$

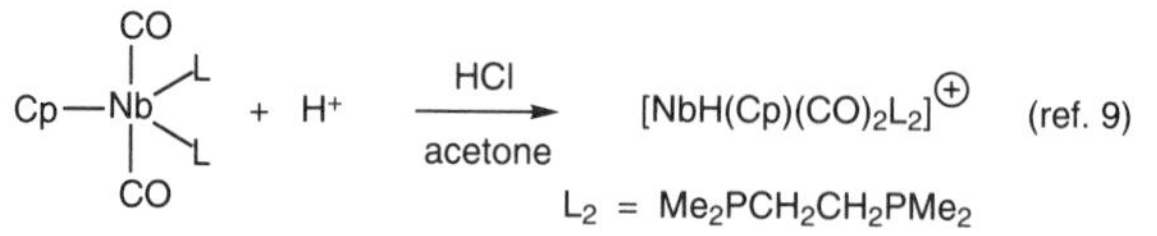

$$Cp\text{-}Nb\begin{matrix}CO\\|\\\\|\\CO\end{matrix}L \;\; + \;\; H^+ \xrightarrow[\text{acetone}]{HCl} [NbH(Cp)(CO)_2L_2]^{\oplus} \quad (\text{ref. 9})$$

$$L_2 = Me_2PCH_2CH_2PMe_2$$

Probably the most common synthesis of hydrides involves reducing a metal complex by a hydride source H^-, usually a borohydride or an aluminium hydride. An intermediate borohydride complex is sometimes observed with $NaBH_4$. The following examples are illustrative:

$$[Fe(Cp)(CO)_2\text{-}Cl] \xrightarrow{NaBH_4} [Fe(Cp)(CO)_2\text{-}H]$$

$$WCl_6 \xrightarrow[NaCp]{NaBH_4} Cp_2WH_2$$

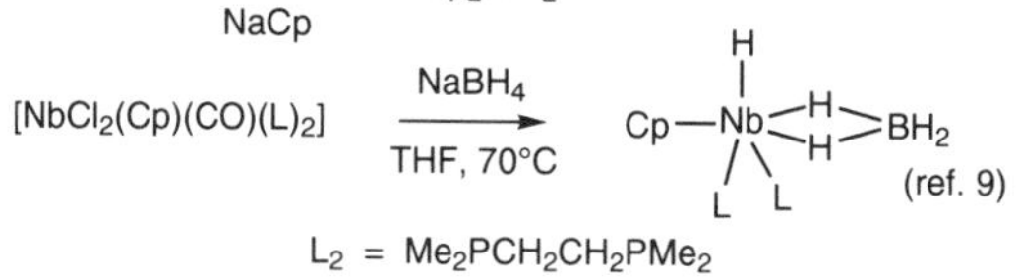

$$[NbCl_2(Cp)(CO)(L)_2] \xrightarrow[\text{THF, 70°C}]{NaBH_4} Cp\text{-}Nb \cdots BH_2 \quad (\text{ref. 9})$$

$$L_2 = Me_2PCH_2CH_2PMe_2$$

$$TaCl_5 \xrightarrow[MeC_5H_4Li]{LiAlH_4} [(\eta^5\text{-}MeC_5H_4)_2TaH_3]$$

$$[(\eta^5\text{-}C_5H_5)_2ZrCl_2] + Na[AlH_2(OCH_2CH_2OMe)_2] \longrightarrow [(\eta^5\text{-}C_5H_5)_2ZrHCl]$$

The zirconium hydride cited in the last example is employed as a stoichiometric reagent for the hydrozirconation of olefins (see Section 4.1).

Finally, numerous coordinated ligands can undergo decomposition to give hydrides: alkoxides, formates, hydroxycarbonyls, alkyls with β-hydrogens, etc., are amongst the most important of these. The following gives typical examples:

$$K_2IrCl_6 \xrightarrow[PPh_3]{EtOH\,/\,H_2O} [IrHCl_2(PPh_3)_3]$$

mechanism : $\quad Me\text{-}C\overset{H}{\underset{H}{|}}\text{-}O\text{-}Ir \longrightarrow MeCHO + H\text{-}Ir$

$$Fe(CO)_5 + OH^{\ominus} \longrightarrow \left[(OC)_4Fe\overset{O}{\overset{\|}{-}}C\text{-}O\text{-}H\right]^{\ominus} \longrightarrow [HFe(CO)_4]^{\ominus} + CO_2$$

$$[PtBr(C_2H_5)(PEt_3)_2] \longrightarrow [PtHBr(PEt_3)_2] + H_2C=CH_2$$

The Detection of Metallic Hydrides

In practice, how can we detect the presence of a metal–hydrogen bond? The problem is not trivial and the existence of these bonds was disputed for many years. The M–H vibration for terminal hydrides appears in the infrared between 1500 and 2300 cm^{-1}, but these weak peaks are often difficult to detect. Until recently, X-rays could not be used to locate hydrogen atoms either: the low electronic density associated with H$^-$ means low scattering power. Fortunately, the X-ray detection of hydrides has now become possible due to improved diffractometer sensitivity but one should note that the M–H distances measured in this way are about 0.1 Å too short. Neutron diffraction, which detects the nuclei rather than electronic density, allows a reliable measurement of these bond lengths.

At present, wherever possible (i.e. for soluble and diamagnetic complexes), the best technique for hydride detection is proton NMR. In the great majority of cases the hydride resonance appears in the range $0 \rightarrow -40$ ppm (with reference to Me$_4$Si), a region that is normally silent. However, it should be noted that hydrides of the d^0 or d^{10} metals resonate at low fields. It should also be emphasized that there is no correlation between the acidity of a hydride and its chemical shift.

Properties of Metallic Hydrides

The M–H bond is of medium stability: thus its dissociation energy varies between 37 and 65 kcal/mol for the series from Mn to Ni. In solution it may be polarized either as M$^{\delta-}$ ... H$^{\delta+}$, or as M$^{\delta+}$ ... H$^{\delta-}$, so that the hydride ligands can be acidic, basic or neutral. Table 3.1 gives some more precise data:

Table 3.1 pK_a values for some transition metal hydrides

Hydride	Solvent	
	water	acetonitrile
H[Cr(CO)$_3$(Cp)]		13.3
H[Mo(CO)$_3$(Cp)]		13.9
H[W(CO)$_3$(Cp)]		16.1
H[Mn(CO)$_5$)]	7.1	15.1
H[Re(CO)$_5$)]		21.0
H$_2$[Fe(CO)$_4$]	4.0 (pK_{a1})	11.4
H$_2$[Ru(CO)$_4$)]		18.7
H$_2$[Os(CO)$_4$]		20.8
H[Fe(CO)$_2$Cp]		19.4
H[Co(CO)$_4$]	strong acid ($\approx$ H$_2$SO$_4$)	8.4
H[Co(CO)$_3$(PPh$_3$)]	6.96	15.4

A comparison of the two last entries in the table shows very clearly the enormous influence which the metal's coligands can exert over hydride acidity. The simple replacement of a CO by a phosphine increases the pK_a by 7 units! This control of the properties of a hydride through variation of the coordinating groups is useful in catalysis (see especially Section 5.3 on the hydroformylation of olefins).

Later we will come across numerous reactions which involve hydrides. Here we will concentrate only on a few essential points. Apart from their acidic or basic characteristics, hydrides also have reducing properties. The reaction with CCl_4 often serves to titrate M−H bonds:

$$M\text{-}H \ + \ CCl_4 \ \longrightarrow \ M\text{-}Cl \ + \ CHCl_3$$

From a utilitarian viewpoint, the majority of 'useful' hydride reactions are insertions such as:

$$M\text{-}H \ + \ CO_2 \ \longrightarrow \ M\text{-}C(O)OH$$

$$M\text{-}H \ + \ CS_2 \ \longrightarrow \ M\text{-}C(S)SH$$

$$M\text{-}H \ + \ \underset{}{>\!\!=\!\!<} \ \longrightarrow \ M\overset{|}{\underset{|}{C}}\text{-}\overset{|}{\underset{|}{C}}\text{-}H$$

Transient formyl complexes are obtained with CO; they are generally unstable and tend to regenerate the hydride. However, in certain cases they have been isolated and characterized:[10]

$$[Re(Cp)(CO)(PPh_3)(NO)]^{\oplus} \xrightarrow[\text{THF}]{\text{Li[BHEt}_3]} [Re(Cp)(CHO)(PPh_3)(NO)]$$

They very probably act as intermediates in the Fischer–Tropsch process for the conversion of $H_2 + CO$ into hydrocarbons.

A Typical Example: Cobalt Tetracarbonyl Hydride HCo(CO)₄

This was one of the first hydrides to be characterized and is one of the most important in terms of industrial application, since it takes part in the Roelen process for olefin hydroformylation (see Section 5.3). It can be synthesized directly from its constituents:

$$Co \ + \ 4\,CO \ + \ 0.5\,H_2 \ \xrightarrow{\text{250 atm}} \ HCo(CO)_4$$

It is also obtained by hydrogenolysis of the metal–metal bond in cobalt carbonyl:

$$Co_2(CO)_8 \ + \ H_2 \ \xrightarrow{\text{150°C, 50 atm}} \ 2\,HCo(CO)_4$$

This process permits the *in situ* generation of $HCo(CO)_4$ during the hydroformylation of olefins. However, in the laboratory it is better obtained by acidification of the tetracarbonylcobaltate anion:

$$[Co(CO)_4]^{\ominus} + H^{\oplus} \ \longrightarrow \ HCo(CO)_4$$

This anion itself is generally prepared from $Co_2(CO)_8$. Using sodium amalgam as the reducing agent, the reaction does not always go to completion and $Hg[Co(CO)_4]_2$ is formed as a byproduct:

$$Co_2(CO)_8 \ + \ 2 \ Na(Hg) \longrightarrow 2 \ Na[Co(CO)_4]$$

The following methods give purer material:

$$Co_2(CO)_8 \ + \ 2 \ KH \longrightarrow 2 \ K[Co(CO)_4] \ + \ H_2$$

$$Co_2(CO)_8 \xrightarrow{\text{concentrated KOH}} 2 \ K[Co(CO)_4]$$

Here, the concentrated aqueous potash is reacted with a benzene solution of cobalt carbonyl by agitation in the presence of a phase transfer agent (generally a quaternary ammonium salt).

The hydride complex is a very volatile and highly toxic pale yellow liquid: m.p. $-26°C$ and b.p. $+47°C$ (with decomposition). It has a trigonal bipyramidal structure wherein the hydride occupies an axial position (Co$-$H $= 1.56$ Å):

$$
\begin{array}{c}
H \\
| \\
OC_{\cdot\cdot} \\
\diagdown Co-CO \\
OC^{\diagup}| \\
CO
\end{array}
$$

The equatorial carbonyls are slightly displaced towards the hydrogen:

$$C_{ax}-Co-C_{eq} \ = 99.7°$$

The Co$-$H bond is of medium strength, at 57 kcal/mol. Spectroscopically, the hydride is characterized by a proton NMR peak at -10.7 ppm upfield from tetramethylsilane. A weak IR stretch for the Co$-$H bond is seen at 1934 cm^{-1} and the carbonyls appear at 2121, 2062 and 2043 cm^{-1}. For comparison, note that the perfect tetrahedral symmetry in $[Co(CO)_4]^-$ means that the carbonyls appear as a single peak at 1887 cm^{-1}. This displacement to lower frequencies indicates a stronger polarization of the C ... O bond than in HCo(CO)$_4$.

The decomposition of HCo(CO)$_4$ is exothermic and occurs even at ambient temperatures:

$$2 \ HCo(CO)_4 \longrightarrow H_2 \ + \ Co_2(CO)_8$$

$$\Delta H = -27.6 \ \text{kJ/mol}$$

This hydride is only slightly soluble in water (5.6×10^{-2} mol/l) but behaves as a strong acid in MeOH: $pK_a \approx -4$. Nevertheless, measurements using X-ray photoelectron spectroscopy indicate a net charge of -0.75 on the hydrogen. Thus, in spite of its acidic behaviour (H$^+$ and Co(CO)$_4{}^-$) in a polar solvent, it can be considered as a true hydride ($H^{\delta-}-Co^{\delta+}(CO)_4$) under apolar or gas phase conditions. Some of its most typical reactions are summarized below:

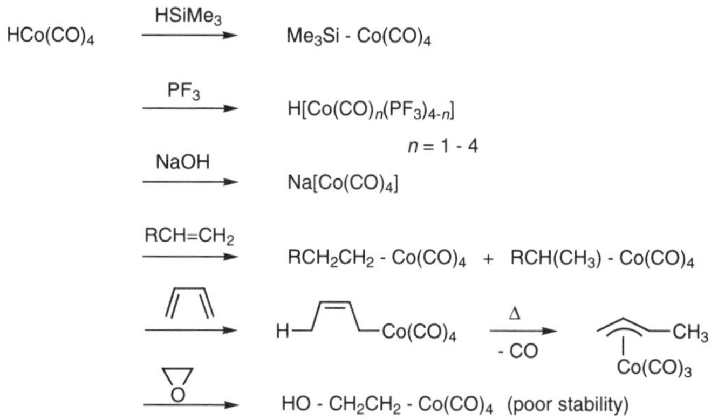

HCo(CO)4
HSiMe3
Me3Si - Co(CO)4
PF3
H[Co(CO)n(PF3)4-n]
n = 1 - 4
NaOH
Na[Co(CO)4]
RCH=CH2
RCH2CH2 - Co(CO)4 + RCH(CH3) - Co(CO)4
H
Co(CO)4
Δ
- CO
CH3
Co(CO)3
O
HO - CH2CH2 - Co(CO)4 (poor stability)

3.2 The Metal Carbonyls

The metal carbonyls were some of the first organometallic compounds to be prepared and are still amongst the most important in terms of chemical versatility and industrial production tonnage. Thus they are well worthy of attention. A list of stable metal carbonyls is given in Table 3.2 (overpage).

Osmium also forms the stable homoleptic carbonyl clusters $Os_5(CO)_{16}$, $Os_6(CO)_{18}$, $Os_7(CO)_{21}$ and $Os_8(CO)_{23}$, upon pyrolysis of $Os_3(CO)_{12}$.[11] Marginally stable carbonyls such as $Ti(CO)_6$, $Ta(CO)_n$, $Pd(CO)_4$, $Pt(CO)_4$, $Cu(CO)_3$, $Ag(CO)_3$, $Au(CO)_2$, $Al(CO)_2$, can be prepared and studied by trapping gaseous metal atoms in a matrix of carbon monoxide.[12] The last of these complexes is the first binary carbonyl to contain a main-group metal. Finally, the recent discovery of stable homoleptic cations such as $[M(CO)_2]^+$ (M = Ag, Au) and $[M(CO)_4]^{2+}$ (M = Pd, Pt) deserves particular mention.[13] Some details on their preparation will be given below.

The Carbon Monoxide Ligand

The following coordination modes are known for CO:

2e
M−C≡O
M=C=O
terminal
μ2-bridging
semi-bridging
μ3-bridging
4e
M=C=O
M
M=C=O→M
6e
η2(μ3C, μ2O)

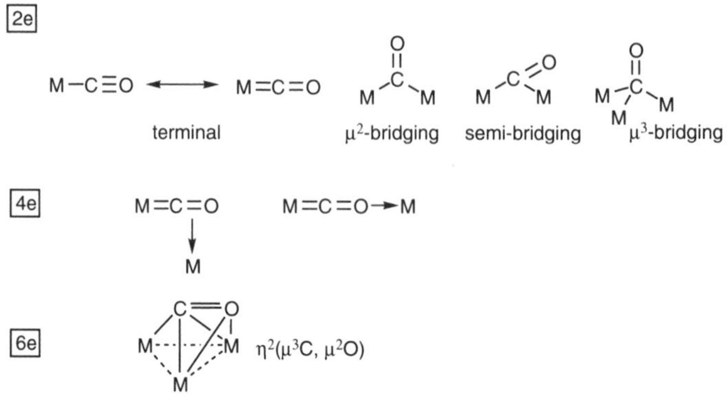

In the vast majority of cases, carbon monoxide plays the role of a two-electron ligand. The four-electron mode was discovered in Australia in 1975[14] and the six-electron mode in Germany in 1981.[15] These higher degrees of coordination are clearly seen by infrared

Table 3.2   Stable binary metal carbonyls

$[Ti(CO)_6]^{2-}$	$V(CO)_6$ $[V(CO)_6]^-$	$Cr(CO)_6$	$Mn_2(CO)_{10}$	$Fe(CO)_5$ $Fe_2(CO)_9$ $Fe_3(CO)_{12}$	$Co_2(CO)_8$ $Co_4(CO)_{12}$ $Co_6(CO)_{16}$	$Ni(CO)_4$
$[Zr(CO)_6]^{2-}$	$[Nb(CO)_6]^-$	$Mo(CO)_6$	$Tc_2(CO)_{10}$ $Tc_3(CO)_{12}$	$Ru(CO)_5$ $Ru_2(CO)_9$ $Ru_3(CO)_{12}$	$Rh_2(CO)_8$ $Rh_4(CO)_{12}$ $Rh_6(CO)_{16}$	
$[Hf(CO)_6]^{2-}$	$[Ta(CO)_6]^-$	$W(CO)_6$	$Re_2(CO)_{10}$	$Os(CO)_5$ $Os_2(CO)_9$ $Os_3(CO)_{12}$	$Ir_2(CO)_8$ $Ir_4(CO)_{12}$ $Ir_6(CO)_{16}$	

spectroscopy, where much reduced IR frequencies $v(CO)$ are observed. A spectacular example is furnished by the value of $v(CO) = 1330$ cm^{-1} for the six-electron carbonyl,[15] which may be compared with the normal 2e carbonyl range of 1850–2100 cm^{-1}. The weakened C...O interaction in the six-electron ligand is also seen in its bond length: $r(C...O) = 1.303$ Å[15] versus 1.10–1.18 Å for a typical 2e carbonyl ligand.

The interaction between CO and a transition metal has already been subjected to a detailed analysis in Section 2.5. The simplified diagram below depicts a classical M—CO bond where the carbon monoxide coordinates through the carbon lone pair as a two-electron ligand:

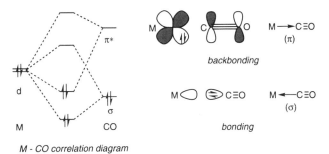

M - CO correlation diagram

The interaction has two components. The first is a ligand-to-metal σ bond which results from transfer of electron density from the carbon lone pair to an empty metal d orbital (OC $\rightarrow$ M σ bond). A second interaction of filled metal d orbitals with the empty π^* orbital of the CO leads to a reverse metal-to-ligand electron transfer which results in a second bond (M $\rightarrow$ CO π backbond). Thus, the CO behaves simultaneously as a σ donor and a π acceptor. The π backbonding causes a filling of the carbonyl π^* level, which induces a weakening of the carbon–oxygen bond. This effect is visible both in the infrared spectrum, where $v(CO)$ passes from 2143 cm^{-1} in free CO to less than 2100 cm^{-1} in a terminal carbonyl, and in the C...O bond length which increases from 1.1282 Å in free CO to about 1.15 Å in a terminal 2e carbonyl ligand. As a curiosity, we should note that $v(CO)$ is situated at 2235.5 cm^{-1} in $[Au(CO)_2]^+$ which implies that coordination of carbon monoxide to gold strengthens the C$\equiv$O bond.[16]

Metal–Carbonyl Synthesis

The most straightforward synthesis of a metal–carbonyl complex employs the direct reaction of CO with a metal, for example:

$$\text{Fe} + 5\,\text{CO} \xrightarrow[\text{200 atm}]{\text{200}^\circ\text{C}} \text{Fe(CO)}_5$$

Nonetheless, the reductive carbonylation of a metal salt in the presence of CO is a much more general method. CO itself may serve as the reducing agent in certain cases. Most of these reactions occur at high CO pressures and are not usually performed in university laboratories for security reasons. This technical problem also partially explains the slow development of metal carbonyl chemistry at the end of the nineteenth century. Typical examples include:

$$\text{Co(O}_2\text{CCH}_3)_2 \xrightarrow[\text{180}^\circ\text{C, 330 atm}]{\text{CO, H}_2} \text{Co}_2\text{(CO)}_8$$

$$\text{MnCl}_2 \xrightarrow[\text{200}^\circ\text{C, 200 atm}]{\text{CO, Na - Ph}_2\text{CO}} \text{Mn}_2\text{(CO)}_{10}$$

$$\text{Re}_2\text{O}_7 \xrightarrow[\text{250}^\circ\text{C, 200 atm}]{\text{CO}} \text{Re}_2\text{(CO)}_{10}$$

$$\text{OsO}_4 \xrightarrow[\text{250}^\circ\text{C, 350 atm}]{\text{CO}} \text{Os(CO)}_5$$

$$\text{RuI}_3 \xrightarrow[\text{175}^\circ\text{C, 250 atm}]{\text{CO, Ag}} \text{Ru(CO)}_5$$

$$\text{WCl}_6 \xrightarrow[\text{100}^\circ\text{C}]{\text{Fe(CO)}_5} \text{W(CO)}_6$$

In addition to neutral species, reductive carbonylation can also be used to prepare stable homoleptic anions or cations:

$$\text{Au(SO}_3\text{F)}_3 \xrightarrow[\text{HSO}_3\text{F}]{\text{CO, 25}^\circ\text{C}} \text{[Au(CO)SO}_3\text{F]} \xrightarrow[\text{SbF}_5]{\text{CO}} \text{[Au(CO)}_2\text{]}^+\,\text{[Sb}_2\text{F}_{11}\text{]}^-$$

$$\text{(ref. 16)}$$

The same strategy allows preparation of the stable cations $[\text{Pd(CO)}_4]^{2+}$ and $[\text{Pt(CO)}_4]^{2+}$.[17]

The traditional synthesis of multinuclear carbonyls consists of the decomposition of a simpler carbonyl by photolysis or thermolysis, for example:

$$\text{Co}_2\text{(CO)}_8 \xrightarrow[\text{70}^\circ\text{C}]{\Delta} \text{Co}_4\text{(CO)}_{12}$$

$$\text{Fe(CO)}_5 \xrightarrow[\text{CH}_3\text{CO}_2\text{H}]{h\nu} \text{Fe}_2\text{(CO)}_9$$

In certain rare cases, it is possible to prepare a metal–carbonyl complex by displacing other ligands. The following is the best laboratory preparation of $\text{Mn}_2\text{(CO)}_{10}$:

$$2 \text{ Cp'Mn(CO)}_3 \ + \ 2 \text{ Na} \ + \ 4 \text{ CO} \xrightarrow[\text{reflux}]{\text{(MeOCH}_2\text{CH}_2)_2\text{O}} \text{Mn}_2\text{(CO)}_{10} \ + \ 2 \text{ Cp'Na}$$

Cp' = ⟨⟩—Me

The Detection and Properties of Metal Carbonyls

The identification of a complex containing a metal–carbonyl bond is usually achieved easily by means of infrared or ^{13}C NMR spectroscopy. The infrared stretching vibrations of the C=O bonds give rise to exceptionally intense absorption bands which are readily detected. Valuable information concerning the coordination modes, the carbon–oxygen force constants and the local symmetry at the metal can be deduced from the position, shape, intensity and multiplicity of these bands. For neutral complexes, terminal CO ligands generally give stretching bands in the range 1900–2100 cm^{-1} whilst bridging COs lie between 1700 and 1850 cm^{-1}. The overall charge of the complex has a significant influence upon the position of the CO stretch, as can be seen in the following series: $[\text{Mn(CO)}_6]^+$ 2090, Cr(CO)_6 2000, $[\text{V(CO)}_6]^-$ 1859 and $[\text{Ti(CO)}_6]^{2-}$ 1750 cm^{-1}. Table 3.3 correlates local symmetry at the metal centre with the theoretical number of carbonyl peaks observed in the infrared spectrum.

Table 3.3 IR-active $\nu(\text{CO})$ modes for some point groups

Complex*	Point group	$\nu(\text{CO})$ modes
Ni(CO)$_2$L$_2$	C$_{2v}$	2 (A$_1$ + B$_1$)
trans-RuX$_2$(CO)$_2$L$_2$	D$_{2h}$	1 (B$_{1u}$)
CpMn(CO)$_3$	C$_{3v}$	2 (A$_1$ + E)
Fe(CO)$_3$L$_2$ (TBP, axial L)	D$_{3h}$	1 (E')
XMn(CO)$_5$	C$_{4v}$	3 (2A$_1$ + E)
trans-Mo(CO)$_4$L$_2$	D$_{4h}$	1 (E$_u$)
Ni(CO)$_4$	T$_d$	1 (T$_2$)
Cr(CO)$_6$	O$_h$	1 (T$_{1u}$)

* For a complete analysis of the IR spectra of M(CO)$_n$L$_{6-n}$, see ref. 18.

A complete theoretical analysis of the IR spectra of metal–carbonyl complexes allows the calculation of the metal–carbon and carbon–oxygen bond force constants.[19] However, the procedure is rather complicated and is generally used only with compounds having reasonably high symmetry. Some results are given below in Table 3.4.

Table 3.4 Force constants for some mono-nuclear metal carbonyls (mdyne Å^{-1})

Carbonyl	k_{CO}	k_{MC}
Cr(CO)$_6$	17.2	2.08
Mo(CO)$_6$	17.3	1.96
W(CO)$_6$	17.2	2.36
Ni(CO)$_4$	17.9	2.02
CO	19.8	

The M–C bond strength, which varies significantly from metal to metal, has a direct influence upon the ease of substitution of the carbonyl ligands. It is generally accepted that the stability of the M–C bond mainly reflects the extent of the π-backbonding interaction.

The second routine method for the detection of M–CO functionalities is ^{13}C NMR spectroscopy,[20] where the carbonyl resonance generally appears between 190 and 230 ppm downfield from tetramethylsilane. It should be noted that the exchange of carbonyl ligands between different sites in a molecule is often fast relative to timescale of the NMR measurement, which means that non-equivalent exchanging carbonyls often give an averaged signal. Therefore room temperature ^{13}C NMR is an unreliable way of defining local symmetries within a metal–carbonyl complex. The CO exchange reactions are slowed down at low temperatures and the appearance of different spectra with changing temperature is a good indication that fluxional processes are operating in solution. It must be added here that the *intrinsic* structures of metal–carbonyl complexes are also often sensitive to temperature, or their physical state. For example, $Co_2(CO)_8$ has two bridging carbonyls in the solid phase, but only terminal carbonyls in solution. These structural interconversions are possible because the energy differences between the isomeric forms are very small. We illustrate this point with the structures of the more common metal–carbonyl complexes.

M = Mn, Tc, Re

solution: $\nu(CO)$ 2069, 2055, 2032 cm^{-1}

solid state: $\nu(CO)$ 2071, 2044, 2042, 1866, 1857 cm^{-1}

bridging COs

M = Ru, Os

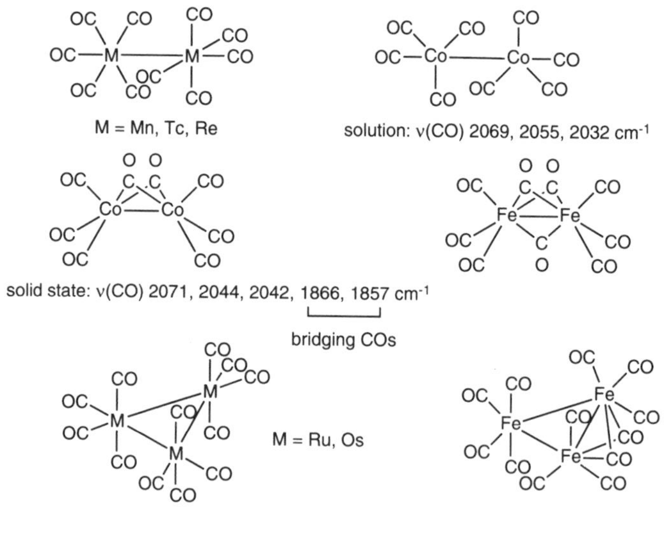

Two Representative Examples

Vanadium Hexacarbonyl V(CO)$_6$

This is the only simple binary metal carbonyl which violates the 18-electron rule: the paramagnetic vanadium centre has only 17 electrons. The preparation is given below:[21]

$$VCl_3 \xrightarrow[\text{[cot]}]{Na\,/\,THF\,/\,CO} [V(CO)_6]^- \xrightarrow{H^+} [HV(CO)_6] \xrightarrow{\Delta} V(CO)_6$$

The active reducing agent is probably the $[cot]^{2-}$ dianion. The vanadium hexacarbonyl hydride intermediate is unstable and spontaneously undergoes homolysis of the H−V bond. The product, the vanadium carbonyl radical, does not dimerize for steric reasons and is obtained in the form of sublimable blue-green pyrophoric crystals. They show a magnetic moment of 1.81 BM (Bohr Magneton) when dissolved in benzene at 20°C, and melt with decomposition at 70°C. The single infrared carbonyl stretch at 1973 cm^{-1} (CH$_2$Cl$_2$ room temperature) indicates that the molecule has perfect O$_h$ symmetry, but the blue colouration of the solid monomer reflects a charge transfer complex formed between neighbouring V(CO)$_6$ molecules. In this regard, it should be noted that vanadium vapour reacts with CO to give a dimer which is stable at low temperature (20 K).

Vanadium carbonyl undergoes dismutation with Lewis bases:

$$3 \text{ V}^0 \longrightarrow \text{V}^{2+} + 2 \text{ V}^-$$

examples include:

$$3 \text{ V(CO)}_6 + 6 \text{ L} \longrightarrow [VL_6]^{2+}, 2 [V(CO)_6]^-$$
$$L = \text{aniline, MeCN, PhCN...}$$

$$3 \text{ V(CO)}_6 + 4 \text{ L} \longrightarrow [VL_4]^{2+}, 2 [V(CO)_6]^-$$
$$L = \text{Ph}_3\text{PO, DMSO...}$$

However, simple substitutions with reagents such as phosphines are possible in non-ionizing solvents; disubstituted derivatives are readily obtained:

$$2 \text{ R}_3\text{P} + \text{V(CO)}_6 \longrightarrow [V(CO)_4(PR_3)_2] \quad R = \text{Et, Pr, Ph...}$$

The use of hexane, or a similar solvent without donor power, is necessary to avoid dismutation. The *trans* structure of the products is easily deduced by infrared spectroscopy, where a single CO peak, reflecting D$_{4h}$ symmetry, is observed. Dimeric compounds are obtained with diphosphines, for example:

X-ray analysis of the molecule indicates a vanadium–vanadium interaction (2.733 Å) which is noticeably shorter than the sum of the covalent radii. The presence of this double bond can easily be predicted from the 18-electron rule. Finally, the reaction with arenes leads to η^6-complexes:

These red, diamagnetic complexes obey the 18-electron rule. Before closing this section we should note that $[Nb(CO)_6]^-$ and $[Ta(CO)_6]^-$, the heavier analogues of the hexacarbonylvanadate anion, have also been prepared.[22]

Iron Pentacarbonyl Fe(CO)₅

This, the most important industrial metal carbonyl, was discovered by Mond and Berthelot in 1891. It is prepared directly from iron and carbon monoxide and takes the form of a pale yellow liquid which is highly toxic, albeit less so than $Ni(CO)_4$. Its physical characteristics include m.p. $-20°C$, b.p. $+103°C$. The IR spectrum shows two stretches at 2022 and 2000 cm^{-1} (C_6H_{12}). The molecular structure is trigonal bipyramidal (TBP), with $Fe-C_{axial} = 1.807$ Å and $Fe-C_{eq} = 1.827$ Å according to a gas phase electron diffraction study. This tbp structure is not stereochemically rigid and the axial and equatorial carbonyls exchange rapidly, even at low temperatures. A single averaged signal is observed by ^{13}C NMR spectra at room temperature, and it is necessary to cool the compound to $-38°C$ *in the solid phase*, before two discrete signals at $+216.0$ (axial CO) and 208.1 ppm (equatorial CO, Me_4Si reference) appear.[23] Berry[24] has proposed a 'pseudorotation' mechanism for these exchanges that is nowadays generally accepted: its activation energy is extremely low (≈ 2 kJ/mol).

The mean dissociation energy of the Fe—CO bond is low (27.7 kcal/mol), which leads to high reactivity.

$Fe(CO)_5$ chemistry is very diverse. Photolytically, it reacts slowly under UV or sunlight to give an insoluble golden-yellow precipitate of $Fe_2(CO)_9$. Under thermolysis at 250°C it decomposes to give finely divided pyrophoric iron. It is attacked by mineral acids in ether to give Fe^{2+} salts. Alkaline bases cause its conversion into a hydride:

$$Fe(CO)_5 \ + \ OH^{\ominus} \longrightarrow \ [FeH(CO)_4]^{\ominus} + \ CO_2$$

It is also easily oxidized:

$$Fe(CO)_5 \ + \ 2\,CuCl_2 \longrightarrow \ FeCl_2 \ + \ 2\,CuCl \ + \ 5\,CO$$

$$Fe(CO)_5 \ + \ I_2 \ \xrightarrow{25°C} \ cis\text{-}\,Fe(CO)_4I_2$$

Amines promote dismutation:

$$2\,Fe(CO)_5 \ + \ 6\,NR_3 \longrightarrow \ [Fe(NR_3)_6]^{2+}\,[Fe(CO)_4]^{2-} \ + \ 6\,CO$$

Reduction with sodium amalgam gives Collman's reagent, $Na_2Fe(CO)_4$, whose applications in organic synthesis will be discussed in detail in Section 4.2. Many substitution reactions are also observed, these include:

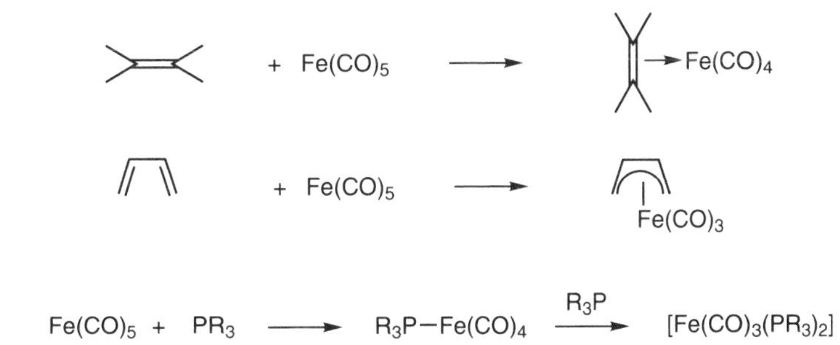

$$Fe(CO)_5 + PR_3 \longrightarrow R_3P{-}Fe(CO)_4 \xrightarrow{R_3P} [Fe(CO)_3(PR_3)_2]$$

The iron tricarbonylbis(phosphine) complex has a *trans* structure, according to an IR spectrum which shows only one carbonyl band (D_{3h} symmetry). If a chelating diphosphine is used, the $P_2Fe(CO)_3$ complex is compelled to adopt a *cis* geometry, which is revealed by three infrared CO stretches. In many cases, a combination of oxidation and substitution reactions is observed, for example:

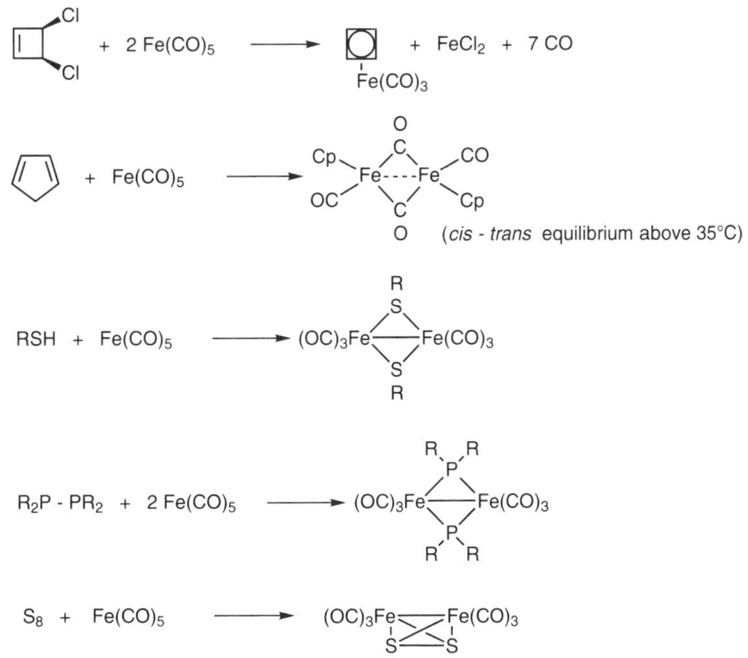

The chemistry of the iron–sulfur cluster is very rich: the S$-$S bond can be selectively cleaved by lithium compounds, hydrides, sodium, etc.[25] The chemistry of iron–sulfur clusters is interesting because they play a fundamental role in biological nitrogen fixation by plants, where Fe_4S_4 groups form an integral part of nitrogenase.

We conclude this highly subjective study of the reactions of $Fe(CO)_5$ by describing its interactions with alkynes. The reaction product depends enormously upon the alkyne substitution pattern, thus:

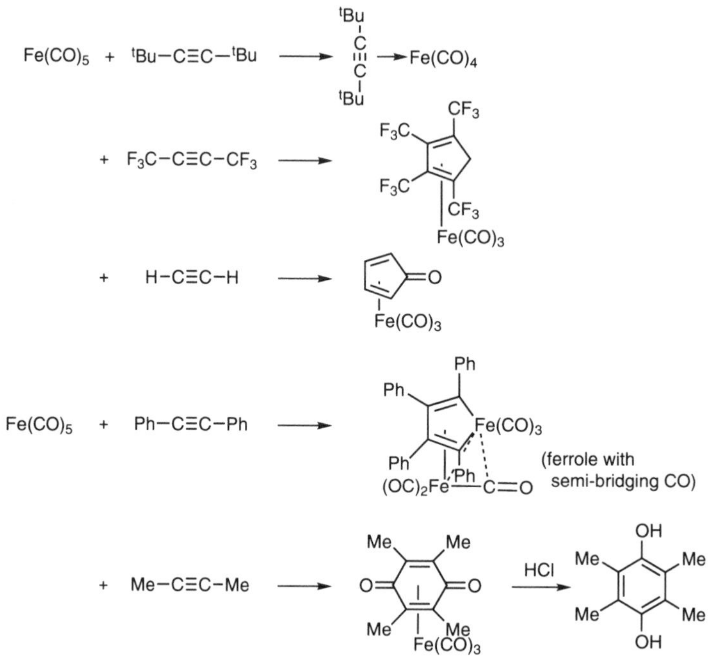

This last reaction forms the basis of the Reppe process for the conversion of acetylenes into hydroquinones in alkaline $Fe(CO)_5$ solution.

The low cost and ready availability of $Fe(CO)_5$ make it a valuable reagent in stoichiometric organic synthesis, where it is often used to effect reduction or carbonylation. A few examples below serve to illustrate its potential:

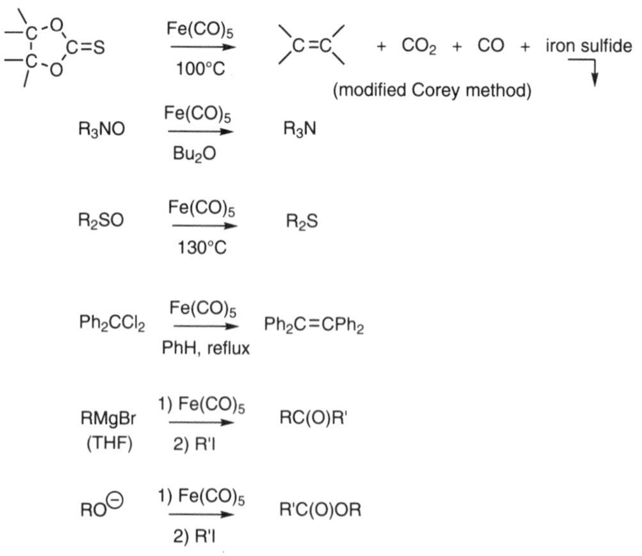

In spite of the wide range of reactions given above, this chapter still gives only a tiny glimpse of the rich and complex chemistry of $Fe(CO)_5$.

3.3 The σ Metal–Carbon Bond

By definition, an organometallic compound contains at least one metal–carbon bond. The parameters which govern the reactivity and thermal or kinetic stability of this bond are the major theme of this book.

Carbon Ligands

The importance of organometallic chemistry stems from the number and diversity of bonding modes which may exist between a transition metal and carbon. These modes may interconvert, thus conferring unique catalytic properties upon many transition metal centres. We briefly review the σ coordination modes here; the π modes will be covered in Section 3.6.

- Alkyl groups: $-CR_3$

$M-CR_3$ (terminal) $M \underset{C R_3}{\overset{}{\diagup \diagdown}} M$ (μ^2-bridge)

The alkyl bridging mode involves a three-centre three-electron bonding scheme which resembles the well known examples in boron hydride chemistry. An aryl group can coordinate similarly, in terminal or bridging mode, in addition to undergoing η^6 complexation through the ring. The bridging mode is frequently found with copper, scandium, yttrium and the lanthanides. Amongst the main group metals, it is most often observed with lithium, beryllium, magnesium and aluminium.

- Alkylidenes : CR_2

$M=CR_2$ (terminal) $M \underset{CR_2}{\overset{}{\diagup \diagdown}} M$ (μ^2-bridge)

At least formally, the terminal mode implies the existence of a metal–carbon double bond. We will see in Section 3.4 that a number of species of very different behaviour are grouped under this general heading.

- Alkylidynes : CR

$M\equiv CR$ (terminal) $M \underset{CR}{\overset{}{\diagup \diagdown}} M$ (μ^2-bridge) $M \overset{CR}{\underset{M}{\diagup | \diagdown}} M$ (μ^3-bridge)

As in the above case, the terminal mode implies a formal triple bond. In the μ^2 mode, an electron pair is delocalized between the carbon and two metal centres. The μ^3 mode allows the formation of clusters.

Synthesis of the Metal–Carbon σ Bond

Route 1

The simplest and most traditional synthetic method consists of reacting a nucleophile R^- with a metallic halide:

$$R^{\ominus} + M-X \longrightarrow R-M + X^{\ominus}$$

The metal counter-ion for the R^- anion can be magnesium, lithium, zinc, aluminium, etc. The following examples serve as illustrations.

$$Cp_2TiCl_2 \xrightarrow{MeLi} Cp_2TiMe_2$$

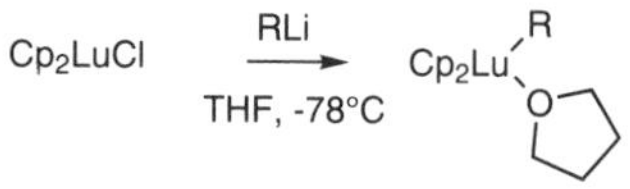

$$Cp_2LuCl \xrightarrow[THF, -78°C]{RLi} Cp_2Lu$$

Organolanthanides having M−C bonds have been discovered recently. Their alkyl groups usually form a bridge between two metal centres, but the monometallic form may be stabilized by the coordination of THF.[26]

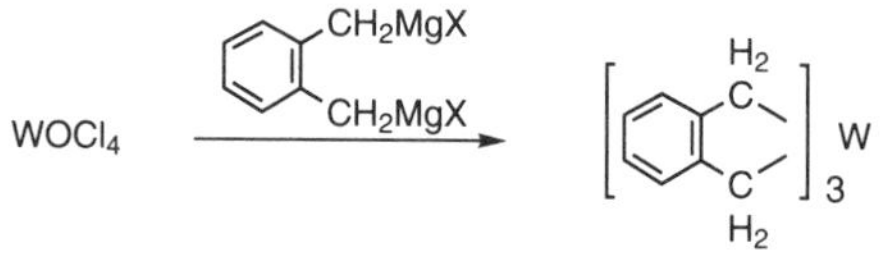

$$WOCl_4 \xrightarrow{} [\quad]_3 W$$

Hexaalkyltungstens are normally unstable; for example, WMe_6 explodes violently. Stabilization by tris-chelation allows these compounds to be handled safely. The displacement of the oxo ligand in the synthesis of such compounds[27] is noteworthy.

$$CpFe(CO)_2I \xrightarrow{HgPh_2} CpFe(CO)_2Ph$$

In the equation above, we see the use of an organomercurial to effect an arylation. It is the preferred reagent here because its weak nucleophilicity means that competing attacks upon the carbonyl ligands are suppressed. Because the nucleophilicity of an alkyl group is lowered upon going from sodium, lithium and magnesium reagents to their aluminium and zinc analogues, these organoaluminiums or organozincs are useful reagents for partial substitution reactions.

$$NbCl_5 \xrightarrow{ZnMe_2} [NbCl_2Me_3]$$

$$TiCl_4 \xrightarrow{Al_2Cl_3Me_3} [TiCl_3Me]$$

Route 2

The attack of a metallic nucleophile upon an organic halide (or any equivalent electrophile), which is merely the reverse of the preceding reaction scheme, is also possible:

$$M^{\ominus} + R-X \longrightarrow R-M + X^{\ominus}$$

The following examples are typical:

$$[Mn(CO)_5]^{\ominus} + MeI \longrightarrow MeMn(CO)_5$$

$$[CpFe(CO)_2]^{\ominus} + [Ph_3S]^{\oplus} \longrightarrow CpFe(CO)_2{-}Ph + Ph_2S$$

This method is much less common because, on the whole, these organometallic anions are more difficult to synthesize than their main-group analogues.

Route 3

A third method involves the insertion of an unsaturated carbon-containing compound into a M−X bond. For an olefin, we have the following general scheme:

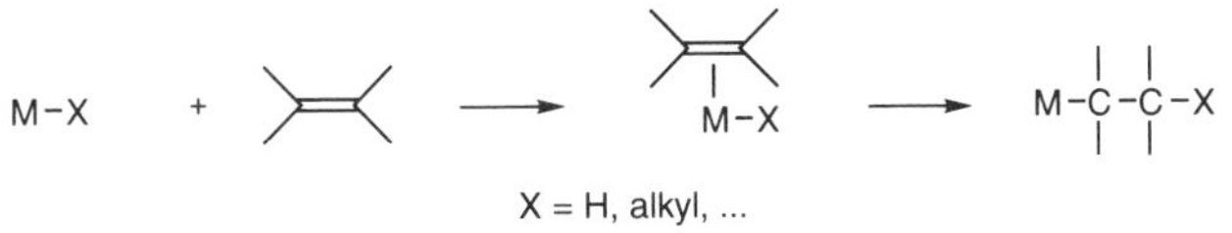

X = H, alkyl, ...

The key step of this insertion is an intramolecular *cis* addition of M−X to the double bond. Usually, this olefin insertion is *reversible* (see the decomposition of metal–alkyl bonds by β-elimination below) and this reversibility plays a crucial role in catalytic olefin hydrogenation, hydroformylation, hydrosilylation, hydrocyanation, etc. by organometallic complexes. For perfluorinated olefins, the product M−R_f bond is very strong and the reaction becomes irreversible. The following examples give an insight into these processes:

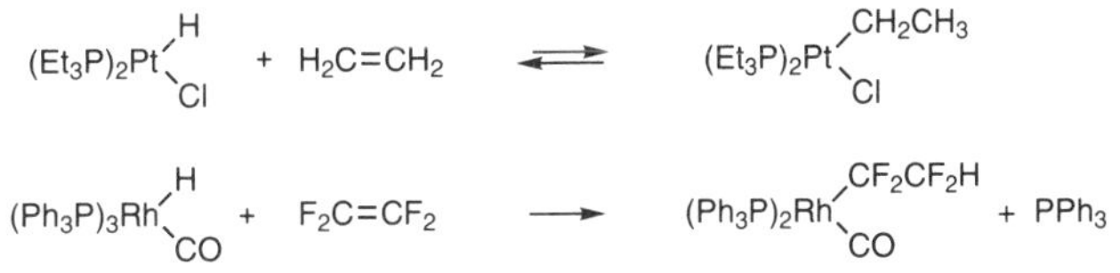

Other unsaturated carbon-containing molecules such as acetylenes and carbenes can be inserted into M−X bonds:

Note that the addition of Zr−H across the alkyne proceeds with a *cis* stereochemistry.

Route 4

The attack of an external nucleophile on a π-bound carbon ligand constitutes a fourth access to M−C bonds. π-coordination of even relatively unreactive olefins to an electron-poor metal centre generally renders them susceptible to nucleophilic attack. If we consider the ligand X in MX as a nucleophile, then this intermolecular process becomes rather similar to the insertion process in route 3 above. It should be noted, however, that the two processes have opposite stereochemistries. Here, for obvious steric reasons, the external nucleophile attacks the face of the π-complexed ligand which is oriented away from the metal, leading to a *trans* addition. The schemes below allow a comparison of the two processes.

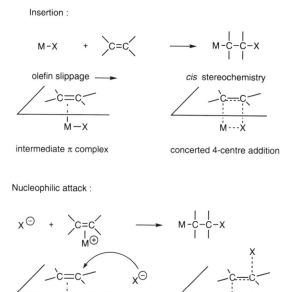

The examples below are representative.

Route 5

M—C bonds can also be created by elimination reactions, as follows:

$$M - U - R \xrightarrow[\text{or } h\nu]{\Delta} M - R + U$$

The following examples give an idea of the potential of this approach:

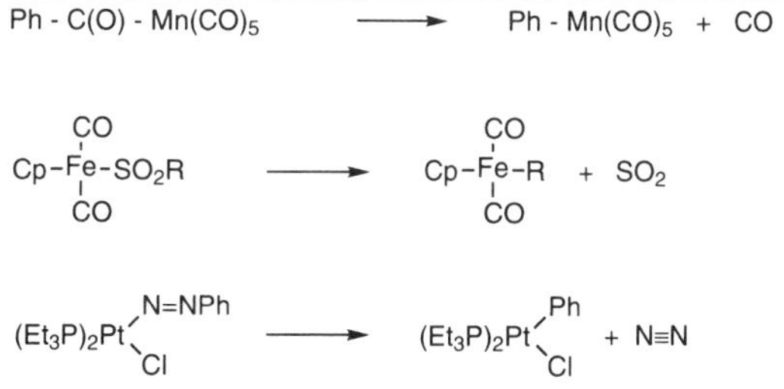

$$Ph\text{-}C(O)\text{-}Mn(CO)_5 \longrightarrow Ph\text{-}Mn(CO)_5 + CO$$

$$Cp\text{-}\underset{CO}{\overset{CO}{Fe}}\text{-}SO_2R \longrightarrow Cp\text{-}\underset{CO}{\overset{CO}{Fe}}\text{-}R + SO_2$$

$$(Et_3P)_2Pt\underset{Cl}{\overset{N=NPh}{\diagup}} \longrightarrow (Et_3P)_2Pt\underset{Cl}{\overset{Ph}{\diagup}} + N\equiv N$$

Route 6

Cyclometallations to create M−C bonds constitute a special class of elimination reaction. The following general scheme:

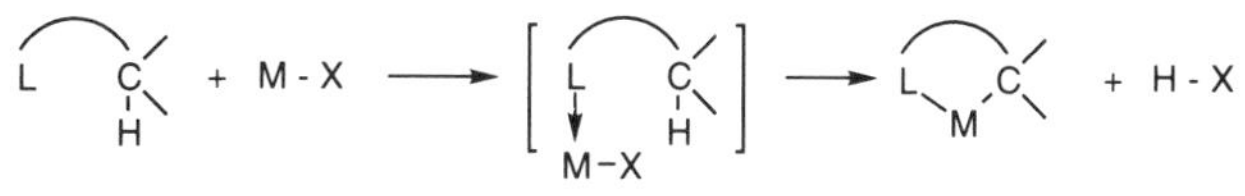

$$L\ \underset{H}{C} + M\text{-}X \longrightarrow \left[L \underset{M-X}{\overset{}{C}} \underset{H}{} \right] \longrightarrow L\text{-}M\text{-}C + H\text{-}X$$

operates when a C−H bond (polarized $C^{-\delta}$−$H^{+\delta}$ or thermodynamically weak) is activated by being brought close to a metal centre. The generation of five-membered heterocycles is favoured for conformational reasons, but three-, four- or six-membered rings are also sometimes generated. Orthometallation of aromatics is observed frequently.

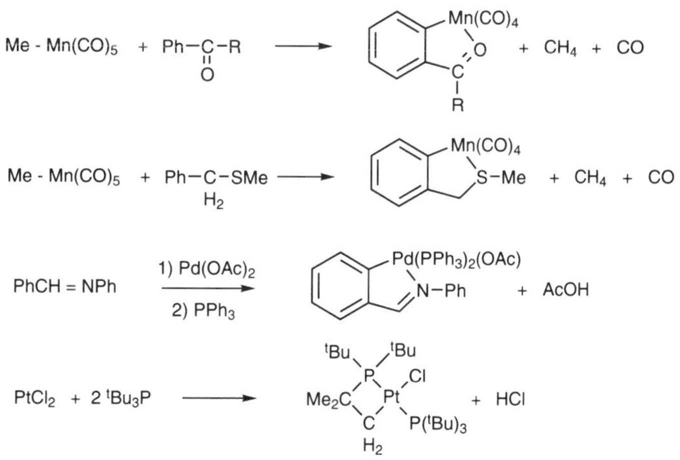

$$Me\text{-}Mn(CO)_5 + Ph\text{-}\underset{O}{\overset{}{C}}\text{-}R \longrightarrow \quad + CH_4 + CO$$

$$Me\text{-}Mn(CO)_5 + Ph\text{-}\underset{H_2}{\overset{}{C}}\text{-}SMe \longrightarrow \quad + CH_4 + CO$$

$$PhCH=NPh \xrightarrow[\text{2) } PPh_3]{\text{1) } Pd(OAc)_2} \quad + AcOH$$

$$PtCl_2 + 2\ {}^tBu_3P \longrightarrow \quad + HCl$$

Route 7

The last important synthesis of M−C bonds involves oxidative addition. In most cases this process is concerted and the first-formed product has a *cis* stereochemistry (see Section 2.3):

$$M + R\text{-}X \longrightarrow M \overset{R}{\underset{X}{\cdots}} \quad (18e)$$
$$(16e)$$

The following examples underline the generality of this method:

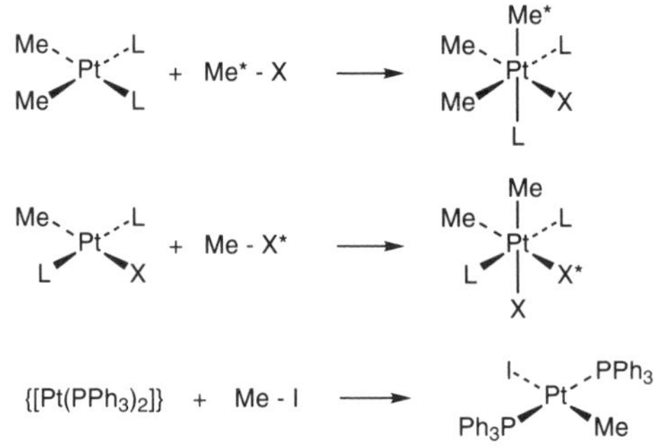

In the last case, the product of *trans* stereochemistry probably results from a reaction which proceeds through an SN_2 mechanism (see Section 2.3).

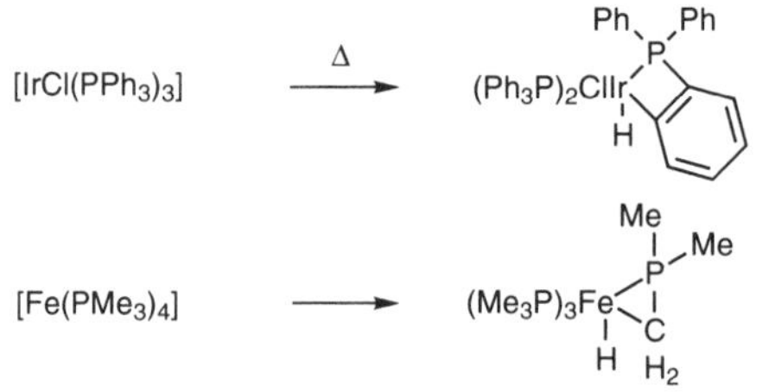

The example showing the insertion of iron into a C−H bond of an alkyl group is an illustration of one of the most important objectives in organometallic chemistry. If an intermolecular version of this process were mastered, the catalytic functionalization of hydrocarbons[29] could be envisaged. The lanthanides and actinides seem to offer one of the best chance of solving this problem, as can be seen below:[30]

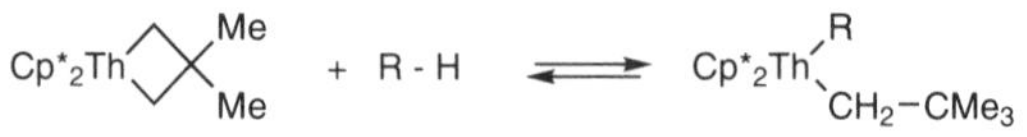

Detection of the Metal–Carbon σ Bond

IR is almost useless for the routine detection of the M−C bond. ^{13}C NMR is useful for diamagnetic complexes, particularly when the metal nucleus has a spin which allows the characteristic $^1J(C-M)$ coupling to be observed. This is the case for the following metals:

Spin 1/2	Isotopic abundance
^{103}Rh	100
^{183}W	14
^{187}Os	1.6
^{195}Pt	34
^{199}Hg	17

^{2}J($\underline{H}$–C–$\underline{M}$) couplings can also be detected for these metals in their proton NMR spectra. As usual, X-ray diffraction studies represent the most foolproof method of characterization.

Properties of the σ M–C Bond

The metal–carbon bond is moderately strong and dissociates somewhere between 30 and 65 kcal/mol. For example, in the manganese series:

Complex	Bond	Dissociation energy (kcal/mol)
PhCH$_2$-Mn(CO)$_5$	Mn-CH$_2$	29
CH$_3$-Mn(CO)$_5$	Mn-CH$_3$	44
CF$_3$-Mn(CO)$_5$	Mn-CF$_3$	49
C$_6$H$_5$-Mn(CO)$_5$	Mn-C$_6$H$_5$	49
H-Mn(CO)$_5$	Mn-H	59

The values come from reference 31 with D(Mn-Mn) = 38 kcal/mol.

The observed trends are quite general and transfer reasonably well to other metals. Thus perfluorination strengthens metal–alkyl bonds, metal–aryl bonds tend to be stronger than the metal–alkyls, and metal–carbon σ linkages are weaker than metal–hydride bonds. These thermodynamic values partially explain why transition metals activate the H–H bond more easily than the C–H bond: the H–H and C–H enthalpies are similar, so the activation depends significantly upon the relative strength of the M–C and M–H bonds.

The instability of the M–C bond which is observed experimentally results more from kinetic than thermodynamic factors. The most important decomposition mode involves β-elimination of an olefin (or a related unsaturated ligand) to give a metal hydride:

This β-elimination operates for alkyl, vinyl (and sometimes aryl) groups because the M–H bond is generally slightly more stable than M–C. In the examples overpage the unsaturated ligand produced in the elimination step remains in the zirconium coordination sphere, whilst the intermediate Zr–H bond reacts further and is lost as RH.

These examples merit some discussion. In two of the three cases (benzyne, cyclohexyne) the coordinated ligands generated by β-elimination are unstable in the free state. This is a beautiful illustration of the capacity of transition metals to stabilize highly reactive species. In the last case, the β-elimination at the sp^2 rather than the sp^3 CH bond probably indicates that the alkyne complex is more stable than the allene complex which could result from β-elimination at the sp^3 carbon. It should also be noted that the chemical properties and strong backbonding in these zirconacycles mean that they are better formulated as three-membered [C₂Zr] rings than π complexes (see Section 2.2).

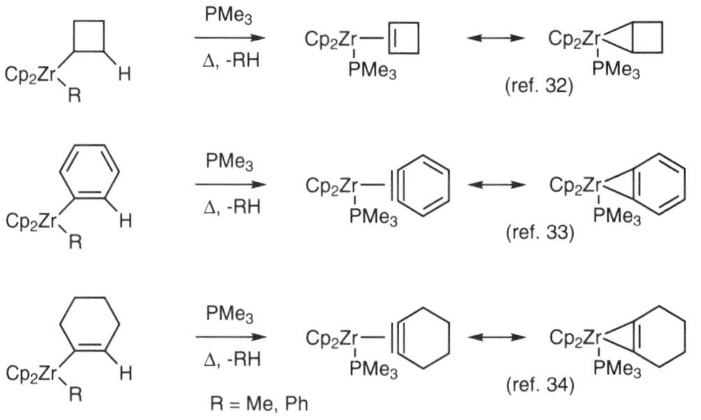

A number of strategies may be used to inhibit the β-elimination process. It is possible to:

(a) Remove the β-hydrogens, as in the case of groups such as CH_3, CH_2Ph, CH_2CMe_3, CH_2SiMe_3, CH_2CF_3, CH_2PR_2.

(b) Use precursors which cannot give stable olefins. According to Bredt's rule, it is impossible to form an olefin at a bridgehead carbon because the planarity necessary for π overlap between olefinic carbons cannot be achieved. So metal-norbornyl complexes, for example, are stable:

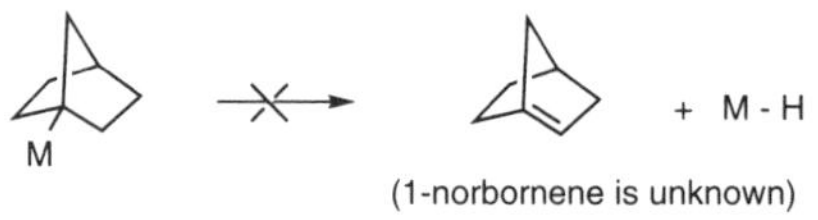

(1-norbornene is unknown)

(c) Prevent the approach of the metal to the β-hydrogen atoms, either by steric hindrance or careful choice of alkyl group geometry:

$Cr(CHMe_2)_4$ $L_2(H)Pt - C \equiv C - H$

(Pt - C≡C - H linear)

(d) Ensure that no vacant sites can become available at the metal centre. The intermediate in the β-elimination process involves a π-coordinated olefin, which means that β-elimination cannot occur if the metal is saturated by strongly bound ligands. Thus:

$CpFe(CO)_2Et$ $CpMo(CO)_3Et$

(18e)   (18e)

Even when β-elimination appears favourable, there are fascinating cases where the reaction is blocked *en route*. In these 'agostic' M...H bonds[35] the J(C...H) coupling constant lies between normal values for ^{1}J(C—H) and ^{2}J(C—M—H), which makes the interaction easy to detect. It is mostly found in d^0 complexes:

$$\text{Me}_2\text{P}\cdots\underset{\overset{\displaystyle|}{\text{Cl}}}{\underset{\overset{\displaystyle|}{\text{P}}\ \text{Me}_2}{\text{Ti}}}\cdots\overset{\text{Cl}}{\underset{\text{H}}{\cdots}}\text{H}\cdots\text{CH}_2 \quad \text{agostic hydrogen}$$

Complete C–H bond rupture is assisted by an electron transfer from an occupied metal d orbital into the vacant σ^* of the C–H bond (cf. π backbonding in a η^2 metal–olefin complex). This transfer, which destabilizes the C–H bond by populating its antibonding orbital, obviously cannot take place in a complex having a d^0 configuration. Consequently, rupture does not occur easily at such centres. It should be noted here that the absence of d electrons defavourizes rather than blocks β-elimination completely, as the above examples with Zr(IV) show.

When hydride β-elimination is impossible, at least two other reaction pathways may cause the cleavage of M–C bonds. They involve the reductive elimination of alkyl or aryl radicals and the α-elimination of alkanes:

$$M\overset{R}{\underset{X}{\cdot}} \longrightarrow [M] + R–X$$

$$R–M–C\overset{H}{<} \;\rightleftharpoons\; \left[\,R–\overset{H}{M}=C<\,\right] \;/\; \left[\,\underset{M=\!\!=C–}{\overset{R\cdots H}{}}\,\right] \;\longrightarrow\; R–H + M=C<$$

The following examples illustrate the two processes:

$$(\text{Ph}_3\text{P})\text{AuMe}_3 \xrightarrow{\ \Delta\ } (\text{Ph}_3\text{P})\text{AuMe} + \text{C}_2\text{H}_6$$

$$(\text{R}_3\text{P})_2\text{Pt}\overset{\text{Me}}{\underset{\text{Me}}{\diamond}} \xrightarrow{\ \Delta\ } [\text{Pt}(\text{PR}_3)_2] + \overset{\text{Me}}{\underset{\text{Me}}{\diamond}} \quad (\text{ref. 36})$$

$$\text{WMe}_6 \xrightarrow{\ \Delta\ } \text{CH}_4 + \text{C}_2\text{H}_6 \ (\text{traces}) + \text{polymers}$$

WMe$_6$ is only properly stable at around $-10°$C. Some ethane is produced by reductive-elimination, but the principal decomposition pathway involves the α-elimination of methane. The equilibrium occurring in the first step of α-elimination has been demonstrated in the following case.[37]

As well as M–C bond rupture, the metal carbon σ bond undergoes insertion reactions.

$$\text{M-R} + \text{U} \longrightarrow \text{M-U-R} \qquad \text{U} = \text{CO, CO}_2\text{, O}_2\text{, SO}_2\text{, SO}_3\text{, allene, RNC...}$$

The most important of these is probably the insertion of carbon monoxide to give a metal–acyl complex:[38]

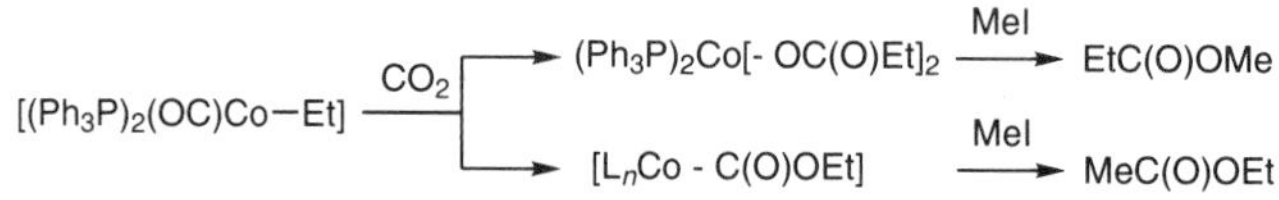

This process is crucial in carbonylation and hydroformylation catalysis (see Section 5.3). The key step is intramolecular, as can be shown clearly with CH_3–$Mn(CO)_5$. The mechanism is as follows:[39]

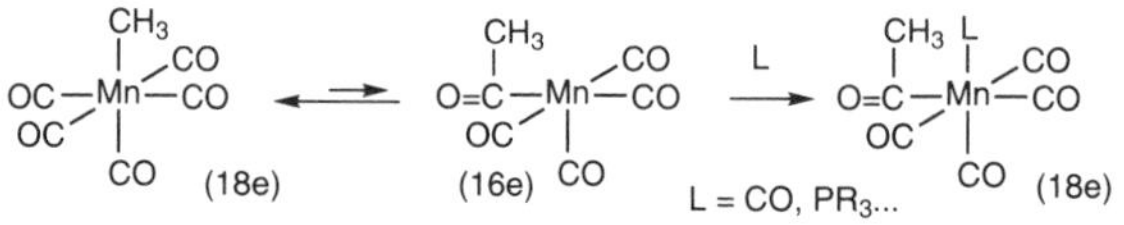

The compound exists in a dynamic equilibrium, where the methyl nucleophile reversibly attacks the electrophilic carbonyl ligands. This *cis*-methyl migration gives an unstable 16-electron intermediate which may be stabilized by coordination of a further ligand, such as external CO. When the CH_3–$Mn(CO)_5$ is allowed to equilibrate under an atmosphere of labelled CO, it is clearly seen that it is a CO *cis* to the methyl which is incorporated into the acyl, rather than the incoming labelled CO. Thus the external CO merely displaces the initial equilibrium, which means that it can be replaced by any good two-electron ligand, such as a phosphine or phosphite. It should also be noted that the insertion generally takes place with retention of carbon stereochemistry at the alkyl group. Also, carbonylation of the very oxophilic metals on the left of the periodic table tends to give η^2-acyls having metal–oxygen coordination rather than η^1-acyls:

$$\text{NaMn(CO)}_5 + \text{MeI} \xrightarrow[\text{(or Me}_2\text{SO}_4)]{} \text{Me - Mn(CO)}_5$$

$$\text{NaMn(CO)}_5 + \text{MeC(O)Cl} \longrightarrow \text{Me - C(O) - Mn(CO)}_5 \xrightarrow[\text{- CO}]{\Delta} \text{Me - Mn(CO)}_5$$

The η^2-acyls may be distinguished from their η^1-analogues by their reduced IR stretching frequencies $\nu(CO)$ and a displacement of the ^{13}C NMR carbonyl signal to very low fields (see below).

Some fascinating cascade insertions of CO have been observed with the lanthanides:[40]

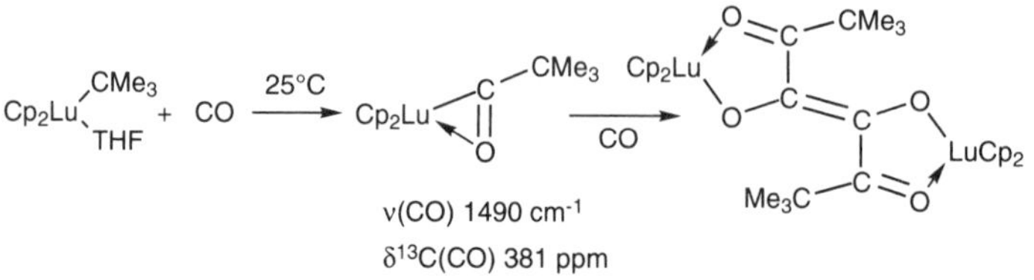

Here we can see the possible beginnings of CO polymerization!

The insertion of small molecules other than CO has been less well studied. Carbon dioxide gives two types of product:[41]

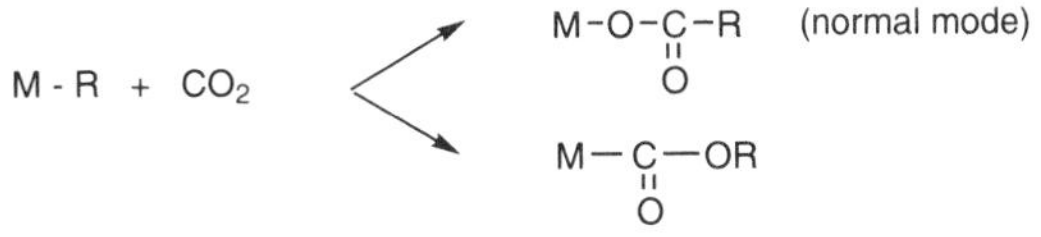

$$M - R + CO_2 \longrightarrow M - O - C - R \quad (normal\ mode)$$

These two modes are seen simultaneously in the following case:

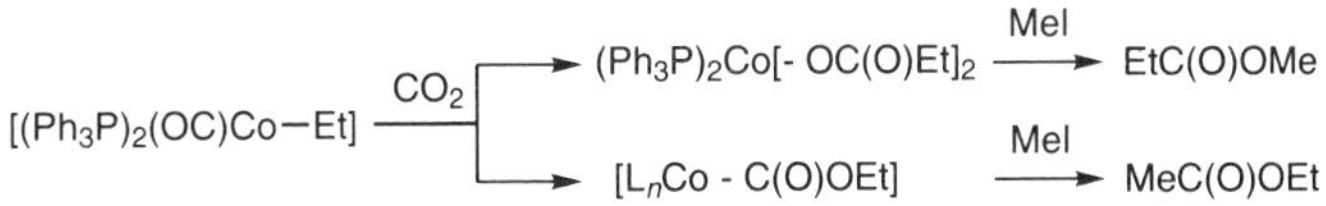

The insertion of SO_2 is an even more complicated process which can lead to a whole series of products.

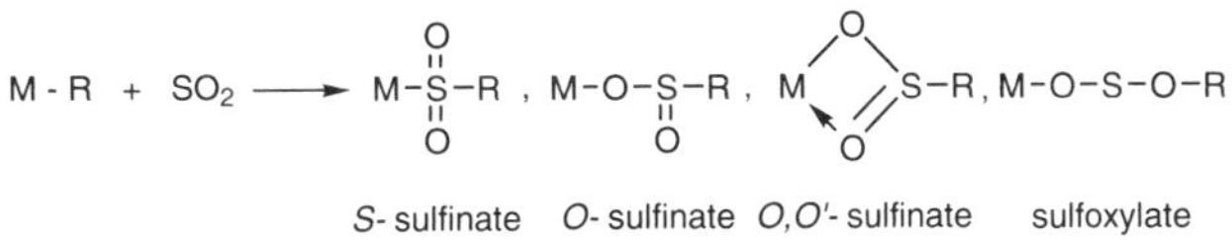

S- sulfinate   O- sulfinate   O,O'- sulfinate   sulfoxylate

We finish with an overview of the principal insertion reactions associated with the three-membered $[C_2Zr]$ rings which we mentioned previously.

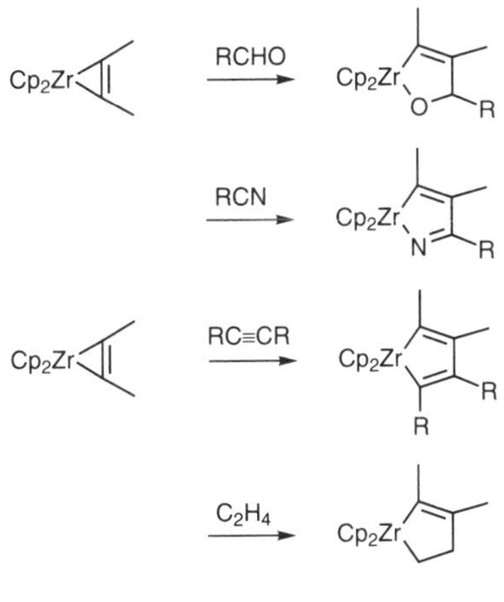

The product metallacycles are valuable sources of organic compounds (especially heterocycles) through protonation, halogenation, or exchange of zirconium for another element. A very rich synthetic chemistry is being developed in this area.[42]

A Typical Example: Methylmanganesepentacarbonyl Me−Mn(CO)₅

This compound was discovered in 1957. Its preparation may be performed as follows:

$$NaMn(CO)_5 + MeI \longrightarrow Me - Mn(CO)_5$$
$$(or\ Me_2SO_4)$$

$$NaMn(CO)_5 + MeC(O)Cl \longrightarrow Me - C(O) - Mn(CO)_5 \xrightarrow[- CO]{\Delta} Me - Mn(CO)_5$$

$NaMn(CO)_5$ is prepared by cleaving the metal–metal bond of $Mn_2(CO)_{10}$ with sodium. Its protonation leads to $HMn(Co)_5$ whose reaction with diazomethane gives the product.

$$HMn(CO)_5 \ + \ CH_2N_2 \longrightarrow Me\text{-}Mn(CO)_5$$

$HMn(CO)_5$ may also be prepared by the protonation of $NaMn(CO)_5$. $Me{-}Mn(CO)_5$ takes the form of colourless sublimable crystals, m.p. 95 °C. Quite stable to air and moisture, it has C_{4v} symmetry, a $Mn{-}CH_3$ distance of 2.185 Å and IR bands at $v(CO) = 2112, 2111$ and $1990 \ cm^{-1}$. Halogens cleave the M−C bond:

$$MeMn(CO)_5 \ + \ X_2 \longrightarrow X\text{-}Mn(CO)_5 \ + \ MeX$$
$$X = Br, I$$

protic acids do likewise:

$$MeMn(CO)_5 \ + \ HX \longrightarrow X\text{-}Mn(CO)_5 \ + \ CH_4$$
$$X = Br, I, CF_3CO_2$$

CO insertion has already been discussed; SO_2 gives an *S*-sulfinate:

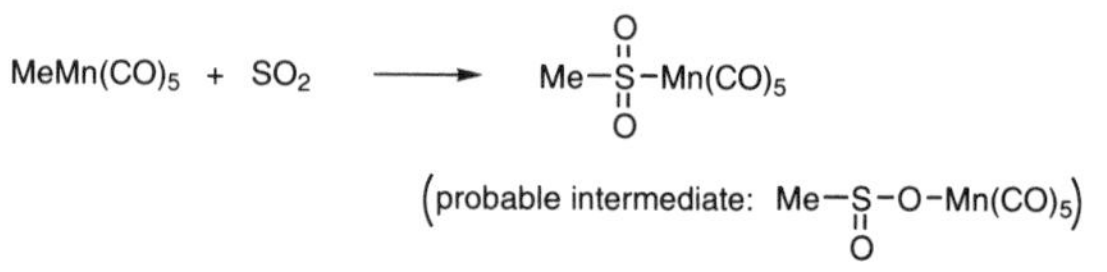

$$MeMn(CO)_5 \ + \ SO_2 \longrightarrow Me\overset{O}{\underset{O}{-\!\!\overset{\|}{\underset{\|}{S}}\!\!-}}Mn(CO)_5$$

$$\left(\text{probable intermediate: } Me{-}\underset{\|}{\overset{}{S}}{-}O{-}Mn(CO)_5\right)$$

Lewis acids such as $AlBr_3$ catalyse the migration of the methyl group to one of the *cis* CO ligands:

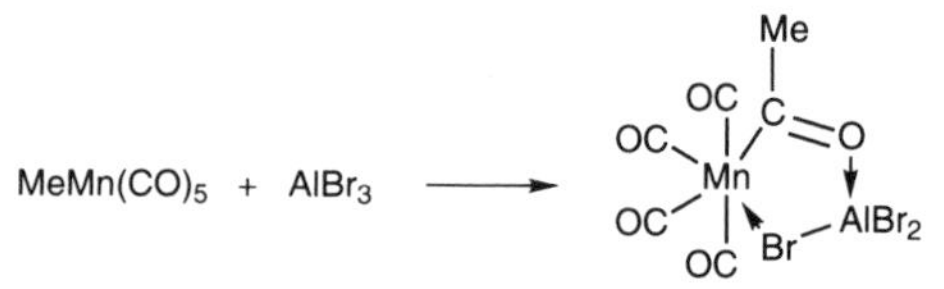

$$MeMn(CO)_5 \ + \ AlBr_3 \longrightarrow$$

(structure established by X-ray[43])

Reaction with CS_2 gives a dithiocarboxylate:

$$MeMn(CO)_5 \ + \ CS_2 \longrightarrow$$

and olefins lead to a wide range of products:

$$MeMn(CO)_5 \ + \ F_2C{=}CF_2 \longrightarrow Me\text{-}CF_2CF_2\text{-}Mn(CO)_5$$

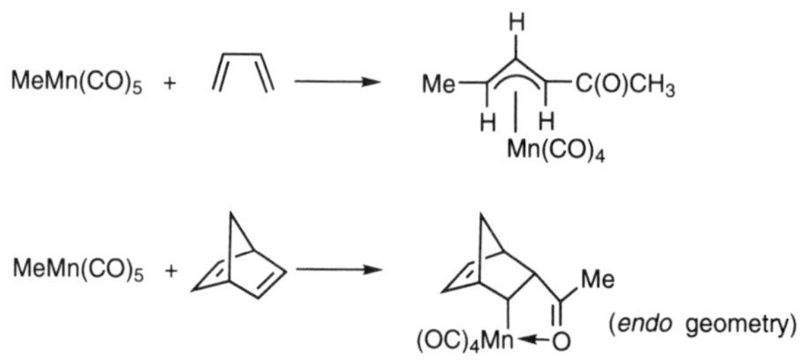

$$MeMn(CO)_5 \ + \ \text{//}\backslash\backslash \longrightarrow$$

$$MeMn(CO)_5 \ + \longrightarrow$$

(*endo* geometry)

We have already mentioned orthometallation and substitution reactions involving two-electron ligands. By way of conclusion, we should note that $Me-Mn(CO)_5$ effects the cleavage of P$-$C bonds in certain cases:[44]

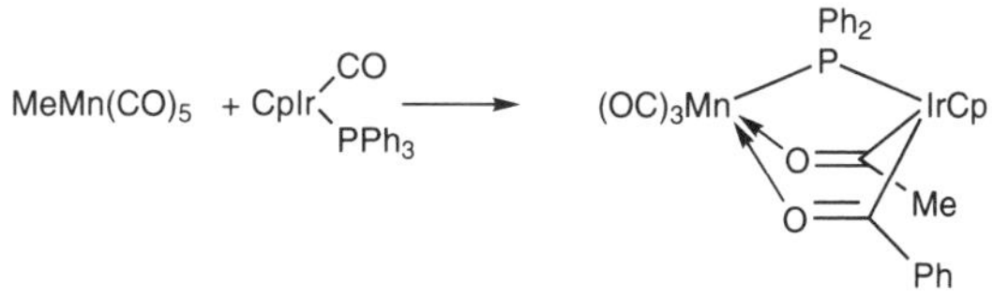

3.4 Carbene Complexes

Fischer's 1964 discovery of carbene complexes was a major breakthrough in organometallic chemistry. It provided the first examples of metal–carbon multiple bonds; their reactivity was expected to be as rich as the carbon–carbon double bond in organic chemistry. This has not proved to be the case, but these carbene complexes are implicated in many crucial processes, such as olefin metathesis and polymerization.

Carbene complexes may usefully be divided into two groups. The Fischer carbenes have heteroatom substituents and show a behaviour typical of electrophiles ($C^{\delta+}$). The Schrock (alkylidene) complexes,[45] which were discovered 10 years later, have carbon substituents which confer nucleophilic properties ($C^{\delta-}$). We will discuss the factors which modulate the polarity of these carbene complexes in more detail below, but suffice it to say here that their complementary reactivity means that the chemistry of carbon–metal double bonds is very rich.

Synthesis of Carbene Complexes

For ease of discussion, we treat the preparation of electrophilic carbene complexes (Fischer) and nucleophilic carbene complexes (Schrock) separately, even if this division is sometimes rather artificial.

Electrophilic Carbene Complexes

The most general synthesis comprises the attack of a nucleophile upon a coordinated carbonyl, or related ligand.

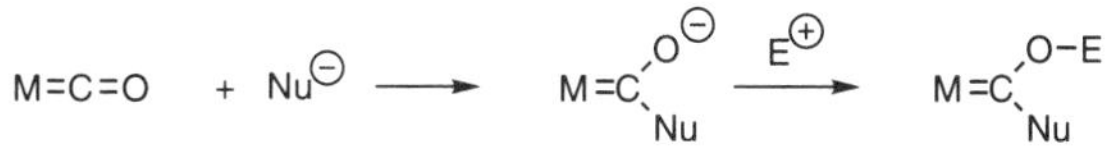

The following examples show the wide variety of metal carbonyls, nucleophiles and electrophiles which can be used to prepare a Fischer carbene complex:

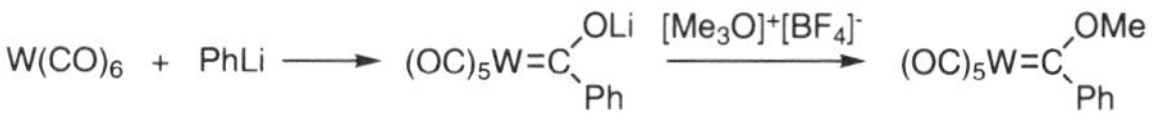

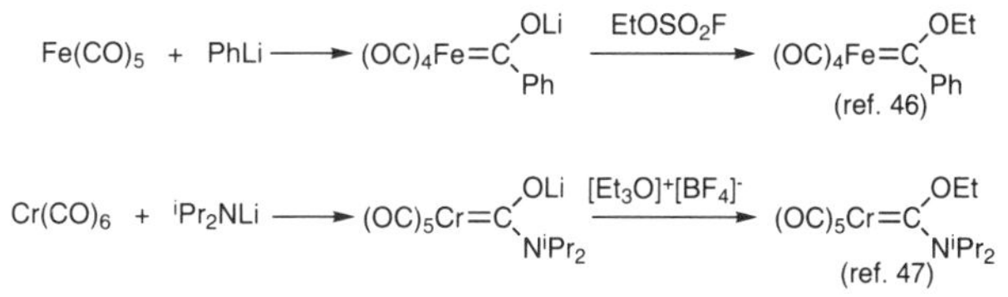

Furthermore, the coordinated carbonyl can be replaced by isoelectronic species such as thiocarbonyls or isonitriles; in these cases the attack is much easier:

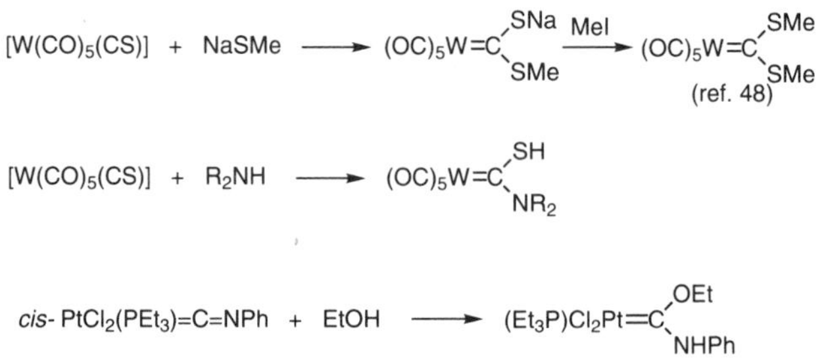

$$[W(CO)_5(CS)] + NaSMe \longrightarrow (OC)_5W=C \underset{SMe}{\overset{SNa}{<}} \xrightarrow{MeI} (OC)_5W=C \underset{SMe}{\overset{SMe}{<}}$$
(ref. 48)

$$[W(CO)_5(CS)] + R_2NH \longrightarrow (OC)_5W=C \underset{NR_2}{\overset{SH}{<}}$$

$$cis\text{-} PtCl_2(PEt_3)=C=NPh + EtOH \longrightarrow (Et_3P)Cl_2Pt=C \underset{NHPh}{\overset{OEt}{<}}$$

The acylate intermediate in the previous synthesis may be viewed as an acyl complex:

$$M=C \underset{Nu}{\overset{O^{\ominus}}{<}} \longleftrightarrow \overset{\ominus}{M}-C \underset{Nu}{\overset{O}{<}}$$

So, it is straightforward to conceive a synthetic path employing the alkylation of an acyl complex at the oxygen atom:

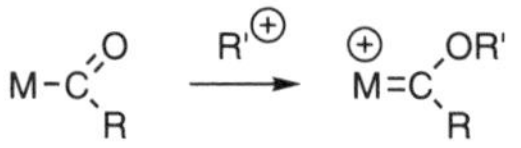

$$M-C \underset{R}{\overset{O}{<}} \xrightarrow{R'^{\oplus}} \overset{\oplus}{M}=C \underset{R}{\overset{OR'}{<}}$$

It is sometimes possible to combine the preparation of the acyl complex with its transformation into the carbene, as in the following method. In this more sophisticated variant, a two-electron ligand (Nu^-) may be used in the first step to promote insertion of CO into the metal–alkyl bond (see Section 3.3).

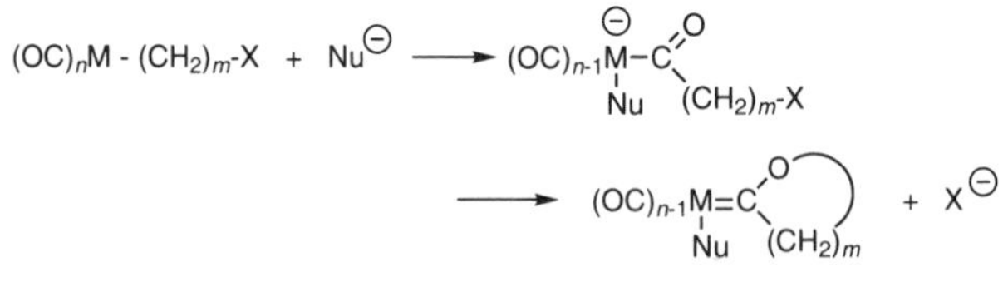

$$(OC)_nM \text{ - } (CH_2)_m\text{-}X + Nu^{\ominus} \longrightarrow (OC)_{n-1}\underset{Nu}{\overset{\ominus}{M}}-C \underset{(CH_2)_m\text{-}X}{\overset{O}{<}}$$

$$\longrightarrow (OC)_{n-1}\underset{Nu}{M}=C \underset{(CH_2)_m}{\overset{O}{<}} \Big) + X^{\ominus}$$

The acylate ion which is obtained then undergoes an entropically favoured intramolecular alkylation, which results in the formation of a ring. A typical example is given below:[49]

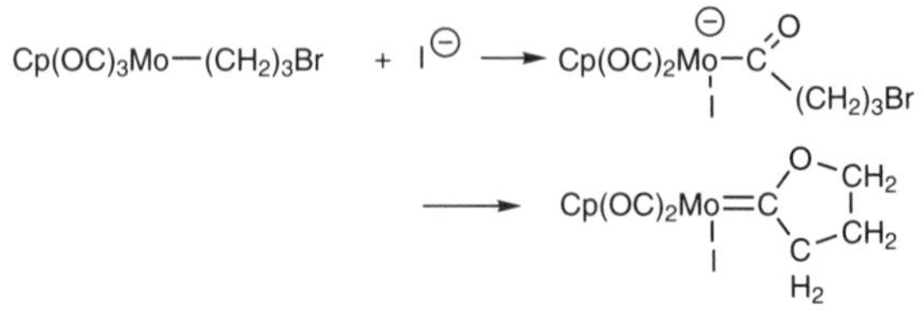

$$Cp(OC)_3Mo-(CH_2)_3Br + I^{\ominus} \longrightarrow Cp(OC)_2\overset{\ominus}{Mo}-C \underset{(CH_2)_3Br}{\overset{O}{<}}$$

$$\longrightarrow Cp(OC)_2Mo=C \underset{\overset{|}{\underset{H_2}{C}}-CH_2}{\overset{O^{\frown}CH_2}{<}}$$

A protonation of another acyl precursor gives a rare example of a stable complexed hydroxycarbene:[50]

$$Cp(NO)(PPh_3)Re-C\underset{Ph}{\overset{O}{\diagdown}} \xrightarrow{\ HOSO_2CF_3\ } \left[Cp(NO)(PPh_3)Re=C\underset{Ph}{\overset{OH}{\diagdown}} \right]^{\oplus}$$

$$\xrightarrow{\ MeOSO_2F\ } \left[Cp(NO)(PPh_3)Re=C\underset{Ph}{\overset{OMe}{\diagdown}} \right]^{\oplus}$$

The electrophilic character of the carbenic carbon in Fischer complexes facilitates an attack by nucleophiles. Consequently, it is easy to convert a heteroatom-substituted (Fischer) carbene complex into an alkylidene (Schrock) compound:

$$M=C\underset{R}{\overset{OR'}{\diagdown}} + Nu^{\ominus} \longrightarrow \left[M-\underset{R}{\overset{OR'}{\underset{|}{C}}}-Nu \right]^{\ominus} \xrightarrow{\ H^{\oplus}\ } M=C\underset{R}{\overset{Nu}{\diagdown}} + R'OH$$

$$Nu = H, R, R_2N, RS$$

The examples below are typical:

$$(OC)_5W=C\underset{Ph}{\overset{OMe}{\diagdown}} \xrightarrow[-78°C]{MeLi} \left[(OC)_5W-\underset{Ph}{\overset{OMe}{\underset{|}{C}}}-Me\right]^{\ominus} \xrightarrow[-78°C]{HCl} (OC)_5W=C\underset{Ph}{\overset{Me}{\diagdown}}$$
(ref. 51)

$$(OC)_5W=C\underset{Ph}{\overset{OMe}{\diagdown}} \xrightarrow[\text{2) } CF_3CO_2H, -78°C]{\text{1) } K[HB(OCHMe_2)_3], -78°C} (OC)_5W=C\underset{Ph}{\overset{H}{\diagdown}}$$
(ref. 52)

These two alkylidene complexes are only stable at low temperature.

The attack of an electrophile upon an acetylide may also give complexed carbenes:

$$M-C\equiv C-R + E^{\oplus} \longrightarrow \left[M-\overset{\oplus}{C}=C\underset{R}{\overset{E}{\diagdown}} \longleftrightarrow M=C=\overset{\oplus}{C}\underset{R}{\overset{E}{\diagdown}} \right] \xrightarrow{AH} M=C\underset{\underset{R}{\overset{|}{C}-H}}{\overset{A}{\diagdown}}\,E$$

example:

$$Cp(OC)(Ph_3P)Fe-C\equiv C-R \xrightarrow[\text{2) MeOH}]{\text{1) } H^{\oplus}} \left[Cp(OC)(Ph_3P)Fe=C\underset{CH_2R}{\overset{OMe}{\diagdown}} \right]^{\oplus}$$
(ref. 53)

The protonation of vinyl complexes follows the same general scheme:

$$Cp(OC)_2Fe-CH=CH_2 \; \underset{\ }{\overset{H^{\oplus}}{\rightleftharpoons}} \; \left[Cp(OC)_2Fe=C\underset{Me}{\overset{H}{\diagdown}} \right]^{\oplus}$$ (ref. 54)

A complexed alkyl functionality having a leaving group α to the metal may be transformed into a carbene complex by abstraction with an electrophile:

$$M-\underset{|}{\overset{|}{C}}-X \xrightarrow{\ E^{\oplus}\ } \left[M=C\diagdown \right]^{\oplus} + EX$$

The two following examples are representative:[55]

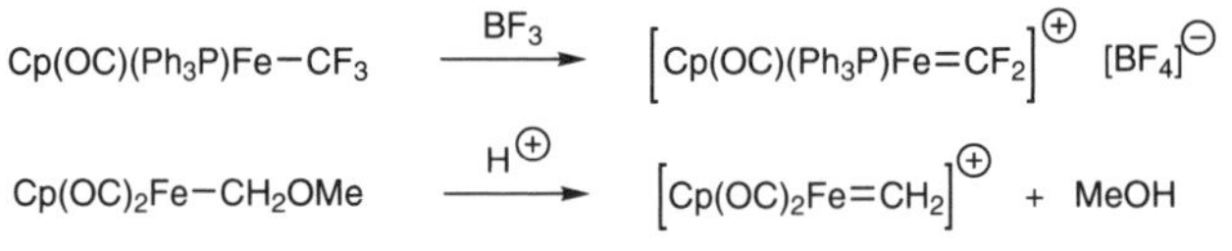

$$Cp(OC)(Ph_3P)Fe-CF_3 \xrightarrow{BF_3} \left[Cp(OC)(Ph_3P)Fe=CF_2\right]^{\oplus} [BF_4]^{\ominus}$$

$$Cp(OC)_2Fe-CH_2OMe \xrightarrow{H^{\oplus}} \left[Cp(OC)_2Fe=CH_2\right]^{\oplus} + MeOH$$

Another interesting application of this method has been developed recently.[56] The synthesis of the initial M—C bond results from the reaction of an organometallic anion with an aldehyde:

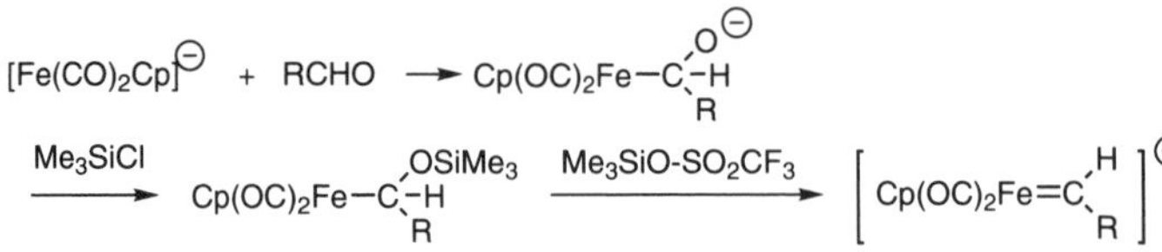

$$[Fe(CO)_2Cp]^{\ominus} + RCHO \longrightarrow Cp(OC)_2Fe-\overset{\overset{\displaystyle O^{\ominus}}{|}}{\underset{\underset{\displaystyle R}{|}}{C}}-H$$

$$\xrightarrow{Me_3SiCl} Cp(OC)_2Fe-\overset{\overset{\displaystyle OSiMe_3}{|}}{\underset{\underset{\displaystyle R}{|}}{C}}-H \xrightarrow{Me_3SiO-SO_2CF_3} \left[Cp(OC)_2Fe=C\overset{H}{\underset{R}{\big\backslash}}\right]^{\oplus}$$

In each of the previous examples, the carbene functionality has been synthesized, stepwise, within the coordination sphere of the metal. However, a preformed free carbene may sometimes be complexed to an unsaturated metal centre. Thus, diazomethanes can serve as precursors of complexed carbenes:

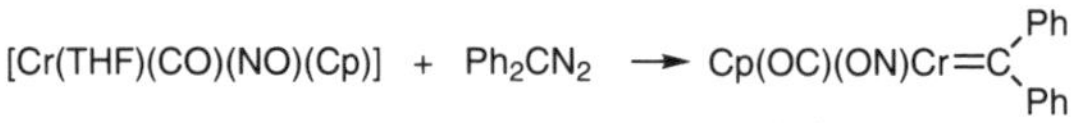

$$[Cr(THF)(CO)(NO)(Cp)] + Ph_2CN_2 \longrightarrow Cp(OC)(ON)Cr=C\overset{Ph}{\underset{Ph}{\big\backslash}}$$

(ref. 57)

This reaction is not particularly general because alkylidene bridges, μ^2-CR_2, are often formed. For example:

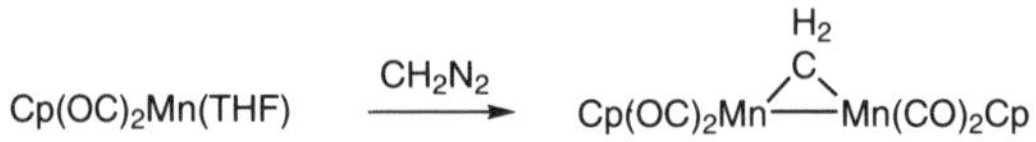

$$Cp(OC)_2Mn(THF) \xrightarrow{CH_2N_2} Cp(OC)_2Mn\overset{\overset{\textstyle H_2}{\overset{\textstyle C}{\diagdown}}}{\underline{\qquad}}Mn(CO)_2Cp$$

Diazoalkane complexes, where nitrogen is retained within the metal coordination sphere, are also frequently isolated.

The recent discovery of stable free carbenes (see later) has allowed the direct synthesis of many carbene complexes which would otherwise be difficult to obtain, thus:

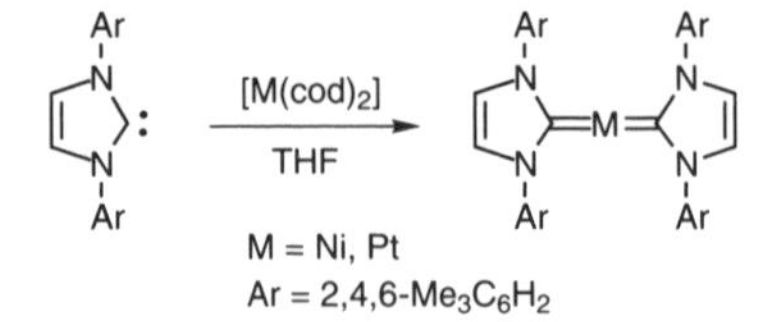

M = Ni, Pt
Ar = 2,4,6-Me$_3$C$_6$H$_2$
(ref. 58)

The stability of these homoleptic 14-electron carbene complexes is remarkable. The metal geometry is linear and the two carbene planes make a dihedral angle of 50°. The first stable rare earth carbene complexes have also been prepared in similar fashion:[59]

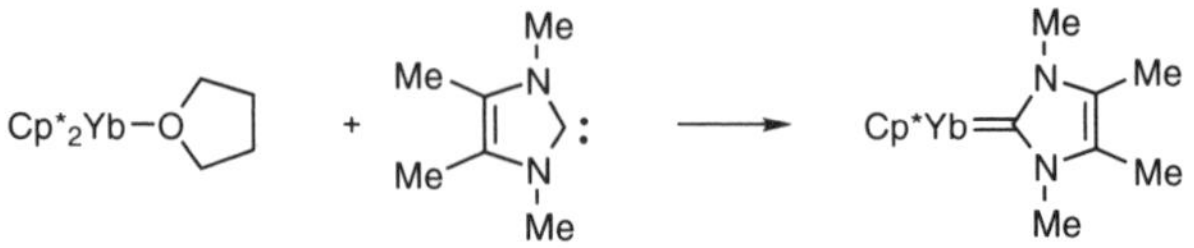

The Yb$-$C bond, at 2.552(4) Å, is rather long, which confirms that these complexes are related to the Fischer class (see later). Using a similar method, Lappert[60] has shown that an electron-rich olefin can be cleaved by an electrophilic organometallic substrate. The reaction scheme is as follows :

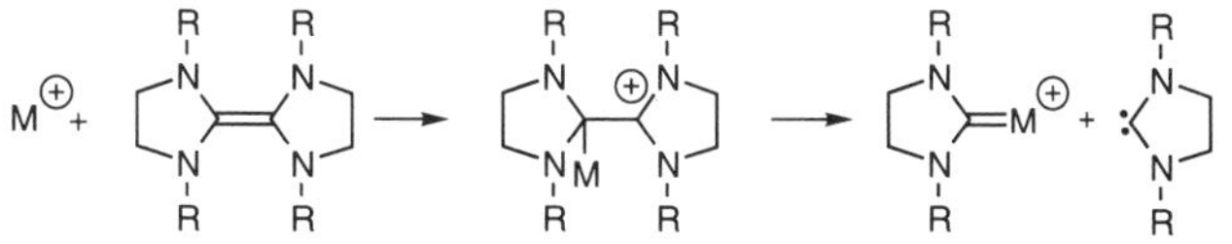

Examples have been described where $M^+ = Mo(CO)_5$ or $Rh(PPh_3)(CO)Cl$.

Finally, the reaction of a *gem*-dihalide with an organometallic dianion may be conceptually simple but, in practice, it has very restricted applications:

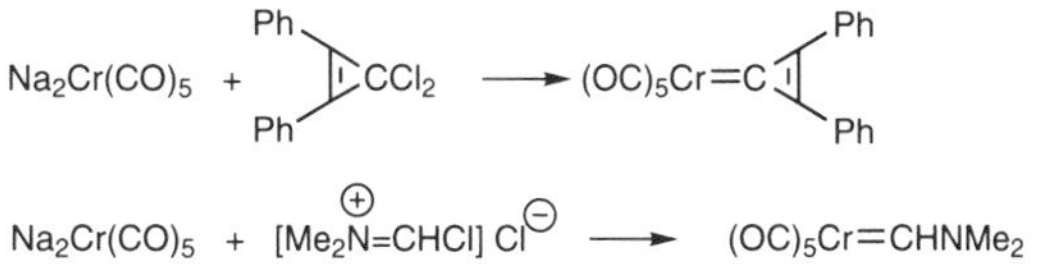

Nonetheless, a related synthesis using an iminium salt can be generalized:

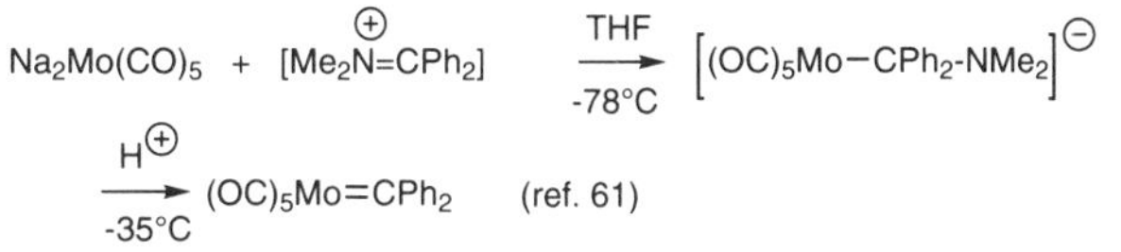

Nucleophilic Carbene Complexes (Alkylidenes)

In general, alkylidene complexes are obtained by α-elimination from a metal dialkyl:

This elimination is favoured by sterically hindered alkyl groups and small, basic phosphine coligands such as PMe_3. The alkyl groups R and CHR_1R_2 must be *cis* in the starting material. This type of elimination only occurs easily with metals in high oxidation states such as Ta(V), W(VI), ... , and involves a first order intramolecular reaction, promoted by an initial interaction between the metal and the α-hydrogen.[62] The following examples are important:

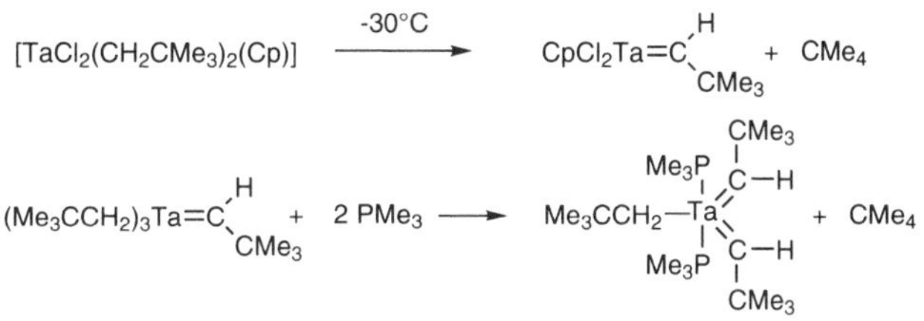

Instead of the thermal elimination of RH, ylids can sometimes be employed to generate such complexes. Overall, the reaction may be regarded as a transylidation, where the phosphorus also acts as a base.

$$[TaCl(CH_2CMe_3)_4] \ + \ Ph_3P{=}CH_2 \ \longrightarrow \ (Me_3CCH_2)_3Ta{=}C\underset{CMe_3}{\overset{H}{<}}$$

$$+ \ [Ph_3P{-}CH_3]^+Cl^-$$

An alkylidene may equally be transferred from one metal to another:

$$(Me_3CO)_4W{=}NPh \ + \ (Me_3P)_2Cl_3Ta{=}CHCMe_3 \ \longrightarrow$$

(with product structure: Cl$_2$W with Me$_3$P, NPh, Me$_3$P, and C–H / CMe$_3$)

The Detection of Carbene Complexes

Without doubt, the most important method for the detection of these complexes is ^{13}C NMR spectroscopy. In general, the carbene carbon resonance appears at extremely low field. It is very sensitive to the type of substituents on the carbon, as is shown in Table 3.5 for complexes of chromium and tungsten.

Table 3.5 Chemical shifts for some carbene carbons in chromium and tungsten complexes

	R^1	R^2	$\delta^{13}\underline{C}$ (ppm)
$(OC)_5Cr{=}C\overset{R^1}{<}_{R^2}$	Me	OMe	362
	Ph	OMe	354
	Me	NH_2	290
$(OC)_5W{=}C\overset{R^1}{<}_{R^2}$	Me	NMe_2	271
	Ph	OMe	324
	Ph	Ph	358

These shifts may be compared with the carbonyl carbon resonances, but $\delta(^{13}\underline{C}O)$ shifts are variable themselves, lying between 217 and 226 ppm in the same chromium complexes. It must be underlined that there is no simple correlation between the positive charge carried by the carbene carbon and these $\delta^{13}C$ values; those requiring proof need only consider the Schrock complexes, which resonate to low field of the carbonyls in spite of a very significant negative charge on the carbene carbon:

$$Cp_2Ta\overset{Me}{<}_{CH_2} \quad \delta(^{13}\underline{C}H_2) = 224 \text{ ppm} \qquad Cp_2Nb\overset{Cl}{<}_{CHCMe_3} \quad \delta(^{13}\underline{C}H) = 299 \text{ ppm}$$

Carbenes: Theoretical and Structural Aspects

The differences between Schrock and Fischer carbenes are difficult to explain using our simplified Hückel orbital approach; it neglects questions of spin, which are fundamental in these compounds. Nonetheless, we will continue to employ the perturbation theory used up to now, which means that our conclusions will be highly qualitative. More than ever, they combine both theoretical and experimental approaches.

The aliphatic carbenes, typified by methylene (CH_2), are very reactive species whose existence is deduced from their reaction products. They cannot be isolated. However, a family of stable non-aliphatic carbenes has recently been characterized by X-rays. Here the carbon is both sterically protected by bulky substituents and electronically stabilized by the lone pairs of the nitrogen atoms, which partly offset its electron deficiency. The structural characteristics of one of these free carbenes are summarized below:[63]

- diamagnetic singlet state

- <NCN angle = 102°

- N-C bond length = 1.37 Å

- $\delta^{13}C$ = 211.4 ppm (Me_4Si)

These parameters fit perfectly with the theoretical data which we will derive. We begin by looking at the well known frontier MOs of the AH_2 group, (A = C, N, O, etc. ...) , presented below:

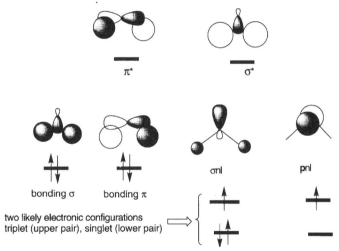

Firstly, we see two bonding orbitals (of σ and π type), at low energy. These are followed by two non-bonding MOs, one of sigma (σn1) and the other of pi symmetry, having similar ionization potentials. As we have six bonding electrons, there are two possible ways of filling the orbitals which are available:

1. $(\sigma \text{ bond})^2 (\pi \text{ bond})^2 (\sigma nl)^1 (\pi nl)^1$, which gives a *triplet* configuration
2. $(\sigma \text{ bond})^2 (\pi \text{ bond})^2 (\sigma nl)^2$, corresponding to a *singlet* configuration.

Calculations show that the ground state configuration is the triplet, which lies approximately 15 kcal/mol below the singlet. The interatomic angle is different for each configuration:

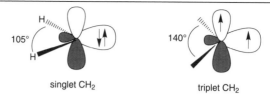

singlet CH$_2$ triplet CH$_2$

Singlet methylene simultaneously shows the characteristics of a base (a lone pair oriented away from the substituents) and a Lewis acid (owing to the empty orbital perpendicular to the atomic plane). This vacant orbital is capable of undergoing conjugation with any π-type MOs localized on the α-substituents (oxygen or nitrogen lone pairs, for example). As we see below, this leads to profound modification of the properties of the carbene.

The Case of a Carbene having a Heteroatomic Substituent in the α Position

As a model, we consider the carbene R−C−OR, whose oxygen atom is oriented to allow an interaction of one of its lone pairs with the vacant orbital on the carbene. This perturbation gives the following:

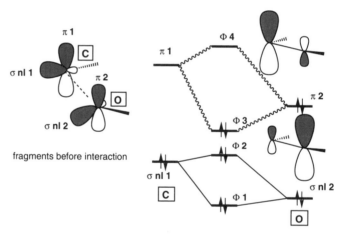

fragments before interaction

fragments after interaction

The σ molecular orbitals (σnl1 and σnl2), mix very little and give rise to Φ1 and Φ2, which are only slightly displaced from their original energies. On the other hand, the π MOs interact strongly and produce a classical combination of bonding and antibonding orbitals, Φ3 and Φ4 respectively. Therefore, the singlet becomes the more stable configuration for the carbene, because the triplet form would require the population of the high-lying antibonding combination Φ4. This is a general rule for heteroatom-substituted carbenes; in HC−OH, for example, the lowest lying triplet state is 17 kcal/mol higher in energy than the singlet.

We have underlined that the electronic state of the carbene depends strongly upon its substituents and is quite difficult to predict with certainty; furthermore, this difficulty is not overcome when we bind our carbene to a metal fragment ML$_n$. We will rationalize the varying results observed for carbene complexes within the framework of the general scheme which we have just discussed. A first, rather interesting, analogy appears when we consider the hypothetical formation of ethylene from two triplet methylene moieties:

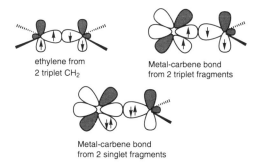

ethylene from
2 triplet CH$_2$

Metal-carbene bond
from 2 triplet fragments

Metal-carbene bond
from 2 singlet fragments

By correctly aligning the spins of the two triplet entities (above), we find an orientation which corresponds to the singlet ethylene molecule. The same scheme applies if a triplet carbene interacts with a triplet metal centre to form a carbene complex. For the complexation of a singlet carbene species, we find a situation which we have already encountered very frequently: a donation of an electron pair from the carbene to the metal, and a corresponding backdonation of metal electron density onto the carbene. We will study these two broad classes of metal–carbene interaction separately, interpreting their structure and reactivity in terms of the concepts above.

Schrock Carbenes

Their bonding may be interpreted as the interaction of a triplet carbene and a triplet ML$_n$ fragment, as for the case of ethylene above. To simplify our diagrams, we treat only the metal orbitals which are involved in bonding to the carbene; these comprise an orbital of σ symmetry, (the d$_{z^2}$), and one of π symmetry, each having one unpaired electron. As in ethylene, the triplet character in each component is lost as the bond is formed. The orbital diagram below gives a *precis* of these interactions and their resulting MOs.

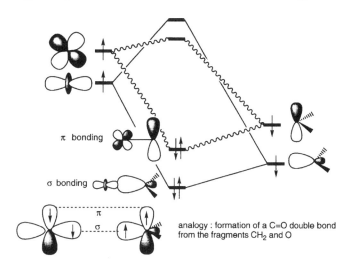

π bonding

σ bonding

π

σ

analogy : formation of a C=O double bond
from the fragments CH$_2$ and O

The most important conclusion to be drawn from this scheme is that we form two MOs, of which the highest, *the HOMO, is localized mainly upon the carbon*. Because the fragments involved have different electronegativities, we can draw an analogy between the orbital interactions in the product with those in formaldehyde, comprising the triplet fragments CH_2 and O. Note, though, that the bond here is polarized in the opposite sense, $M^{\delta+}CH_2^{\delta-}$; consequently the *LUMO is localized on the metal*.

Fischer Carbenes

Their description, from the HC$-$OH entity whose orbitals were described above, is given in the following diagram:

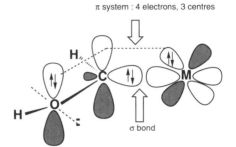

The second oxygen lone pair, which does not interact with the M$-$C bond, is represented by two bold dots for simplicity. The frontier orbitals comprise two MOs of σ symmetry (one on carbon, the other on the metal containing a total of two electrons), and three MOs of π symmetry (one on carbon, one on oxygen, and one on the metal, containing a total of four electrons). In the general perturbation diagram, we see the following:

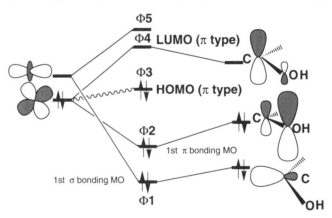

As above, the σ type MOs give a pattern typical of a classical single bond. However, the π system is comprised of three MOs in an allyl-like arrangement: one bonding ($\Phi2$), one non-bonding ($\Phi3$) and one antibonding ($\Phi4$). The antibonding *LUMO of the carbenic system is localized on the carbon, whilst the HOMO resides mainly on the metal*, which is precisely the reverse of the situation in Schrock carbenes. The two Fischer carbene frontier orbitals have the following form:

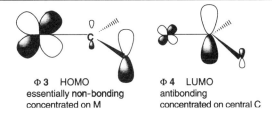

$\Phi 3$ HOMO
essentially **non-bonding**
concentrated on M

$\Phi 4$ LUMO
antibonding
concentrated on central C

The electronic properties of each type are summarized in the following table:

Schrock carbenes	Fischer carbenes
analogy with CH_2O	analogy with the allyl anion
2 electrons in the π system	4 electrons in the π system
HOMO centred on C	HOMO centred on M
LUMO centred on M	LUMO centred on C
nucleophilic carbon	electrophilic carbon

Again, these qualitative results are in good agreement with the energies obtained from more sophisticated calculations in ref. 64:

$$(OC)_5Mo{=}CHOH \quad \text{overall bond energy} \quad Mo{=}C = 60 \text{ kcal/mol}$$

$$CpCl_2Nb{=}CH_2 \qquad\qquad\qquad\qquad\quad Nb{=}C = 74 \text{ kcal/mol}$$

(Note that the $C{=}C$ double bond has an energy of 163 kcal/mol).

The double bond character of the metal–carbene interaction is more accentuated in the Schrock case; this is seen clearly from the barrier to rotation about the metal–carbon bond, a useful indicator of the energy of the π component:

$$(OC)_5Cr{=}CHOH \;:\; \text{barrier of 0.41 kcal/mol}^{65}$$

$$CpCl_2Nb{=}CH_2 \;:\; \text{barrier of 14.6 kcal/mol}^{64}$$

This lowered barrier in the Fischer case is because one of the π-bonding orbitals ($\Phi3$) has no significant localization on the carbon. Additionally, the delocalization of the other ($\Phi2$) over three centres results in a weaker overlap between the metal and carbon than in the Schrock case. This may be explained in more familiar terminology as a mesomeric hybrid of the two forms:

Y = OR, NR$_2$, ...

This rationale is confirmed by structural studies, which show multiple bond character in the C—Y interaction (Y = a substituent having a lone pair):

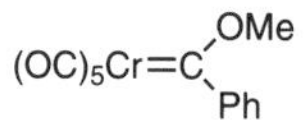

$$(OC)_5Cr=C\begin{smallmatrix}OMe\\Ph\end{smallmatrix}$$

Cr=C 2.04 Å   C-OMe 1.33 Å

Cr-CO *trans* 1.87 Å   O-C sp³ 1.41 Å

Cr-CO *cis* 1.89 Å

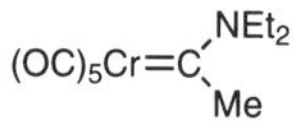

$$(OC)_5Cr=C\begin{smallmatrix}NEt_2\\Me\end{smallmatrix}$$

Cr=C 2.16 Å

C-N   1.31 Å

rotation barrier around C-N *ca* 25 kcal/mol

A typical single Cr—C bond length is approximately 2.21 Å; as expected, the percentage shortening is much greater in the Schrock case below.

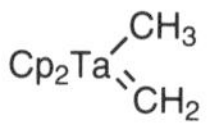

$$Cp_2Ta\begin{smallmatrix}CH_3\\CH_2\end{smallmatrix}$$

Ta-CH₃ 2.246 Å

Ta=CH₂ 2.026 Å

Chemical Properties of Carbene Complexes

Because the electrophilic Fischer complexes were discovered first and are more easily prepared and handled than their Schrock analogues, they have been studied in much more detail. We look at their chemistry first.

Electrophilic Carbene Complexes

The thermal stability of these compounds is often low; their decomposition by heat usually gives olefins, by formal dimerization of the carbene:

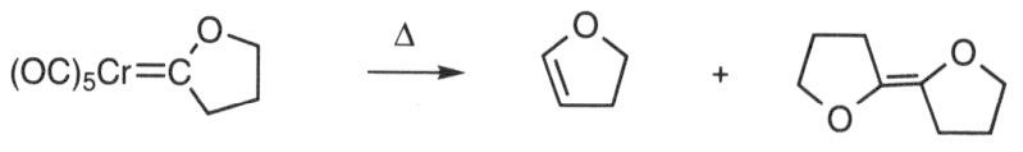

However, the absence of cyclobutanone in the product mixture means that the reaction pathway cannot involve the free carbene itself. Oxidation generally gives carbonyl compounds:

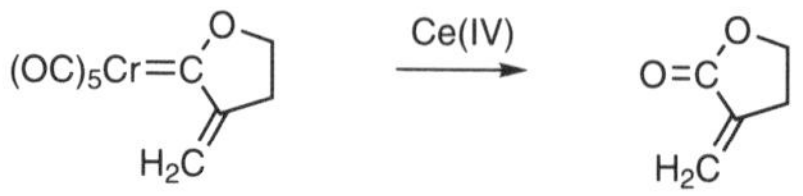

This reaction is useful for the characterization of carbene complexes; suitable oxidants include Ce(IV), DMSO, air, etc. Protons attached to a carbon α to the carbenic centre are rather acidic; this allows the synthesis of substituted carbene complexes through their abstraction by a base:

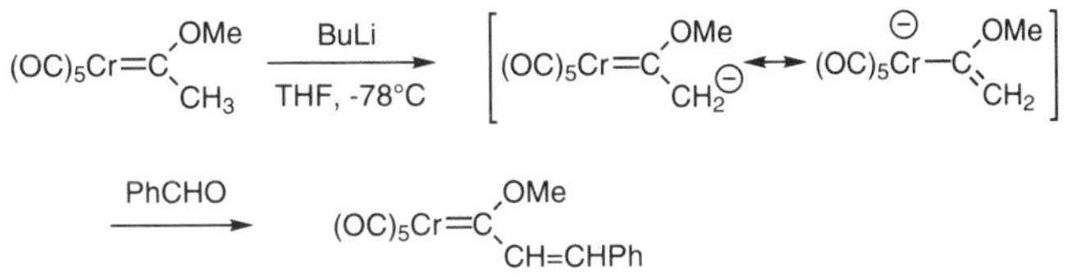

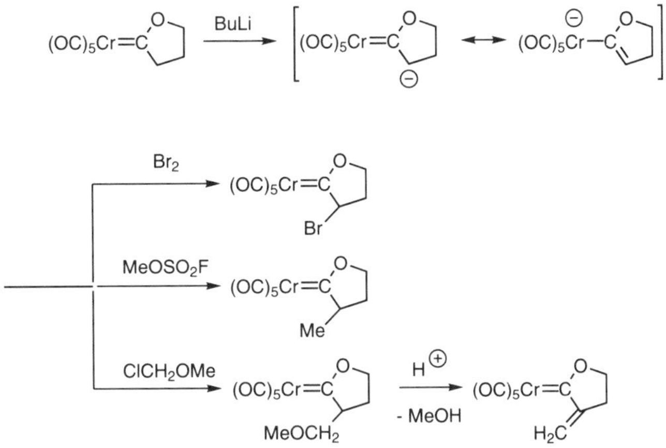

$$(OC)_5Cr=C \overset{OMe}{\underset{CH_3}{}} \xrightarrow[\text{THF, -78°C}]{\text{BuLi}} \left[(OC)_5Cr=C \overset{OMe}{\underset{CH_2^{\ominus}}{}} \longleftrightarrow (OC)_5Cr-C \overset{OMe}{\underset{CH_2}{}} \right]$$

$$\xrightarrow{\text{PhCHO}} (OC)_5Cr=C \overset{OMe}{\underset{CH=CHPh}{}}$$

$$(OC)_5Cr=C\langle \text{(oxolane ring with O)} \xrightarrow{\text{BuLi}} \left[(OC)_5Cr=C\langle \text{(ring, }\ominus\text{)} \longleftrightarrow (OC)_5Cr-C\langle \text{(ring, }\ominus\text{)} \right]$$

$$\xrightarrow{\text{Br}_2} (OC)_5Cr=C\langle \text{(ring with Br)}$$

$$\xrightarrow{\text{MeOSO}_2\text{F}} (OC)_5Cr=C\langle \text{(ring with Me)}$$

$$\xrightarrow{\text{ClCH}_2\text{OMe}} (OC)_5Cr=C\langle \text{(ring with MeOCH}_2) \xrightarrow[-\text{MeOH}]{\text{H}^{\oplus}} (OC)_5Cr=C\langle \text{(ring with H}_2\text{C)}$$

We have already mentioned nucleophilic attacks on the carbenic carbon, but will add a number of examples here:

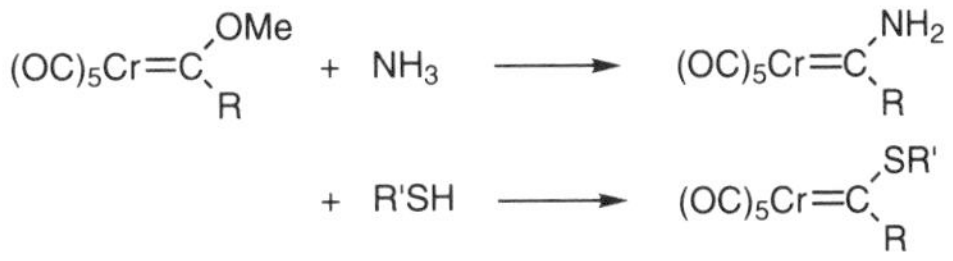

$$(OC)_5Cr=C \overset{OMe}{\underset{R}{}} + NH_3 \longrightarrow (OC)_5Cr=C \overset{NH_2}{\underset{R}{}}$$

$$+ R'SH \longrightarrow (OC)_5Cr=C \overset{SR'}{\underset{R}{}}$$

Phosphines attack similarly, to give reactive ylides which may evolve through several pathways:

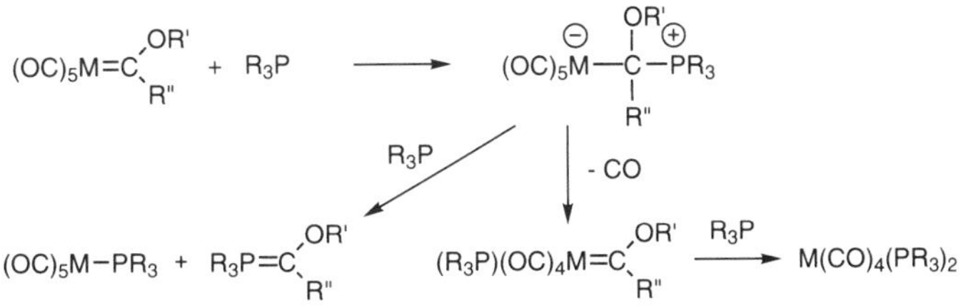

$$(OC)_5M=C \overset{OR'}{\underset{R''}{}} + R_3P \longrightarrow (OC)_5M-\overset{OR'}{\underset{R''}{\overset{|}{\underset{|}{C}}}}-PR_3 \; (\ominus / \oplus)$$

$$\overset{R_3P}{\swarrow} \qquad \overset{-CO}{\downarrow}$$

$$(OC)_5M-PR_3 + R_3P=C \overset{OR'}{\underset{R''}{}} \qquad (R_3P)(OC)_4M=C \overset{OR'}{\underset{R''}{}} \xrightarrow{R_3P} M(CO)_4(PR_3)_2$$

Phosphorus ylides give olefins by coupling reactions with one equivalent of a carbene complex:

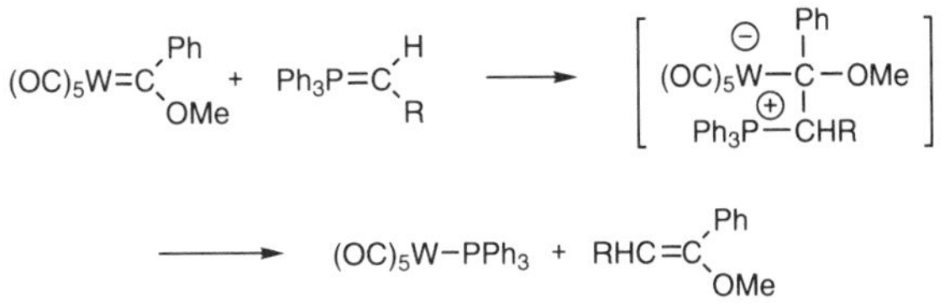

$$(OC)_5W=C \overset{Ph}{\underset{OMe}{}} + Ph_3P=C \overset{H}{\underset{R}{}} \longrightarrow \left[(OC)_5W-\overset{Ph}{\underset{Ph_3P-CHR}{\overset{|}{\underset{|}{C}}}}-OMe \; (\ominus / \oplus) \right]$$

$$\longrightarrow (OC)_5W-PPh_3 + RHC=C \overset{Ph}{\underset{OMe}{}}$$

We can consider the reaction with olefins as an unusual form of nucleophilic attack. Back attack on the carbene leads to a cyclopropane, through nucleophilic abstraction.[66]

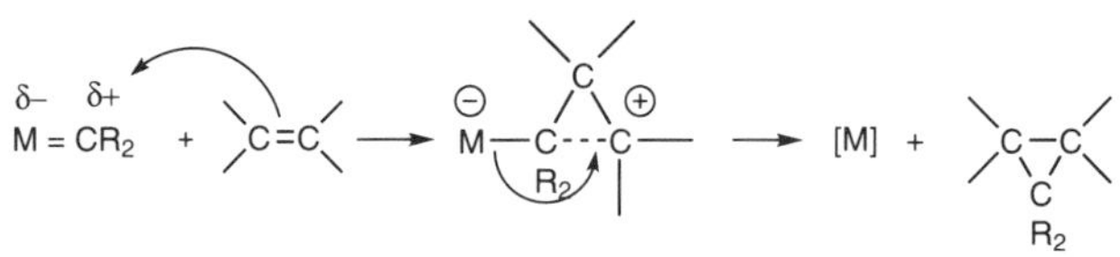

This cyclopropanation procedure can only be made to work stoichiometrically, because no carbene complex can be regenerated from the products (see Section 4.3). However, the formal $[2+2]$ cycloaddition of a carbene complex with an olefin gives a metalla-cyclobutane which may evolve to give a new carbene complex and a different olefin. This reaction (metathesis) is necessarily reversible and may be used as the key step in catalysis (see Section 5.6). The following examples illustrate the two different processes:

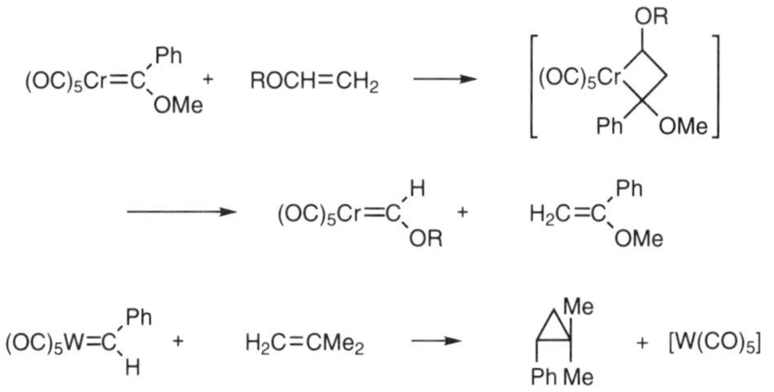

In a related reaction, alkynes give aromatic cyclopropenium cations:

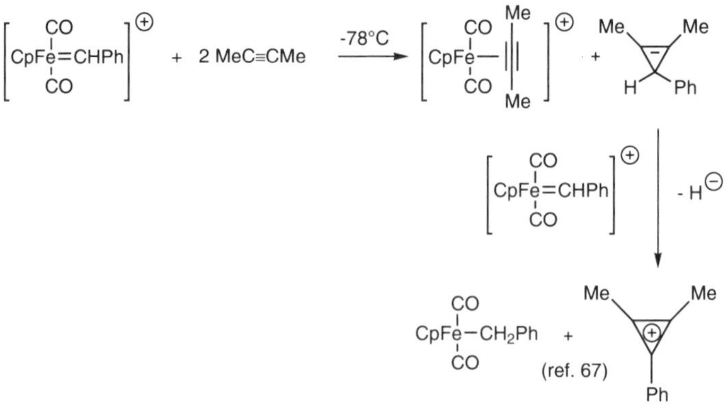

When the carbene has an aromatic or vinylic substituent, alkynes react to form six-membered rings (three atoms from the vinylic carbene, two from the alkyne and one from the CO), through a formal cycloaddition discovered by Dötz.

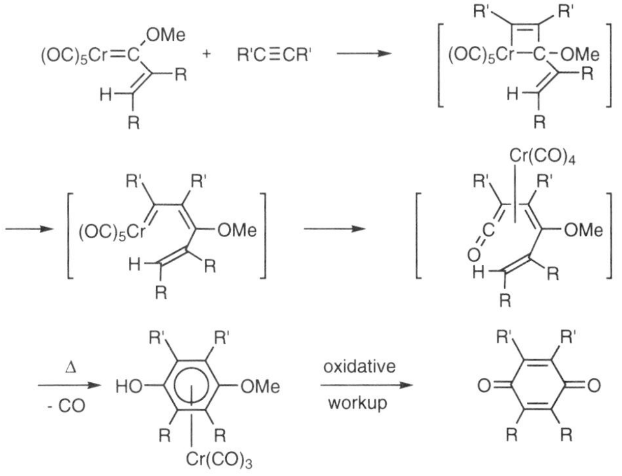

Therefore, carbenes bearing aromatic groups may be used as naphthalene precursors:[68]

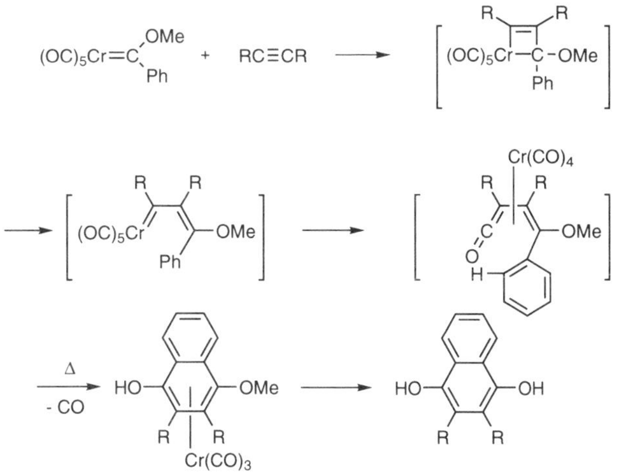

This reaction of carbene complexes with alkynes has appeared in many synthetic guises: for example, it may be applied to the synthesis of pyrroles.[69]

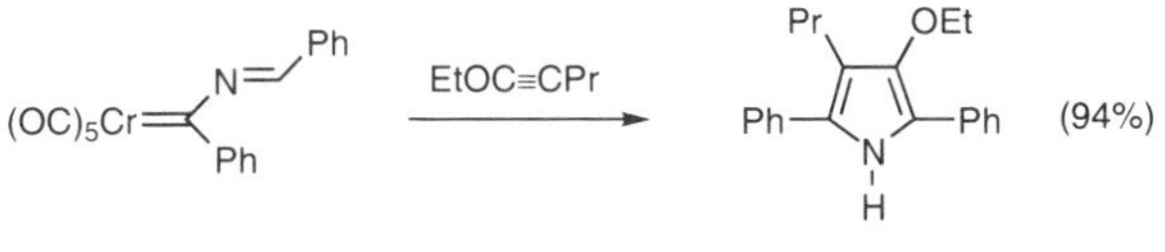

The intermediacy of ketenes in the Dötz reaction is supported by studies which show an equilibrium between ketene and carbene complexes under UV irradiation. The ketene complexes may be trapped by nucleophiles:

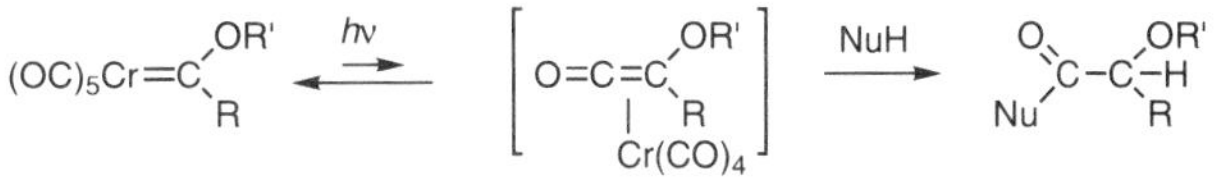

The nucleophile may be a functional group present within the carbene complex, as in the following intramolecular synthesis:

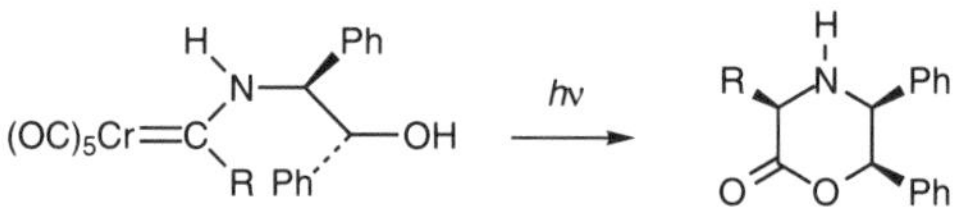

We close by noting that, today, the synthetic utility of carbene complexes is so extensive[70] that it would be impossible to list each of their applications here.

Nucleophilic Carbene Complexes

The reactivity of Schrock carbene complexes is governed by the high nucleophilicity of the carbenic carbon. Thus they behave as bases in reactions with Lewis acids

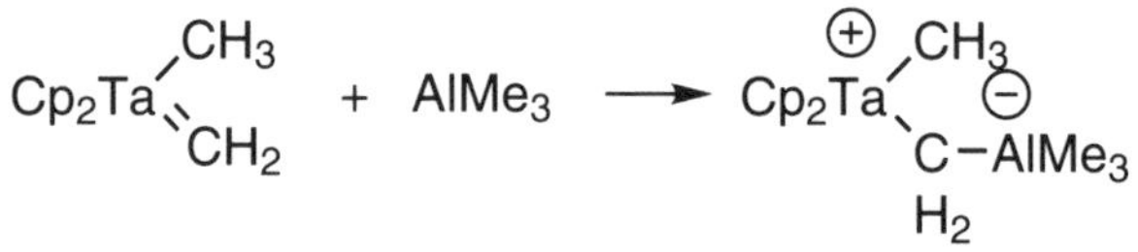

and resemble phosphorus ylides in their reactions with ketones. This allows Wittig-like chemistry.

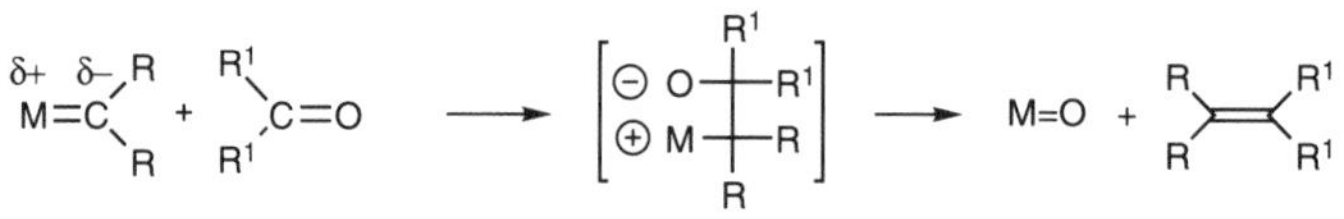

They may also be employed for the functionalization of esters, a transformation which is not possible with the corresponding phosphorus reagents:

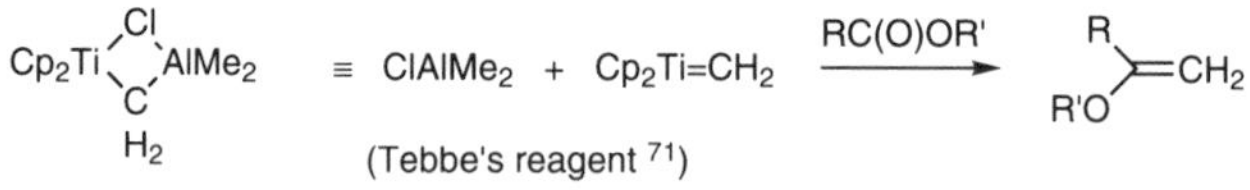

With alkenes, Schrock carbenes show the same behaviour as Fischer carbene complexes: transient metallacyclobutanes capable of catalysing metathesis are formed.

We underline once more that the 'Fischer' and 'Schrock' definitions of carbene complexes are merely extreme limits of a continuum of behaviour; complexes which show characteristics between the nucleophile and electrophile limiting definitions are known. As an example, $[H_2C=OsCl(NO)(PPh_3)_2]$ reacts with both the electrophile SO_2 and the nucleophile CO:[55]

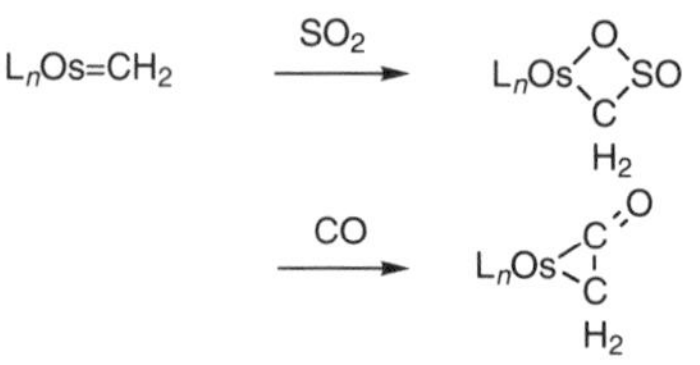

Two Representative Examples

(Diphenylcarbene)pentacarbonyltungsten $[(OC)_5W{=}CPh_2]$[73]

The compound is prepared as follows:

$$(OC)_5W{=}C\overset{OMe}{\underset{Ph}{\diagdown}} \xrightarrow[\substack{-78°C \\ ether}]{PhLi} \left[(OC)_5W{-}C\overset{OMe}{\underset{Ph}{\overset{|}{-}Ph}}\right]^{\ominus} \xrightarrow[-78°C]{HCl} (OC)_5W{=}C\overset{Ph}{\underset{Ph}{\diagdown}}$$

It may be purified by chromatography on silica and crystallized from pentane at around − 20°C. The complex is a red, air stable, crystalline material (m.p. 65–66°C), which decomposes slowly in solution. Its most important spectroscopic characteristics include $v(CO)$ 2070, 1971 and 1963 cm^{-1} in heptane, which may be compared with $[(OC)_5W{=}C(Ph)OMe]$ $v(CO)$ 2070, 1958 and 1945 cm^{-1}. The weaker and more polarized $(C^{\delta+}{-}O^{\delta-})$ bonds in the latter show that the presence of the oxygen lone pair induces a higher σ-donor character in C(OMe)Ph than in CPh$_2$. The carbene resonance in $[(OC)_5W{=}CPh_2]$ appears at 358 ppm (vs TMS), 34 ppm to low field of its analogue in $[(OC)_5W{=}C(OMe)Ph]$, which suggests a more positively charged carbon in the former case, possibly because the oxygen lone pair serves to diminish the charge on the carbene atom in the Fischer complex. The carbenic carbon of $[(OC)_5W{=}CPh_2]$ has classical, planar, sp^2 hybridization and lies 2.15 Å from the tungsten; this distance may be compared with the mean W−CO separation of 2.02 Å. Some important reactions are outlined below:

$$(OC)_5W{=}C\overset{Ph}{\underset{Ph}{\diagdown}} \xrightarrow[\substack{98°C, reflux}]{\Delta, heptane} Ph_2CH_2 \ (10\%) + Ph_2C{=}CPh_2 \ (35\%) + W(CO)_6 \ (19\%)$$

$$\xrightarrow{DMSO} Ph_2C{=}O \ (88\%)$$

$$\xrightarrow[\substack{100°C, 20\ h}]{H_2,\ 69\ atm} Ph_2CH_2 \ (49\%)$$

$$\xrightarrow[-78°C]{Me_3P} (OC)_5W{-}\overset{\oplus}{\underset{Ph}{\overset{Ph}{C}}}{-}PMe_3 \quad ^{\ominus}$$

$$\xrightarrow[20°C]{Ph_3P} Ph_3P{=}CPh_2$$

$$\xrightarrow{Ph_3P{=}CH_2} Ph_2C{=}CH_2 \ + \ Ph_3P{\longrightarrow}W(CO)_5$$

$$\xrightarrow[37°C]{H_2C{=}CHOEt} \underset{EtO\quad Ph}{\triangle}\overset{Ph}{|} \ (65\%)$$

$$(OC)_5W{=}C\overset{Ph}{\underset{Ph}{\diagdown}} + H_2C{=}C\overset{Me}{\underset{Me}{\diagdown}} \longrightarrow \underset{Me\quad\ \ \ Ph}{\overset{Me\qquad Ph}{\triangle}} \ (10\%) + H_2C{=}C\overset{Ph}{\underset{Ph}{\diagdown}} \ (76\%)$$

$$+ \overset{Me}{\underset{H}{\diagdown}}C{=}C\overset{H}{\underset{Me}{\diagdown}} \longrightarrow Ph_2C{=}CHMe \ (54\%)$$

$$+ H_2C{=}C\overset{OMe}{\underset{Ph}{\diagdown}} \longrightarrow Ph_2C{=}CH_2 \ + \ (OC)_5W{=}C\overset{OMe}{\underset{Ph}{\diagdown}}$$

In the last case, the choice of olefin determines whether metathesis, cyclopropanation or a combination of both are observed.

(Neopentylidene)(tris-neopentyl)tantalum [Me₃CCH=Ta(CH₂CMe₃)₃][73]

Two syntheses have been described:

$$TaCl_5 + 5\ Me_3C\text{-}CH_2\text{-}MgCl \xrightarrow[25°C]{ether} (Me_3C\text{-}CH_2)_3Ta=CH\text{-}CMe_3 + CMe_4$$
$$(50\text{-}80\%)$$

$$TaCl_5 + 1.5\ (Me_3C\text{-}CH_2)_2Zn \longrightarrow (Me_3C\text{-}CH_2)_3TaCl_2$$

$$\xrightarrow[\text{pentane, 25°C}]{2\ LiCH_2\text{-}CMe_3} (Me_3C\text{-}CH_2)_3Ta=CH\text{-}CMe_3 + CMe_4$$
$$(100\%)$$

Although the second route is more complicated, it gives a purer product and a better yield. A stable pentasubstituted complex is obviously disfavoured by the steric demands of the neopentyl groups. The orange crystalline product (m.p. 71°C) may be sublimed, or recrystallized from pentane at −30 °C. It reacts very rapidly with air, water, protic solvents, carbonyl compounds, nitriles and olefins, but is stable for several months if stored under nitrogen at 25 °C. Its spectroscopic properties are as follows:

- ^{1}H NMR: δ (Ta=C$\underline{H}$) = 1.91 ppm. The abnormal shielding of this proton partly results from the (Ta$^{\delta+}$=CH$^{\delta-}$) polarization of the double bond.
- ^{13}C NMR: δ (Ta=$\underline{C}$H) = 250 ppm, ^{1}J(C−H) = 90 Hz. The very low ^{1}J(C−H) coupling probably reflects an agostic interaction between the CH bond and the metal:

This interaction also contributes to the shielding observed in the ^{1}H NMR spectrum. The reactivity of the complex is dominated by the extreme nucleophilicity of the carbenic carbon, which engenders the properties of a 'super-Wittig' reagent.

$$(Me_3C\text{-}CH_2)_3Ta=CH\text{-}CMe_3 \xrightarrow{Me_2CO} Me_2C=CHCMe_3 \quad (80\%)$$

$$\xrightarrow{PhCHO} \underset{(35\%)}{\overset{Ph}{\underset{H}{>}}C=C\overset{H}{\underset{CMe_3}{<}}} + \underset{(65\%)}{\overset{Ph}{\underset{H}{>}}C=C\overset{CMe_3}{\underset{H}{<}}}$$

(the major product is *cis* in spite of its greater steric hindrance)

$$\xrightarrow{RC(O)OEt} \overset{EtO}{\underset{R}{>}}C=CHCMe_3 \quad R = H\ (90\%),\ R = Me\ (60\%)$$

$$\xrightarrow{HC(O)NMe_2} \overset{Me_2N}{\underset{H}{>}}C=CHCMe_3 \quad (77\%)$$

$$\xrightarrow{CO_2} Me_3CCH=C=CHCMe_3 \quad (> 50\%)$$

In some cases, tantalum enolates are formed:

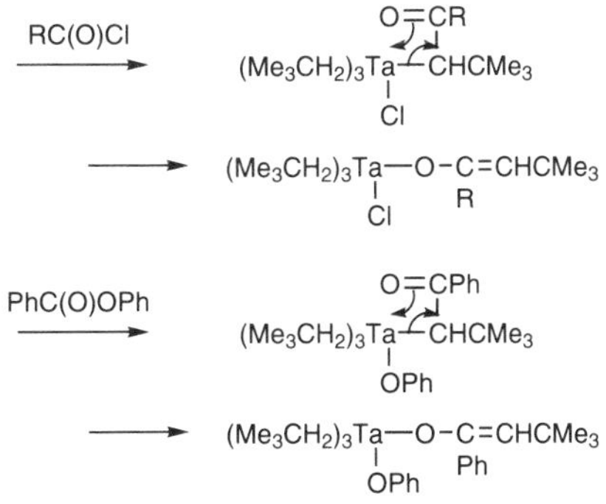

$$\xrightarrow{RC(O)Cl} \quad \underset{\underset{Cl}{|}}{(Me_3CH_2)_3Ta} \overset{O \mathbin{=\hspace{-3pt}=} CR}{\underset{}{\diagdown\hspace{-6pt}\diagup CHCMe_3}}$$

$$\longrightarrow \quad \underset{\underset{Cl}{|}}{(Me_3CH_2)_3Ta} - O - \underset{R}{C} = CHCMe_3$$

$$\xrightarrow{PhC(O)OPh} \quad \underset{\underset{OPh}{|}}{(Me_3CH_2)_3Ta} \overset{O \mathbin{=\hspace{-3pt}=} CPh}{\underset{}{\diagdown\hspace{-6pt}\diagup CHCMe_3}}$$

$$\longrightarrow \quad \underset{\underset{OPh}{|}}{(Me_3CH_2)_3Ta} - O - \underset{Ph}{C} = CHCMe_3$$

Nitriles react analogously:

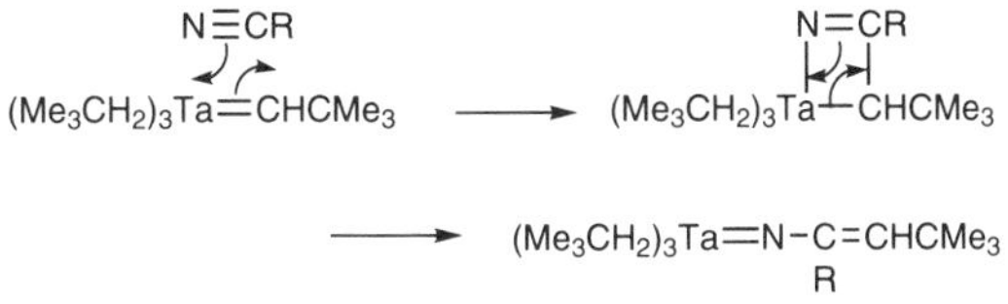

$$\underset{}{(Me_3CH_2)_3Ta} \overset{N \equiv CR}{=\hspace{-3pt}=\hspace{-3pt} CHCMe_3} \quad \longrightarrow \quad \underset{}{(Me_3CH_2)_3Ta} \overset{N = CR}{\underset{}{\diagdown\hspace{-6pt}\diagup CHCMe_3}}$$

$$\longrightarrow \quad \underset{R}{(Me_3CH_2)_3Ta} \equiv N - C = CHCMe_3$$

Hydrochloric acid gives a new carbene complex:

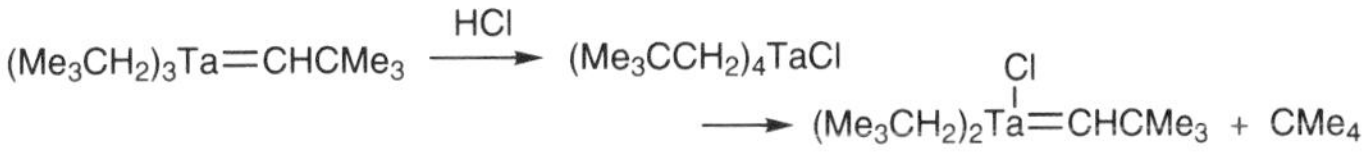

$$(Me_3CH_2)_3Ta = CHCMe_3 \xrightarrow{HCl} (Me_3CCH_2)_4TaCl$$

$$\longrightarrow \underset{\underset{}{}}{(Me_3CH_2)_2}\overset{Cl}{\underset{}{Ta}} = CHCMe_3 \; + \; CMe_4$$

However, no appreciable reactivity is observed towards MeI, $PhCH_2Cl$ or ethylene oxide.

3.5   The Carbyne Complexes

The discovery of carbyne complexes by Fischer in 1973[74] completed the formal analogy between organic and organometallic chemistry. However, the metal–carbon triple bond remained a laboratory curiosity until Schrock showed that carbyne complexes could catalyse alkyne metathesis.[75] From then on, the area underwent a rapid development which merits some study.

Synthesis of Carbyne Complexes

Fischer's original synthesis is still the principal route to carbyne complexes. It involves the attack of an electrophile on a Fischer carbene complex:

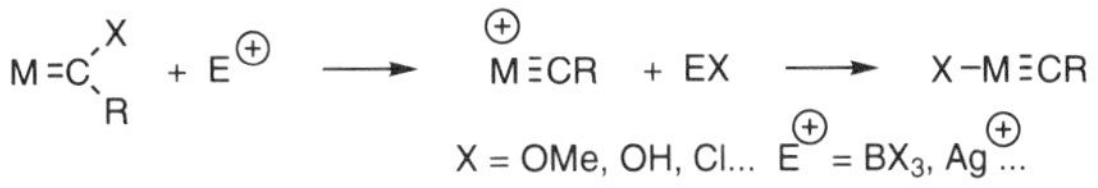

$$M = C \overset{X}{\underset{R}{\diagup}} + E^{\oplus} \longrightarrow \overset{\oplus}{M} \equiv CR \; + \; EX \longrightarrow X - M \equiv CR$$

$$X = OMe, OH, Cl... \quad E^{\oplus} = BX_3, Ag^{\oplus}...$$

for example:

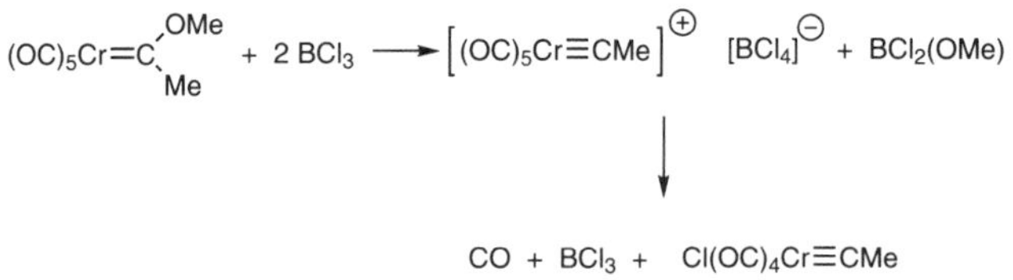

$$(OC)_5Cr=C \begin{smallmatrix} OMe \\ Me \end{smallmatrix} \ + \ 2 \ BCl_3 \ \longrightarrow \ \left[(OC)_5Cr\equiv CMe\right]^{\oplus} \ [BCl_4]^{\ominus} \ + \ BCl_2(OMe)$$

$$\downarrow$$

$$CO \ + \ BCl_3 \ + \ Cl(OC)_4Cr\equiv CMe$$

An extension of the classical Schrock carbene synthesis provides a second route to complexed carbynes. If the steric hindrance around the metal precursor is very severe, it is possible to effect two successive α-hydrogen eliminations:

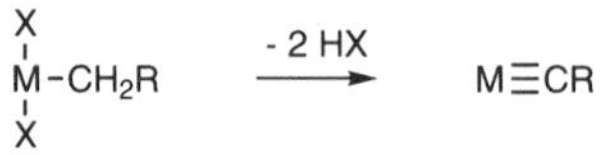

$$\begin{array}{c} X \\ | \\ M-CH_2R \\ | \\ X \end{array} \quad \xrightarrow{\ - \ 2 \ HX\ } \quad M\equiv CR$$

A typical example of this approach has been described by Wilkinson:[76]

$$WCl_6 \ + \ 6 \ LiCH_2SiMe_3 \ \longrightarrow \ (Me_3SiCH_2)_3W\equiv CSiMe_3 \ + \ 6 \ LiCl \ + \ 2 \ Me_4Si$$
$$(\ X = Me_3SiCH_2)$$

A third method which is regularly used involves the deprotonation of a carbene complex by a base:

$$\left[M=CHR\right]^{\oplus} \ \xrightarrow{\ -H^{\oplus}\ } \ M\equiv CR$$

for example:

$$(Me_3CH_2)_3Ta=CHCMe_3 \ \xrightarrow{\ BuLi\ } \ \left[\ (Me_3CH_2)_3Ta\equiv CCMe_3\ \right]^{\ominus}$$

Other less straightforward methods such as the metathesis between a metal–metal triple bond and a carbon–carbon triple bond have been described:

$$L_nM\equiv ML_n \ + \ RC\equiv CR \ \longrightarrow \ \begin{array}{c} L_nM=ML_n \\ | \quad | \\ RC=CR \end{array} \ \longrightarrow \ L_nM\equiv CR$$
$$M = Mo, \ W \ (ref. \ 77)$$

The Detection of Carbyne Complexes

As with carbene complexes, the principal method for the detection of carbyne carbons is ^{13}C NMR, where they lie between 200 and 350 ppm to low field of Me_4Si. The IR absorption for the $W\equiv C$ stretch appears in the range 910–920 cm^{-1} according to very recent work[78] and not between 1300 and 1400 cm^{-1} as previously thought. This means that the $W\equiv C$ triple bond is much weaker than expected: $k \ (W\equiv C) = 5.9$ mdyne/Å rather than the 7.0–7.4 given by previous estimates.

Theoretical and Structural Aspects

Initially, we will have to study the electronics of free carbynes if we are to understand the nature of the M≡C triple bond. The three carbyne electrons which remain after the formation of the CR bond need to be shared between a low energy sp hybrid orbital oriented along the C−R bond axis and two p orbitals that are perpendicular to it. (This result is obvious if we treat the carbyne as a half-alkyne). The M≡C triple bond may be formed from the free carbyne and the metal in two different ways. In case (A), two electrons fill the sp hybrid and the remaining unpaired electron occupies one of the p orbitals of the carbyne. Then, the carbyne lone pair interacts with the metal to create a σ bond and electron density is recovered from the metal by the vacant p orbital (π backdonation). Finally, the second π bond is created by pairing the single electron in the remaining carbyne π orbital with a metal d electron. Thus, the carbyne can be considered as a juxtaposition of a neutral two-electron ligand (L, the sp component) and an anionic one-electron ligand (X, the initially half-filled p orbital). This description is a good representation of the so-called Fischer carbynes (metal in a low oxidation state, carbon substituent carrying a lone pair).

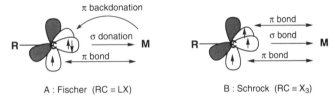

In the second case (B), the three unpaired carbyne electrons interact with three metal d electrons to form a σ bond and two π bonds, as is found in an alkyne. The carbyne ligand thus behaves as the sum of three ligands X. This description represents the so-called Schrock carbynes (metal in a high oxidation state). As with carbene complexes, there is a continuum of intermediate states between these two limiting cases.

The triple M≡C bond is, predictably, very short, for example Ta≡C ≈ 1.74 Å; W≡C ≈ 1.78–1.80 Å. The shortening relative to a single M−C bond is therefore 0.4–0.5 Å. The following complex reported by Schrock allows an excellent comparison:

$$
\begin{array}{l}
\text{W}{\equiv}\text{C} \quad 1.785(8)\ \text{Å} \\
\text{W}{=}\text{C} \quad 1.942(9)\ \text{Å} \\
\text{W -C} \quad 2.258(9)\ \text{Å}
\end{array}
$$

The substituents bound to the carbyne influence the M≡C bond length, just as in the carbene complexes. Substituents having a lone pair tend to reduce the multiple nature of this bond:

$$
Y{-}C{\equiv}M \quad \longleftrightarrow \quad \overset{\oplus}{Y}{=}C{=}\overset{\ominus}{M}
$$

This effect is much less accentuated than in carbene complexes, as is clear from the following data:

$$
\begin{array}{ll}
\text{Br(OC)}_4\text{Cr}{\equiv}\text{CPh} & \text{Br(OC)}_4\text{Cr}{\equiv}\text{C}{-}\text{NEt}_2 \\
\text{Cr}{\equiv}\text{C} \quad 1.68(2)\ \text{Å} & \text{Cr}{\equiv}\text{C} \quad 1.72(2)\ \text{Å}
\end{array}
$$

Finally, it should be noted that the M≡C–R system is not necessarily strictly linear:

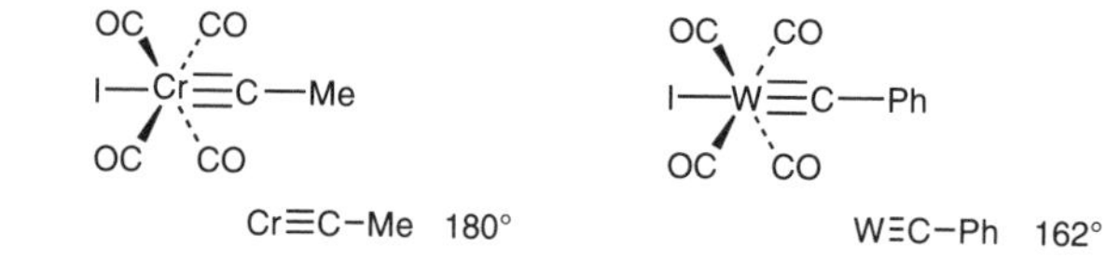

$Cr≡C–Me$ 180° $W≡C–Ph$ 162°

Chemical Properties of Carbyne Complexes

Overall, the reactions of carbyne complexes with electrophiles are better known than those with nucleophiles. Electrophiles which have been used include the proton, sulfur, selenium, tellurium, SO_2, halogens, electron-deficient metal fragments, etc. The reaction with protic acids requires UV irradiation and is thought to involve a transient excited state having a bent carbyne.[79] The following scheme is proposed:

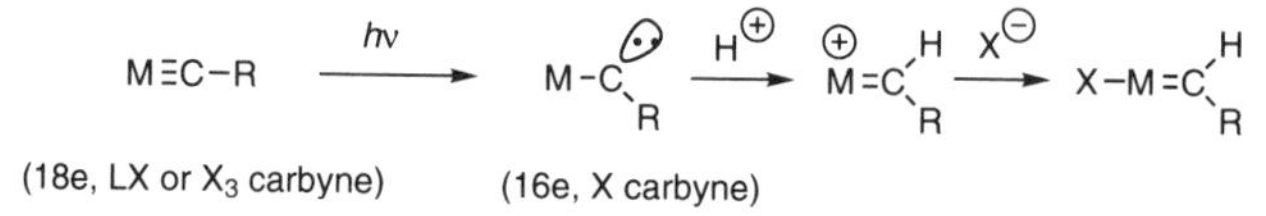

(18e, LX or X_3 carbyne)   (16e, X carbyne)

These bent complexes have not been isolated to date. Sulfur, SO_2 and PtL_2 give cyclic adducts:

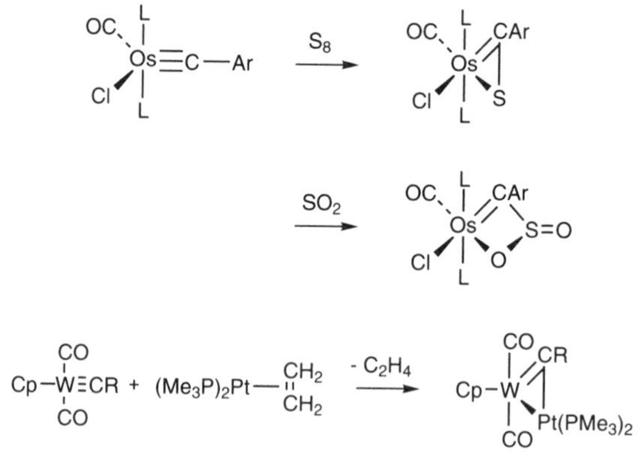

Carbyne complexes tend to dimerize when the steric hindrance around the metal is limited:

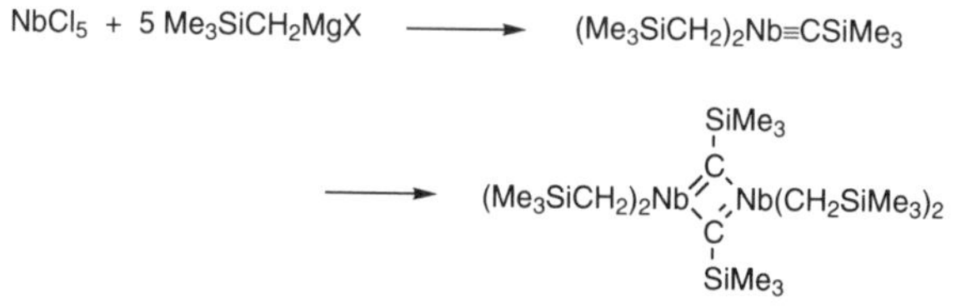

$$NbCl_5 + 5\ Me_3SiCH_2MgX \longrightarrow (Me_3SiCH_2)_2Nb≡CSiMe_3$$

The tendency of carbyne complexes to give metallacyclobutadienes is extremely important, because it is a key step in alkyne metathesis:

$$Cl_3(Et_3PO)W\equiv C-CMe_3 \quad + \quad PhC\equiv CPh \quad \xrightarrow[\text{70°C, 1h}]{\text{toluene}} \quad \left[\begin{array}{c} Cl_3(Et_3PO)W=C-CMe_3 \\ | \\ PhC=CPh \end{array} \right]$$

$$\xrightarrow{} \quad Cl_3(Et_3PO)W\equiv CPh \quad + \quad PhC\equiv C-CMe_3$$

Finally, it has been shown very recently that it is possible to couple two carbynes in a metal coordination sphere to create a carbon–carbon triple bond,[80] for example:

$$\left[Cp^*(EtNC)W \begin{array}{c} \diagup\hspace{-2pt}C\diagdown NEt_2 \\ \diagdown\hspace{-2pt}C\diagup NEt_2 \end{array} \right]^{\oplus} \quad \xrightarrow[\text{- 80°C}]{Br_2,\ CH_2Cl_2} \quad \left[Cp^*(EtNC)Br_2W - \begin{array}{c} NEt_2 \\ | \\ C \\ ||| \\ C \\ | \\ NEt_2 \end{array} \right]^{\oplus}$$

In spite of the data above, our knowledge of carbyne complexes is still rather underdeveloped when compared with the chemistry of coordinated carbenes.

3.6 π Coordination, Theoretical Aspects

One of the major differences between organic and transition metal chemistry lies in the nature of their bonding interactions. In organic chemistry, if we exclude certain solvation effects involving electrostatic interactions, only two-centre two-electron (2c2e) bonds are found with any regularity. Such bonds are also widespread in organic chemistry, as we saw in Chapter 1. However, we have also seen that an alkene and a transition metal may combine through a different bonding scheme which involves a three-centre bond. This is a general phenomenon which we find throughout transition metal chemistry, because d- and f-block metals are capable of simultaneously forming bonds to several centres in an unsaturated molecule or a conjugated radical. The π-type MOs, one of which is available on each unsaturated carbon atom, interact with the metal orbitals to form these bonds. This metal–polyene interaction may be further controlled to some degree by changing the other (σ) ligands attached to the metal, and this ability to modulate reactivity is crucial in catalysis. Here, however, for simplicity, we assume that these other ligands serve merely to confer a precise shape and energy upon the frontier orbitals of the metal bound to the polyene. Then, we can look at the analogies and differences between simple 2c2e bonds and the more complicated interactions found in metal π complexes.

Classical diatomic σ or π bonds consist of a stable, low-energy, bonding MO and a corresponding antibonding MO, usually at much higher energy. The former is populated by two electrons and the latter is empty. The precise nature and energy of these two MOs is defined by the orbital overlaps of the two bonding centres, but the rule $S_\sigma > S_\pi$ generally applies. However, in butadiene or the allyl cation, even simple Hückel theory shows that the 2c2e bond description begins to break down. For example, we find a partial π bond between carbons 2 and 3 of butadiene and a delocalization of the π electron pair over all three centres in the allyl cation. Thus, neither of these species has bonding modes which can be properly described using one solid line per bond with two electrons per line.

This delocalization plays a crucial role in many aspects of chemical reactivity and also results in considerable modification of interatomic distances and angles. Similar

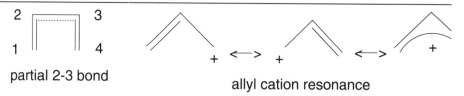

partial 2-3 bond   allyl cation resonance

delocalized bonding schemes exist in organometallic chemistry, where even more profound electronic changes result. To introduce this area, we will examine the characteristics of a typical multicentre bond in detail, before moving on to study its interactions with a transition metal.

Comparison between C_2H_4/H^+, C_2H_4/Br^+ and C_2H_4/ML_n

Let's look at a proton which bridges the double bond of an ethylene molecule. The product ion has an isosceles geometry which, for reasons of symmetry, can form only one three-centre bonding MO. This orbital is symmetrical with respect to the plane bisecting the two CH_2 groups.

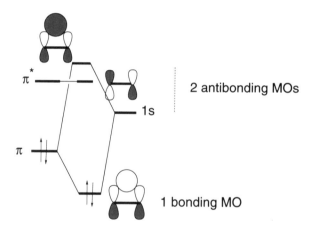

It follows that if any interaction is to occur, the two π electrons have to be spread over the three-centre bonding MO, as was the case in the allyl cation. The resulting stabilization is weak because the overlap is poor, *but energy is gained relative to the isolated reagents* because the newly created MO is more stable than the combined energies of the naked proton and the original π MO. This conclusion is consistent with the Woodward-Hoffmann allowed 1–2 suprafacial/suprafacial proton shifts in such systems, and has been corroborated by more recent theoretical work which shows that the bridging configuration is slightly more stable than the open resonance forms.

The problem is different in the case where Br^+ bridges the π bond because π-type occupied AOs are available on the bromonium ion. A more complicated bonding scheme is involved:

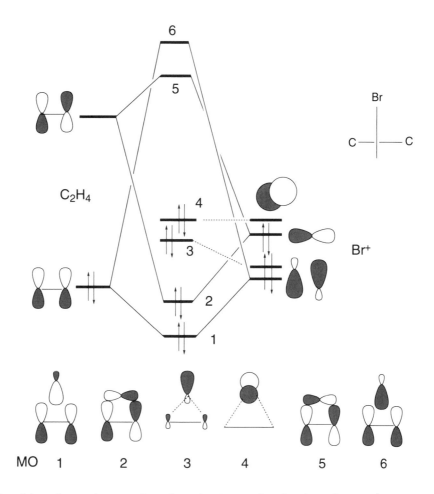

The eight valence electrons (two from the C_2H_4 π bond and six from Br^+) are shared between two bonding and two non-bonding levels. The stability of the ensemble is noticeably higher than that of the fragments because the number of bonding MOs has increased by one. Thus, the orbital scheme confirms the classical experimental result that bromonium ions are relatively stable intermediates.

The situation is rather similar in the case of a transition metal, but many more MOs are available for interaction between the metal centre and the organic molecule. The general principles are, nonetheless, the same and symmetry considerations are fundamentally important.

In the preceding chapters we have developed a number of ideas concerning the nature of electron distribution as a function of symmetry: we revise them here, because they can easily be generalized to any type of polyene. The essential points are outlined in the following diagram. We will start by looking at two metal-centred MO combinations S and A which are symmetric and antisymmetric respectively to the median plane of our olefin.

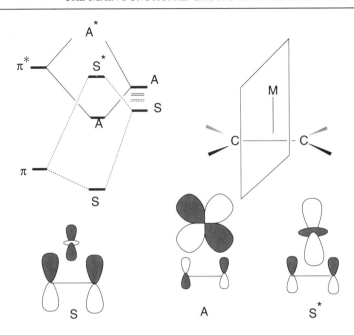

The interaction of the metal with the C−C π bond gives rise to two bonding MOs, S and A, along with two antibonding MOs, S* and A*. Only the first three orbital combinations need be considered, because A* lies at very high energy and is never populated. The bonding interactions which result from this scheme depend upon the number of electrons which are available. A number of cases are possible:

1. *Only the lowest level, S, is populated (two electrons):* this involves classical 2e donation from the ligand to the metal.
2. *S and A are both populated (four electrons):* a very stable configuration is obtained, where the ligand-to-metal transfer (S) is counterbalanced by a 'synergic' metal-to-ligand backbonding interaction (A).
3. *S, A, and S* are all populated (six electrons):* two essentially equal and opposite transfers for the symmetrical (S and S*) combinations exist; this means that these symmetrical components induce no polarization and make no net bonding contribution. Thus, only the backdonation associated with the antisymmetric bonding orbital A maintains the stability of the complex.

   We now consider the electronic effects which occur upon changing the geometry of the alkene; this will show us the relationship between the alkene geometry and the stability of the complex.

The Influence of Alkene Geometry

Let's assume that both S levels and the lower A level are occupied. Then, to enhance the stability of the complex, we clearly have to optimize two positive overlaps whilst minimizing a negative one.

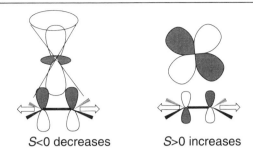

S<0 decreases   *S*>0 increases

We start by stretching the C−C bond. This produces two important effects. Firstly, the antibonding overlap S* decreases because the more separated carbon π lobes interact less strongly with the d_{z^2} orbital (the cone enclosing this orbital makes an angle at the metal of 109° 28′). Secondly, the bonding overlap A increases because the π lobes of the olefin move into a region where the corresponding d MO of the metal is very strongly localized. Thus C−C bond lengthening leads to a better overall metal–ligand overlap. It should be noted that the (*cis* or *trans*) configuration of the alkene is almost invariably conserved during π complexation because the structural reorganization of the olefin upon coordination is relatively small; the precise lengthening relative to the mean alkene C−C distance (1.35 Å) varies with numerous electronic and structural factors that we will not cover here. Nonetheless, as an illustration, we give a case where the modification of the olefin geometry is rather striking:

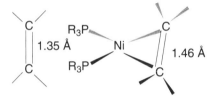

(The d^{10} Ni(0) configuration implies a filled S* level in this complex).

π Coordination of Polyenes

An enormous variety of polyenes, whether they be neutral or ionic, cyclic or linear, odd or even in their number of carbon atoms, give isolatable complexes with the transition metals. The following are merely representatives of the innumerable complexes known:

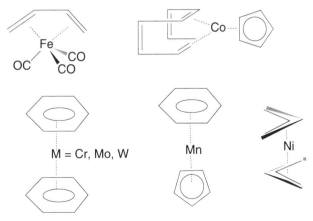

The depiction of the metal–π bonds found in these compounds presents some difficulties because we may use either one single line or, equally, as many lines as there are electron pairs. The latter representation corresponds to traditional 2e bonds. A dashed line translates as *it holds together but if you really want to understand, you'll need a theoretician*. This is often the standard approach, probably because these bonds are not encountered in more classical compounds.

Metal complexes exist which have aromatic ligands, ligands which would be antiaromatic if isolated in the free state, complexes of ions, etc., and obviously we require a bonding scheme which is capable of describing them all. It is easy to evaluate the number of electrons contributed by the ligand in simple cases (benzene or butadiene), but ligands such as allyl and cyclopentadienyl, for instance, pose more problems. We will study a number of common π-bonding modes, to explain the electronic properties of these unsaturated ligands.

We invariably use the same two-step approach: firstly, we obtain the molecular orbitals in the complex by combining the metal fragment with its σ-bonding ligands (see the ML_n structures given in the appendix); subsequently, we combine them them with the π type MOs of the polyene ligand. It is essential to define the symmetry axes clearly for each case and to have reliable data for the fragment MOs. The symmetries and energies for the polyenes are easily obtained, either by direct analysis, or through the Coulson formulae given below.

The MOs of organometallic fragments are less readily available and simple calculations are sometimes necessary to obtain them. Here, we use model metal centres surrounded by CO ligands, and change only the electron count when going from one metal to another. This means using general MO schemes which are applicable to any metal. Finally, we make the approximation that the ligands 'L' in the organometallic fragments are pure σ-type donors, imposing a local field at the metal which is unchanged by the subsequent complexation of the π ligand. Thus, in our approximation, the π ligands adapt to the available MOs. This approach becomes clearer when treating a real example, as we will show in the following paragraphs.

MOs of Polyenes and Annulenes: Coulson Formulae

We need to begin by distinguishing between unbranched linear conjugated polyenes and their cyclic analogues.

Polyene (Linear) Case

In the general case, a polyene comprises n homoatomic conjugated centres (see the diagram above). According to classical Hückel calculations, α is the energy associated with a π-type AO in the isolated atom and β is the resonance integral for adjacent atoms. This means that, using the classical definition of the matrix elements, $H_{rr} = \alpha$ and $H_{rs} = \beta$ when the atoms r and s are close neighbours, otherwise $H_{rs} = 0$. With these constraints, the energy of the pth molecular orbital, E_p, depends only upon p, whilst the normalized coefficient of the atom c_{pr} in the pth order MO depends both upon p and r:

$$E_p = \alpha + 2\beta \cos\frac{p\pi}{n+1} \quad \text{and} \quad c_{pr} = \sqrt{\frac{2}{n+1}} \sin\frac{pr\pi}{n+1}$$

Annulene (Cyclic) Case

With the same rules as the preceding section, the energies and coefficients are given by:

$$E_p = \alpha + 2\beta \cos\frac{2p\pi}{n} \quad \text{and} \quad c_{pr} = \sqrt{\frac{1}{n}} \exp\left\{\frac{2\pi i(r-1)p}{n}\right\}$$

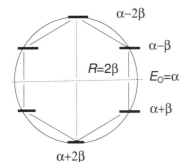

$\alpha - 2\beta$

$\alpha - \beta$

$R = 2\beta$ $E_0 = \alpha$

$\alpha + \beta$

$\alpha + 2\beta$

The MO energies are given by the actual positions of the apices, as a function of β. The lowest apex always has the energy $\alpha + 2\beta$

(The general scheme above allows the orbital energies to be obtained for any regular annulene.)

Before moving on to study a complex, a last word concerning the hybridization of metal MOs of s, p and d symmetry is necessary because, in many cases, the precise form of a hybrid MO depends upon the geometry of the complex.

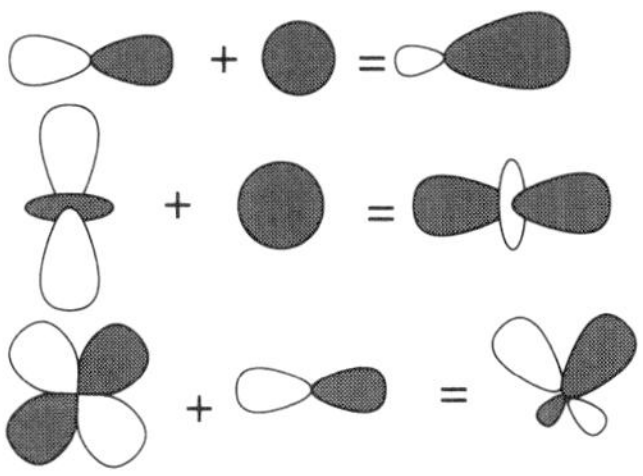

Just as for the s + p type hybrids in organic chemistry, the different lobes are mixed using the simple rule which states that lobes of the same shade (phase) are added and lobes of different phase are subtracted. A number of examples of hybrid orbitals are given above.

Study of $(\Phi_3P)_2Ni(C_2H_4)$

As always, we begin by assigning our symmetry axes, then superimpose the frontier orbitals of the two interacting fragments and, finally, examine their overlaps as a function of symmetry. As the individual molecular orbitals form discrete sets of different symmetry, we can study each of them in isolation.

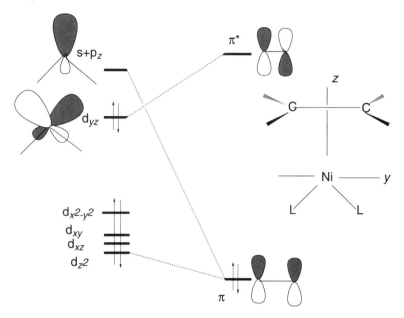

The S interactions, which are symmetrical relative to the xz plane, lead to only a slight overall destabilization. This is because the repulsion between d_{z^2} and π is almost compensated for by the interactions with the higher energy vacant empty MOs $s + p_z$. The A interaction, which is antisymmetric with respect to the xz plane, is strongly stabilizing which means that a stable structure is observed. The interaction of the alkene with the metal centre does not change the d^{10} electronic configuration and the product, which has $10 + (3 \times 2) = 16$ electrons, is unsaturated. If the C_2H_4 rotates by 90° around the Ni–olefin bond, then the original overlap of the olefin π^* orbital with d_{yz} will be replaced by an interaction with d_{xz}. This orbital is lower in energy than d_{yz}, and so the interaction is slightly less effective than in the coplanar configuration. Nonetheless, the relatively small energy gap between the two forms of this compound means that rotation of the complexed ethylene ligand is facile.

We have already seen that the C–C bond is highly elongated in this complex: it lengthens from 1.34 to 1.46 Å upon coordination to the metal.

Comparative Study of the Fe(CO)₃ Complexes of Butadiene and Cyclobutadiene

Butadiene has a finite resonance energy, unlike its unstable antiaromatic congener cyclobutadiene. We can show how complex formation considerably modifies their configurations. Once again, we analyse the fragment MOs as a function of symmetry. The butadiene MOs are known and the threefold rotational axis of the $Fe(CO)_3$ fragment leads to a set of degenerate frontier metal orbitals. Firstly we analyse a butadiene complex:

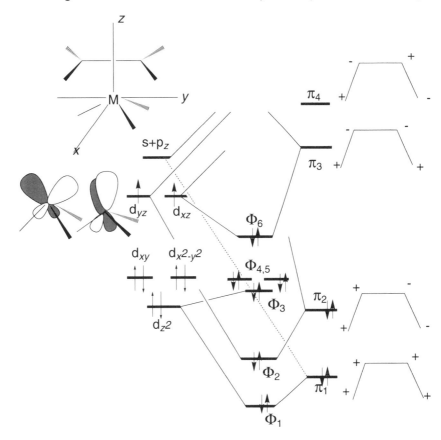

This figure is complicated because it gives every occupied frontier orbital. We will go through them according to their symmetry. π_1, d_{z^2} and the empty orbital $s + p_z$ give two occupied orbitals, Φ_1 and Φ_3, which are completely analogous to their counterparts in $(\Phi_3 P)_2 Ni(C_2 H_4)$. This may initially appear surprising, but the rotational symmetry about the z-axis means that they are unaffected by the number of ligands bound to the metal. The π_2 MO, which has d_{yz} symmetry, is stabilized by its transformation into Φ_2. The energy of d_{z^2} is raised slightly by incorporation into the doubly filled MO Φ_3 whilst d_{xy} and $d_{x^2-y^2}$ remain largely unchanged as the fully occupied orbitals Φ_4 and Φ_5. At higher energy, mixing of the vacant butadiene antibonding orbital π_3 (of d_{xz} symmetry) with the corresponding metal d orbital gives the highly stabilized MO Φ_6, which is doubly occupied in the complex. Overall, we find that six electrons (in Φ_1, Φ_2, Φ_6) are stabilized by complexation, whilst two (in Φ_3) are slightly destabilized. Because the stabilizing influences enormously outweigh the repulsions, the complexation process is highly favoured in this geometry.

The complex is Fe^0, d^8; it therefore has $8 + (3 \times 2) + 4 = 18$ electrons and is saturated. A pictorial diagram of the interactions of the ML_3 fragment with the butadiene orbitals is given below, to show how they interact in space.

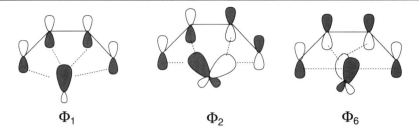

Φ_1 Φ_2 Φ_6

In the case of the cyclobutadiene complex, we focus upon the $d_{xz,\ yz}$ and $s + p_z$ orbitals, which are crucial to the coordination of the ligand. Cyclobutadiene has a fourfold symmetry axis:

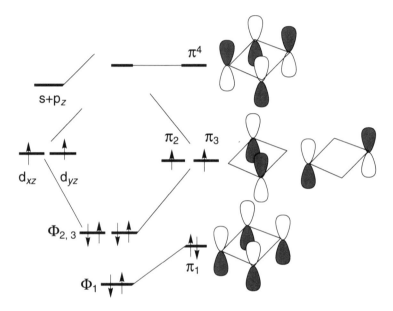

The degenerate interactions involving d_{xz} with π_2 and d_{yz} with π_3 induce a significant stabilization because the two electrons which confer the high spin (triplet) configuration in free cyclobutadiene are transferred into highly stable bonding orbitals in the complex. Thus, the biradical triplet cyclobutadiene ground state is transformed into a singlet, spin-paired, electronic configuration upon coordination. In form, the MOs Φ_1, Φ_2 and Φ_3 strongly resemble their counterparts in butadiene:

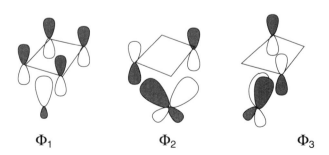

Φ_1 Φ_2 Φ_3

The metal undergoes a d^8 to d^6 conversion during complexation to the cyclobutadiene, because two of its electrons are transferred into energetically low-lying ligand-centred orbitals. As in the previous case the complex has 18 valence electrons and is saturated and stable. This iron complex is important historically, because it was the first example of a stable molecule containing the parent cyclobutadiene and it clearly underlines the potential of metal centres to modify the electronic properties of an organic moiety. This is one of the fundamental principles of catalysis: the presence of the metal permits the generation of intermediates or compounds which would otherwise have too high an energy (i.e. be too unstable) to exist.

The Cyclopentadienyl Ligand

This ligand, encountered in many complexes, may be regarded as $(CH)_5$. Using this formalism, we can visualize a cyclic radical having a π atomic orbital lying orthogonal to each of the five carbon atoms. The MOs of the ring are then:

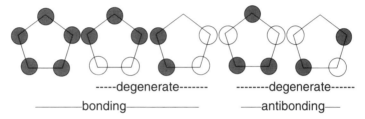

This radical is not stable in the free form. Nonetheless, in the presence of a metal, it can abstract an electron to give the corresponding cyclopentadienyl anion, whose six electrons engender a strongly aromatic character, reminiscent of the isoelectronic benzene. It is, therefore, considered as a *monoanionic ligand* because its coordination requires the formal transfer of one electron from the metal into its π system. We will explore this concept by looking at the molecular orbitals in the compound $(Cp)Co(CO)_2$, where Cp represents the cyclopentadienyl group.

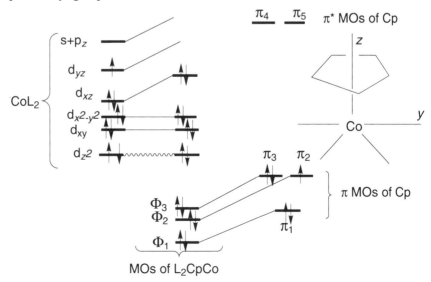

The formation of the bonding orbitals $\Phi_{1, 2, 3}$, induces the transfer of the metal electron into the low-lying molecular orbitals which are localized mainly on the ligand. Thus, the cyclopentadienyl residue gains an electron and the metal is oxidized by one unit. In the approximate orbital diagram above, we again see that the d_{z^2} level is not significantly destabilized because it interacts with the π_1 and $s + p_z$ combinations; the mixture of these three interactions leaves its electronic energy largely unchanged. This reflects the empirical observation that when three orbitals overlap we tend to see a bonding, an antibonding and an essentially non-bonding combination. This 'rule' has little mathematical rigour, but it is easy to demonstrate that it operates in many cases. Here, with d_{z^2} as the median orbital, it applies well. The three orbitals which result from this combination are given below:

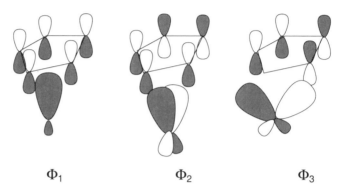

$$\Phi_1 \qquad\qquad \Phi_2 \qquad\qquad \Phi_3$$

The two previous figures give only simplified descriptions of the cyclopentadienyl molecular orbitals because they do not represent the relative sizes of the π lobes. To obtain a true representation of their spatial extension or an analytical solution to a real problem, more rigorous calculations would be necessary.

In the figure above, we note that the orbitals of the cyclopentadienyl ligand bear strong similarities to those of butadiene and cyclobutadiene. It is important to underline that *if a ligand has six valence electrons which are all used to bond to a metal, then three molecular orbitals will be formed. Each of these orbitals will be filled by two electrons, and will give rise to a π bond.* Similarly, butadiene employs two bonding MOs which generate two π bonds. These overlaps may be difficult to imagine in terms of classical bonds, but become clear when we look at the molecular orbital diagrams. The simplest way to state these results is that *we have one π-type bond for every two polyenic π electrons which are bound to the metal,* even if the precise description of the orbitals is complicated by electron delocalization between the metal and the ligand. These conventions at least allow a simple representation of a complicated polycentric delocalized bond. In this light, note that the structure of cyclopentadienyl ligands is a little more complex than we explained previously, and the exact description for any case will only come from a thorough MO calculation which treats all of the observable electronic features. Our six electrons giving rise to the 'aromaticity' of the cyclopentadienyl are in low-lying orbitals generally attributed to the ligands but, in reality, they have some localization on the metal; they even penetrate its supporting ligands, which are defined purely as passive spectators in our first approximation. It goes without saying that all models are rather qualitative and have limits, whence the possibility of classifying certain ligands according to either ionic or covalent models (see Chapter 1).

Thus we may write:

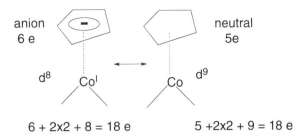

$$6 + 2 \times 2 + 8 = 18 \text{ e} \qquad 5 + 2 \times 2 + 9 = 18 \text{ e}$$

and think in terms of mesomerism between these two contributing forms. For simplicity, we usually adopt the ionic formalism, because the metal uses MOs which we associate with a d^8 centre. Although the covalent (neutral) formalism gives a much more satisfactory indication of the overall polarity of the complex, experimental results show clearly that the metal centre is, in fact, d^8 and not d^9; the complex is diamagnetic and therefore cannot contain the unpaired electron required by the d^9 configuration.

The Allyl Ligand

We see an important difference between the allyl and cyclopentadienyl ligands, because the allyl anion (4πe) and the allyl radical (3πe) have the same energy, at least at the EHT level. This reflects the non-bonding character of the second lowest orbital; it makes little difference to either the total energy or the resonance energy if one or two electrons are found in this non-bonding combination (at higher levels of sophistication, some differences appear). To understand this difference between the allyl and cyclopentadienyl ligands, we will study the allyl ligand in the neutral compound (allyl)Co(CO)$_3$.

For (allyl)Co(CO)$_3$ Φ_1 and Φ_2 have a pronounced 'ligand' character, with very minor metal contributions; however, Φ_3 has a dominant metal contribution and may be visualized as a backacceptor orbital. Here, it is extremely difficult to predict the polarity of the real metal–ligand interaction nor to say whether the metal really transfers its electron to the ligand. We are, therefore, at the limits of our model. If we adopt an ionic formalism, we effectively say that the metal transfers an electron to Φ_3, to give an anionic allyl ligand.

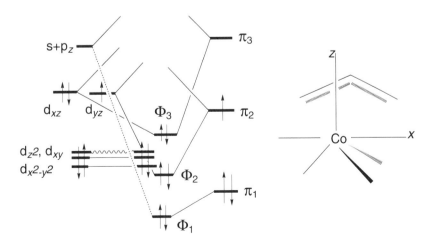

However, this does not reflect the exact nature of the ligand. It is much more reasonable to suggest that the ligand uses the electron found in Φ_3 to form a covalent bond with the metal. Both representations have a certain validity and many authors still write the allyl as a four-electron ligand, thus retaining the description allyl$^-$.

The orbitals Φ_1, Φ_2 and Φ_3 have the following structures:

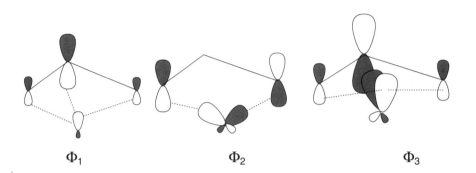

$$\Phi_1 \qquad\qquad\qquad \Phi_2 \qquad\qquad\qquad \Phi_3$$

We conclude this short study by summarizing the principal points of our analyses. Initially, and most importantly, careful calculation and description of the frontier orbitals in the interacting fragments is vital. Then, generally, the simple addition or subtraction of these orbitals is sufficient to give qualitatively reasonable results. To a first approximation, the supporting ligands on the metal may be modelled as simple σ donors, to give reasonable working MOs of suitable symmetry. Finally, it should be noted that, although we have obtained reasonable results by taking only the spatial and symmetrical aspects of the molecule into account, *precise* physical and chemical descriptions of a molecule require much more sophisticated calculations than those which we have employed here.

3.7   The Metal–Metal Bond

Introduction

A general description of a bond comprises two MOs: a bonding orbital, which is filled by a pair of electrons, and a corresponding empty antibonding orbital at much higher energy. In organic chemistry, a line is invariably used to represent this bond, irrespective of whether it is due to σ or π overlap. The use of a few supplementary symbols such as dashes, dotted lines, circles, etc. then allows the representation of almost all of the bonds found in innumerable compounds. *Almost* because there are well known classical compounds whose molecular electron distribution cannot be adequately described by such conventions. The O_2 molecule, which has two electrons in π^* orbitals, and NO (one electron in π^*) are simple cases where conventional representations are in poor agreement with reality. We have just seen that the interpretation of transition metal complexes in terms of 2c2e bonds is often impossible and that the 18-electron rule, which is a very efficient general concept to describe the electronic properties of complexes, does nothing to specify the bonding mode of the ligands. Thus we are obliged to search for a different bonding scheme with new representations and valence rules. Returning briefly to the well studied case of a metal bound to a cyclopentadienyl group, we know that the C_5H_5 ring has three bonding MOs, involving six electrons, which should therefore form three bonds to the metal. Our problem is how to describe these bonds.

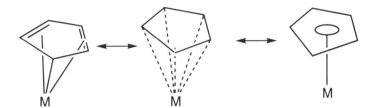

Chemists find the structural representation on the right more convenient, not only because it is better, but because it is simpler too. However, one should note that the vertical line does not have the same implications as the line which we traditionally use to describe a simple σ bond such as M–H. We underline this point here because, unfortunately, such ambiguities and unusual conventions are encountered rather frequently when describing metal–metal bonds.

In this chapter, we will deal only with complexes containing linkages between two metal atoms. A great variety of such compounds, having single, double, triple or quadruple M–M bonds are known from X-ray structure determinations. They may be classified by two important approaches: through structural data, above all the M–M bond length, but also by their electronic properties and information about the filling of their MOs. To characterize these M–M bonds, we talk of *bond order*, which is defined as $1/2(n_1 - n_2)$ where n_1 is the number of electrons in the bonding MOs and n_2 the number of electrons in the antibonding MOs. As in organic chemistry, short bonds tend to be associated with high bond orders, but quantitative attempts to relate bond order directly to internuclear distance are only likely to be successful when very similar structures (both in terms of metals and ligands) are compared. To give a simple example, the presence of bridging ligands can shorten a M–M bond without the bond order being affected; equally, a comparison of Fe–Fe distances with W–W separations has no meaning. Therefore, a double analysis of both structural and electronic characteristics is required to define a precise bonding scheme. For convenience, we will treat the general electronic criteria first and then analyse numerous examples of single and multiple bonds. We cite the work of Cotton and Walton[81] before progressing further: their review treats each of these aspects in depth and is indispensable for those requiring a detailed understanding of the area.

General Electronic Aspects

Consider two metals, having the general electronic configuration M[nd, $(n + 1)$s,p], which are separated by a distance $2\ \text{Å} \leqslant R \leqslant 3.5\ \text{Å}$. This covers the majority of the M–M lengths that are experimentally observed. The s and p atomic orbitals situated on each metal combine through σ and π overlaps. These are analogous to the bonds formed by the light elements (C, N, etc.) but it should be noted that the metal orbitals are larger and much more diffuse. The nd type AOs give three distinct types of overlap which are described in the figure below:

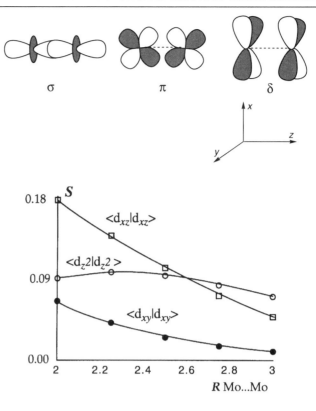

Adopting a convenient set of axes, we have:

- a σ-type overlap, corresponding to $S_\sigma = \langle d_{z^2} \mid d_{z^2} \rangle$
- two π-type overlaps, corresponding to $S_\pi = \langle d_{xz} \mid d_{xz} \rangle$ and $\langle d_{yz} \mid d_{yz} \rangle$
- two overlaps of a new δ type, produced by the degenerate system $S_\delta = \langle d_{xy} \mid d_{xy} \rangle$ and $\langle d_{x^2-y^2} \mid d_{x^2-y^2} \rangle$.

The variation of these d-type overlaps with the intermetallic distance R has been calculated by the EHT method and the results for the case of two Mo atoms are given in the preceding figure.[82] For short separations, the π overlap is stronger than the σ-bonding combination, but the opposite is seen as the M–M distance increases. The $\langle d_{z^2} \mid d_{z^2} \rangle$ (σ) overlap passes through a maximum with increasing M–M separation because at short distances the 'major lobe' of an AO interacts quite strongly with the 'minor lobe' of the other; this destabilizing interaction makes S_σ drop at short internuclear distance, as shown in the figure:

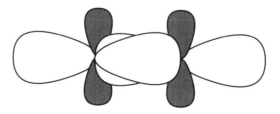

schematic view of the $\langle d_{z^2} \mid d_{z^2} \rangle$ overlap at short M-M distance

Although their relative levels vary, the σ and π interactions are always of comparable strength. It is clear that the δ overlap is the weakest at all distances. Put another way, in every case we observe the sequence:

$$\langle (n+1)s \mid (n+1)s \rangle \approx \langle (n+1)p \mid (n+1)p \rangle \gg \langle nd_{z^2} \mid nd_{z^2} \rangle \approx \langle nd_{xz} \mid nd_{xz} \rangle \gg \langle nd_{xy} \mid nd_{xy} \rangle$$

It should be noted that the $(n+1)s$ and $(n+1)p$ orbitals play only a secondary role in the M–M interaction. This is because their AOs are generally involved in bonds to the ligands and the MOs which result from these interactions are generally strongly metal–metal antibonding. They are therefore found at high energy, far removed from the nd set.[83] (See for example the MOs of the species ML_4, ML_5 and ML_6 in the appendix.)

The principles outlined above are nicely demonstrated in dinickel Ni_2 described in the figure below. With a Ni–Ni distance $R = 2.5$ Å, an EHT calculation of the 3d MOs gives the following:

Qualitatively, as we will see in the following section, this orbital ordering is valid in the majority of cases, even when ligands are present. However, *the δ-type MOs are frequently non-degenerate* in contrast to the case shown in the figure above. Generally, one of the two MO sets, for example the one corresponding to the $d_{x^2-y^2}$-type AO (with the choice of axes adopted here), is displaced to high energy by the ligand electron density. The d_{xy}-type MOs are much less sensitive to the ligand environment, as we will see through studying the following model structures.

Theoretical Study of the Structures $(MCl_4)^{2-}$ and (M_2Cl_8)[84]

The study of these complexes (with M = Cr, Mo, W) is instructive because it leads to important general rules concerning metal–metal multiple bonds. The MOs and the electronic factors which form the basis of our analysis were obtained with the aid of an EHT calculation with M = Mo.[84] The structures of these dimers, which are known with precision from studies of their salts,[85] may be visualized as a combination of two ML_4

fragments of essentially square-planar geometry arranged in either eclipsed (D_{4h}) or staggered (D_{4d}) configurations. These are shown in the following figure:

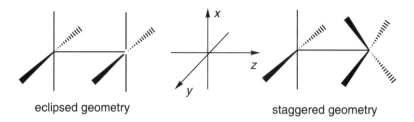

eclipsed geometry   staggered geometry

These different orientations result in only minor electronic differences and the resulting MO scheme is very similar in both cases. In each conformation, the σ and σ^* MOs and the π and π^* degenerate sets have cylindrical symmetry, just as with acetylene or N_2. These σ and π energies are essentially insensitive to rotation about the M−M bond, although small perturbations resulting from secondary interactions between ligands in the metal coordination spheres may occur. The only major effect of changing configuration is seen in the δ and δ^* MOs. In the D_{4h} geometry these orbitals, having irreducible representations B_{2g} for δ and B_{1u} for δ^*, are separated by a small energy gap which results from the weak overlap $\langle d_{xy} \mid d_{xy} \rangle$. In the D_{4d} geometry, the degenerate δ and δ^* set (irreducible representation E_2) is situated approximately half-way between B_{2g} and B_{1u}. As a consequence, if two electrons are available to the δ and δ^* combination, then the transition from the eclipsed to the staggered form is weakly endothermic, but if δ and δ^* are populated by four electrons, it becomes practically zero. An overview of these changes is given in the following:

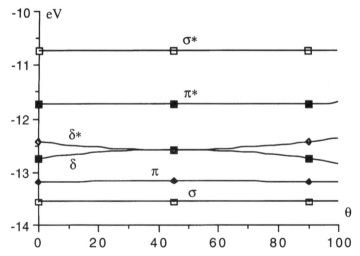

Variation of the σ, π, δ, δ^*, π^*, σ^* levels in $(FeH_4)_2$ as a function of the twist angle (θ), for an Fe-Fe separation of 2.5 Å

We will now move on to a more detailed treatment of the eclipsed geometry because it is found in a wide variety of dimers. We have already discussed the ML_4 fragment which forms the basis of our analysis; it appears below, along with the diagram and treatment of its corresponding eclipsed dimer.

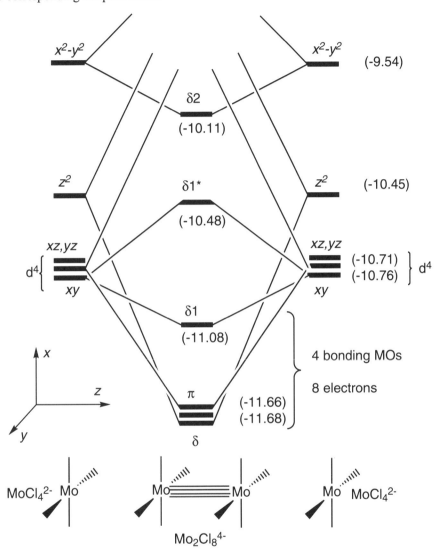

The frontier orbitals of each square-planar $MoCL_4^{2-}$ species[86] comprise a set of three nearly degenerate MOs: d_{xy} (-10.76 eV), d_{xz} and d_{yz} (-10.71). At higher energy, we find d_{z^2} (-10.45) and, finally, the antibonding combination $d_{x^2-y^2}$ (-9.54). Interaction of two of these $MoCl_4^{2-}$ units by combination of MOs of matching symmetry leads to the $Mo_2Cl_8^{4-}$ species, whose orbitals are, in order of decreasing stability:

- a σ MO formed from $d_{z^2} + d_{z^2}$ (-11.68);
- two π MOs formed from $d_{xz} + d_{xz}$ and $d_{yz} + d_{yz}$, which lie at almost the same energy (-11.66);

- a $\delta1$ MO formed from $d_{xy} + d_{xy}$ (-11.08);
- followed by its associated antibonding combination $\delta1^*$ (-10.48);
- finally the bonding combination $\delta2$ composed of $d_{x^2-y^2} + d_{x^2-y^2}$ (-10.11).

The formation of each σ and π MO results in a stabilization of almost 1 eV whilst the more weakly bonding δ MO contributes only 0.7 eV. The metal–metal antibonding orbital $\delta1^*$ is found at lower energy than $\delta2$ because the $d_{x^2-y^2}$ MOs lie at high energy in the parent $MoCl_4^{2-}$ subunits and are only weakly stabilized when they combine as $\delta2$ (0.54 eV). Their gain in stabilization is weaker than in $\delta1$ (0.7 eV) because of important ligand contributions to the $d_{x^2-y^2}$ MO, which dilute the influence of the metal d AO in the $MoCl_4^{2-}$ subunit. This phenomenon appears to a much lesser extent for d_{xy}.

From this analysis we see that the frontier orbitals of the dimer comprise a set of four bonding MOs $(\sigma, \pi, \delta1)$ followed by the weakly antibonding MO $\delta1^*$ at slightly higher energy. This means that if ligands are present, a maximum of four metal–metal bonds can be formed, depending on the nature of the metals involved and the number of electrons that are present. For instance, a quadruple metal–metal bond is found in $Mo_2Cl_8^{4-}$, because each Mo^{II} monomer has four electrons to contribute to the metal–metal framework. Referring to the EHT diagram above, we may represent this intermetallic orbital population by the formula $(\sigma)^2(\pi)^4(\delta)^2$. This molecular orbital scheme is, of course, valid for other electron counts and a very interesting question arises when there is a pool of 10 electrons to share between the two metals. Here, filling of the five lowest MOs leads to a population $(\sigma)^2(\pi)^4(\delta)^2(\delta^*)^2$. When we ask how many formal bonds there are, the reply is simple: *if we populate both a bonding orbital and its antibonding counterpart, no bond will be created.* However, our triply bonded $(\sigma)^2(\pi)^4(\delta)^2(\delta^*)^2$ configuration results in slight electronic repulsions within the system that are accommodated by the reorganization of other levels. These repulsions are small because the $(\delta^*)^2$ overlap is weak; we may calculate the electronic energies involved simply from the EHT diagram above: $-2(11.08) - 2(10.48) + 4(10.76) = -0.08$ eV (this diagram presents MOs for a real system, and therefore includes the effect of the coligands bound to the metals). Note here that we have just verified the simple rule (stated above) that the formal number of metal–metal bonds may be determined from the bond order: $+1$ counted when a bonding MO is filled and -1 for a filled antibonding MO. This bond order, which may be non-integral when an odd number of electrons is present, comprises a practical shorthand for describing the metal–metal interaction. However, like metal oxidation state, it is a useful conceptual tool rather than a thorough description of the magnetic properties or the real electronic states of a complex. It is obvious that care must be taken in the application of bond order criteria, which rely upon the assumptions concerning the MOs of the complex under consideration. In practice, experimental measurement of metal–metal distances and elaborate quantum-mechanical calculations of the bonding index corroborate this scheme, but it must be admitted that it remains rather qualitative.

Theoretical Study of the $MoCl_5^{3-}$ and $Mo_2Cl_{10}^{6-}$ Moieties

The MOs of the $MoCl_5^{3-}$ monomer are known (see Appendix), and we merely adjust the charge to obtain the two $Mo^{II}(d^4)$ structures. The MOs of $MoCl_5^{3-}$ closely resemble those of $MoCl_4^{2-}$ which were analysed in the previous paragraph. The only major difference is caused by a repulsion between the fifth halide ligand and the d_{z^2} orbital, which is consequently displaced to higher energy. This means that the $d_{z^2} + d_{z^2}$ MO should also lie

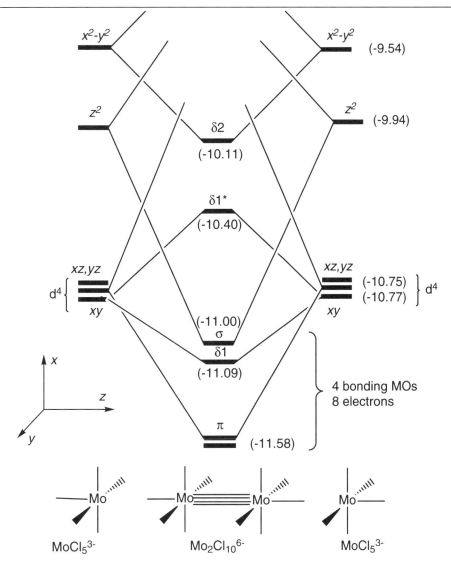

at high energy in the dimer and, accordingly, it appears above the π set comprised of $d_{xz} + d_{xz}$ and $d_{yz} + d_{yz}$. Otherwise there is little difference between the diagrams for $Mo_2Cl_8^{4-}$ and $Mo_2Cl_{10}^{6-}$, because the fifth halide does not interact significantly with the other metal-centred orbitals. From the diagram, we see that a maximum of four bonds may be formed (corresponding to an eight-electron core), and that fewer bonds will again be produced if we introduce more than these eight electrons.

It is useful to comment further on the limitations of the formal bond order method of attributing bonds in these dimers. The use of a similar convention in organic chemistry would lead us to write the incorrect representation [:O=O:] for O_2, which clearly underlines the potential pitfalls of the method if it is used carelessly. A further instructive example is the Fe^{IV}–O structure of porphyrins, which has often been written as Fe=O in recent publications:

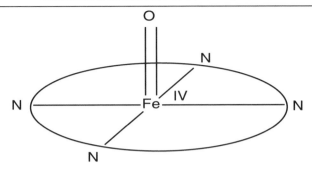

This formula incorrectly implies the properties of a true double bond. In fact, the FeO interaction is comprised of a normal σ bond together with π and π^* MOs resulting from degenerate combinations of metal d_{xz} and d_{yz} orbitals with the $2p_x$, and $2p_y$ functions of the oxygen atom. This means that the real electronic structure of the FeO bond is $(\sigma)^2(\pi)^4(\pi^*)^2$ (giving $3 - 1 = 2$ bonds), a configuration entirely analogous to that of O_2 itself.

General Case

The sections above have showed us that the highest bond order that we can expect to find for complexes between two metal centres is four. A vast number of experimental studies upon well identified complexes have confirmed this hypothesis.

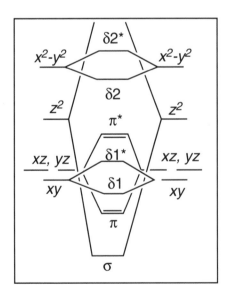

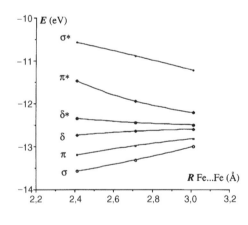

Clearly, since the diagrams in the preceding figures were obtained for complexes containing Cl ligands and quite a short Mo–Mo distance, they can not necessarily be applied in every situation. A diagram applicable in a more general case is given above. It defines a hypothetical compound which has two H_4Fe^{IV} (Fe^{IV}, d^4) structures in an eclipsed geometry, and demonstrates the changes which occur upon elongation of the Fe–Fe bond from 2.4 Å to 2.7 Å and 3.0 Å. The hydrides are both simple to calculate and are satisfactory models for ligands which are poor backdonors. The variation of the energy levels of each orbital with the internuclear distance can be easily explained by consideration of the relative size and orientation of the orbitals involved. In passing, we note that the presence of a fifth ligand on each metal would considerably perturb the position of the σ and $\sigma*$ levels by significantly distorting the energy of the original d_{z^2} orbital, but we will not deal with such considerations in detail.

n	electron count	bond type		
		σ	π	δ
2	$(\sigma)^2$	1	0	0
4	$(\sigma)^2(\pi)^2$	1	1	0
6	$(\sigma)^2(\pi)^4$	1	2	0
8	$(\sigma)^2(\pi)^4(\delta)^2$	1	2	1
10	$(\sigma)^2(\pi)^4(\delta)^2(\delta*)^2$	1	2	0
12	$(\sigma)^2(\pi)^4(\delta)^2(\delta*)^2(\pi*)^2$	1	1	0
14	$(\sigma)^2(\pi)^4(\delta)^2(\delta*)^2(\pi*)^4$	1	0	0

n.b. the relative ordering of the $\delta*$ and $\pi*$ levels may be inverted in some complexes (see previous figure)

It should be noted that the filling of antibonding levels, particularly the $\pi*$ orbitals, results in a significant electronic destabilization which is generally attenuated by increasing the M–M distance. A good example of this phenomenon is given by $Mn_2(CO)_{10}$: the table indicates a single Mn–Mn bond for this $2d^n = 14$-electron configuration and the filling of the $\delta*$ and the two $\pi*$ orbitals results in an extremely long Mn–Mn distance of 2.92 Å.[87]

So many dimeric species are known that we cannot hope to treat each of the structural and electronic types in detail. We will therefore proceed by studying specific cases of single and multiply bonded dimers, and explaining their characteristics in terms of the concepts that we have just developed.

Metal–Metal Single Bonds

There are few cases of single σ-bond formation if only two electrons ($d^1 + d^1$) are present. However $Ta^{IV}–Ta^{IV}$ bonds are known, as we see below:[88]

At the other extreme, numerous systems containing 14 electrons ($d^7 + d^7$) are known; these lead to an electron distribution of $(\sigma)^2(\pi)^4(\delta)^2(\delta*)^2(\pi*)^4$. We give a few examples in the following figure:

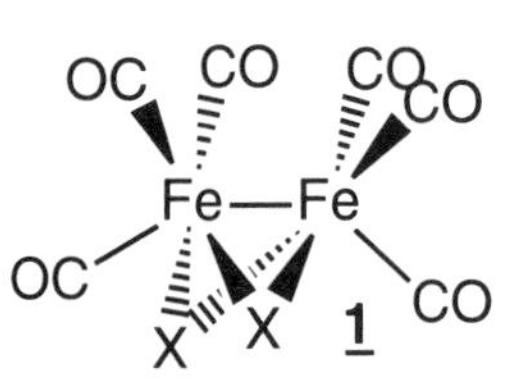

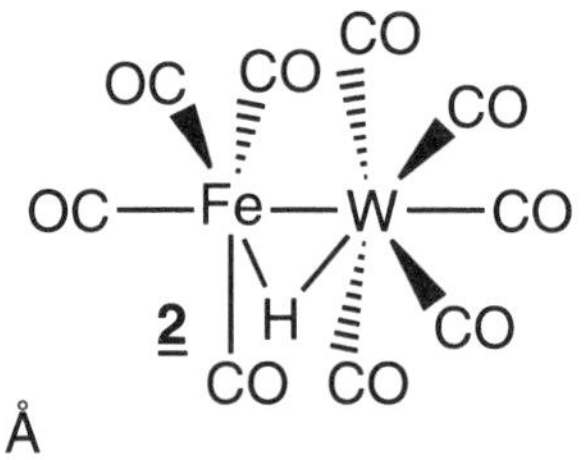

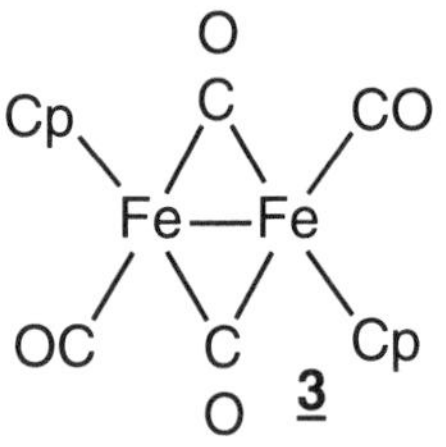

X = μ N(CH$_3$)$_2$ Fe-Fe 2.496 Å
X = μ S   Fe-Fe 2.552 Å   Fe-W 2.989 Å
X = μ PH$_2$ Fe-Fe 2.619 Å   Fe-Fe 2.531 Å

The compounds of type **1** have short Fe—Fe bonds owing to the presence of bridging ligands: they vary from 2.402 Å for μ-NH$_2$ to 2.619 Å for μ-PH$_2$.[89] In the mixed-metal structure **2**, the Fe—W bond is noticeably longer, 2.989 Å because the hydride ligand does not confer the same stabilization as the amido and phosphido bridges.[90] This is verified by the type **3** compounds whose bridging COs reduce the Fe—Fe bond length to 2.531 Å.[91] The Re—Re system is isolobal with Mn—Mn and also features a single bond.[92]

Certain electron-rich complexes also have metal–metal single bonds; examples include the compounds described in the figure below.[93]

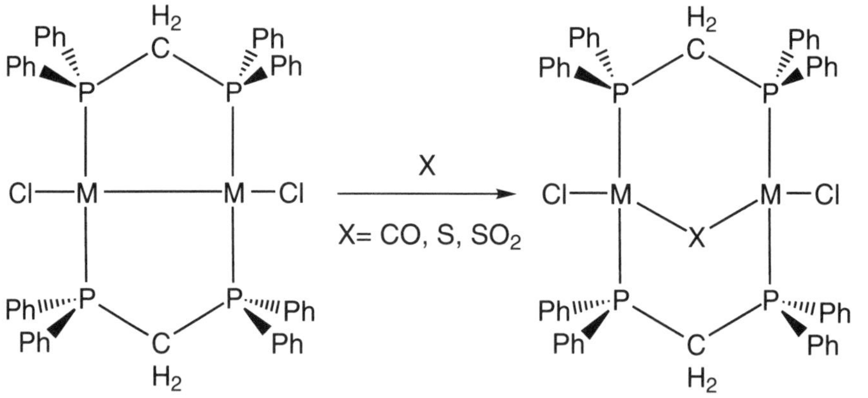

X = CO, S, SO$_2$

Pt-Pt 2.652 Å
Pd-Pd 2.55-2.80 Å

X = SO$_2$, Pd-Pd 3.383 Å
X = S, Pd-Pd 3.25 3 Å

Despite its geometrical simplicity, the explanation of the d electron density in this dimer is rather complex. The MOs of each (d^9) Cl(PR$_3$)Pt subunit are depicted below.

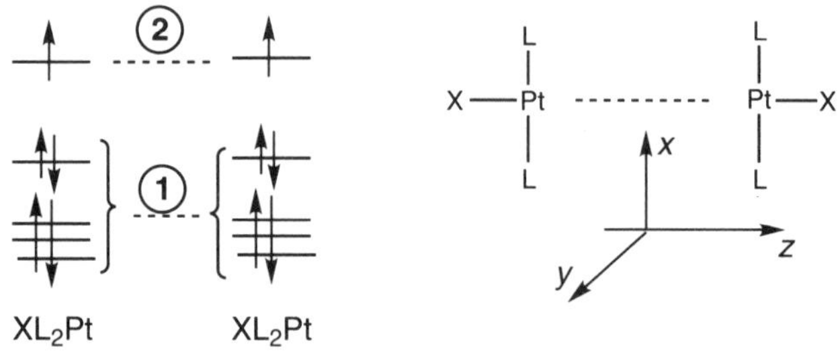

XL$_2$Pt   XL$_2$Pt

In ascending order, we find the low-lying filled d_{xy}, d_{xz} and d_{yz} MOs, which are essentially unperturbed by the ligands. These are followed by the doubly filled orbitals d_{z^2} and $d_{x^2-y^2}$. At still higher energy we reach the HOMO, which comprises a singly occupied hybrid comprised of d_{z^2}, $d_{x^2-y^2}$ and p_z, orbitals pointed towards the remote metal centre. The MOs resulting from fusion of the two structures are complex. Qualitatively, the entire set of doubly occupied levels (marked **1** in the figure) cannot give rise to any formal bonds because combination of any of these orbitals will bring four-electron (bonding and the corresponding antibonding) interactions into play. Only the in-phase combination of the two high energy MOs, (marked **2**) gives a doubly occupied bonding MO without the simultaneous population of its antibonding counterpart. Thus the overall scheme involves only a single formal bond and, once again, it is the bridging ligands which are largely responsible for the stability of the compound.

Amongst singly bonded compounds, $Rh_2(O_2CR)_4L_2$ complexes deserve a particular mention. As shown in the following figure, the carboxylate ligand plays an important role in their structure.[94]

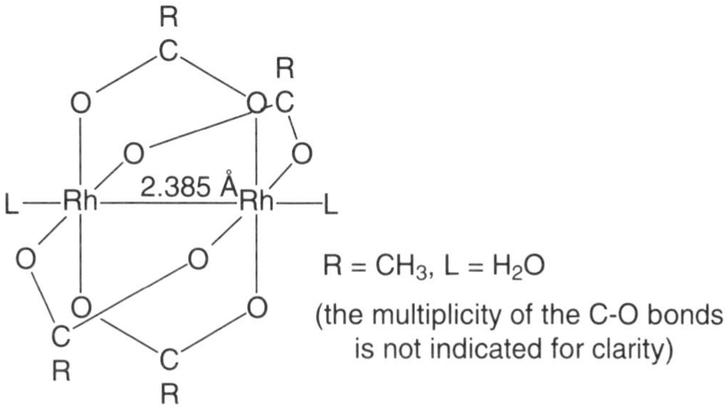

R = CH₃, L = H₂O

(the multiplicity of the C-O bonds is not indicated for clarity)

Once again, the electron distribution $(\sigma)^2(\pi)^4(\delta)^2(\delta*)^2(\pi*)^4$, characteristic of a single bond, is observed.[94]

We conclude this study by underlining that these single bonds are rather complicated, much more so than their representation by a single line might suggest. From even this brief survey, it should be clear that they result from complex electronic combinations which are difficult to generalize. The fact that many bonding interactions are balanced by antibonding contributions explains why bridging ligands are often needed to maintain the stability of the system. An accurate knowledge of geometrical criteria is imperative if we are to understand these orbitals which may or may not give rise to bonding interactions and, therefore, precise structural determinations remain invaluable for an analysis of such systems.

M=M Double Bonds

Structures which contain double bonds pose problems of analysis which are generally at least as difficult as the singly bonded compounds which we have just discussed. For example, a particularly useful illustration of these double bonds is given in certain rhenium aggregates, such as the historically important $Cs_3Re_3Cl_{12}$.[95] Here, the structures may be

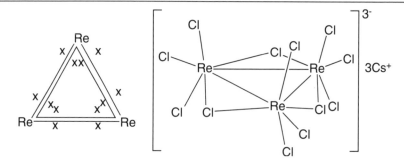

analysed as $3\,Cs^+ + Re_3^{9+} + 12\,Cl^-$, where the core of three rhenium d^4 centres contains three double bonds.

As the analysis of these seminal compounds is rather complex, we merely cite them and pass on to an interpretation of simpler bimetallic species. In general, analysis is simplest for complexes which contain the motif $(d^2 + d^2)$. However, a difficulty arises immediately because we have a conventional orbital ordering (σ, π, δ) in the dimer and π levels which are quasi-degenerate. The filling of the resulting $(\sigma)^2(\pi)^2$ system then leads to a *triplet* electron configuration having a pronounced biradical character. (Remember that a low-spin configuration would require $(\sigma)^2(\pi)^4$). Nevertheless, we talk in terms of a formal double bond for the sake of simplicity. Structural effects in compounds containing M=M double bonds have been extensively studied and, for example, the internuclear distances in the Nb^{III} (and Ta^{III}) complexes such as $Cs_3Nb_2X_9$ and $Rb_3Nb_2X_9$ with $X = Cl$, Br, vary with the metal and the nature of the halide, e.g.:[96]

X	Cl	Br	I
Nb=Nb	2.70 Å	2.77 Å	2.96 Å

The following structure shows the homologues of the complex $M_2X_6(THT)_3$, where the tetrahydrothiophene (THT) serves as a classical two-electron ligand:[97]

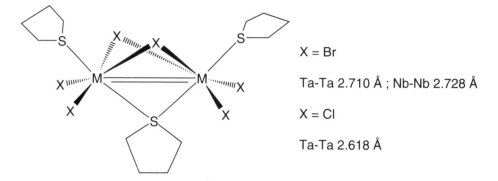

X = Br

Ta-Ta 2.710 Å ; Nb-Nb 2.728 Å

X = Cl

Ta-Ta 2.618 Å

Dimers of the $(d^2 + d^2)$ type are found in $(Mo^{IV})_2$ and $(W^{IV})_2$ complexes such as the compound overpage:[98]

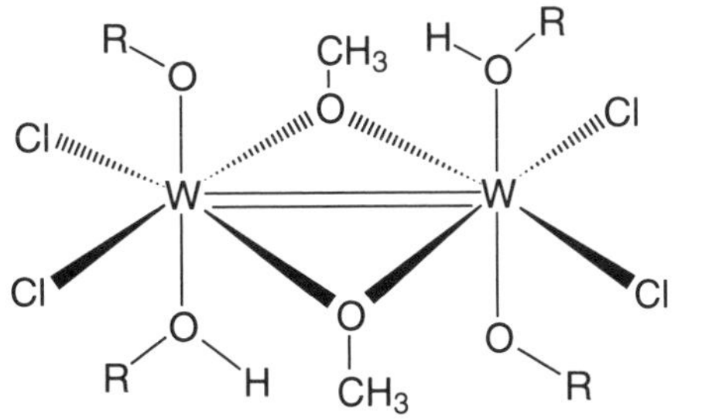

Many other dimers, for instance with V=V double bonds, are known. The example of $(\eta^5\text{-}C_5H_5)_2V_2(CO)_5$, whose V=V distance is 2.462 Å, is interesting because the nature of the bonding in this species was the subject of some controversy.[99] The complex $Fe_2(CO)_4(DBTA)_2$, whose ditertiary-butylacetylene (DBTA) ligand has the doubly bridging configuration frequently encountered in such compounds, is a further example. The Fe−Fe distance is 2.215 Å.[100]

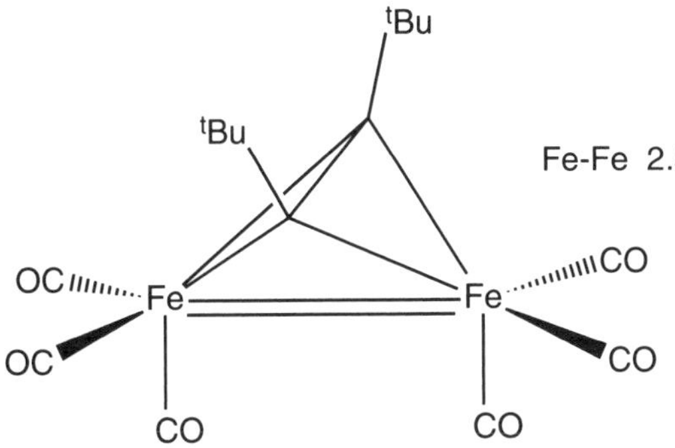

This compound can be formally analysed as follows: each Fe atom forms two bonds with the acetylene and may, therefore, be considered d^6. This leads to an electronic population $(\sigma)^2(\pi)^4(\delta)^2(\delta^*)^2(\pi^*)^2$, which requires a double bond between the two metal centres. In the same way, two formally d^6 centres are combined in the Co, Rh, Ir dimers of the type $(\eta^5\text{-}C_5H_5)_2Co_2(\mu CO)_2$, if we employ a model which treats the bridging COs as four-electron ligands.

To finish, we cite the work of Collman and co-workers[101] who have recently described some fascinating double bonds. These form between Ru or Os atoms incorporated into porphyrin ligands which are constrained to bond face to face:

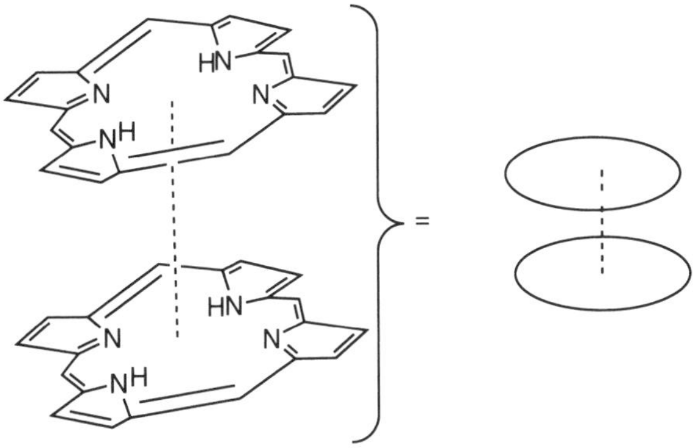

A compound containing the two square-planar porphyrin structures may adopt either an eclipsed or staggered geometry. The eclipsed form has the MOs previously described for the Fe_2 case; this implies a biradical electron configuration $(\sigma)^2(\pi)^4(\delta)^2(\delta*)^2(\pi*)^2$. It is noteworthy that the pattern $(\sigma)^2(\pi)^4(\pi*)^2$ is the same as in O_2 and that the $(\delta)^2(\delta*)^2$ set gives rise to no bond formation. As a result, a double bond is formally obtained between the two metal centres. In the staggered form the δ and $\delta*$ levels are degenerate and non-bonding; here the electron population becomes $(\sigma)^2(\pi)^4(\delta)^4(\pi*)^2$ which does not bring about a noticeable change. With two structures $(d^6 + d^6)$, it is thus possible to obtain dimers with the structure described in the following figure:

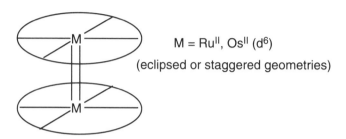

$M = Ru^{II}, Os^{II} (d^6)$

(eclipsed or staggered geometries)

Triple M≡M Bonds

It is practical to distinguish between complexes having six electrons arranged as $(\sigma)^2(\pi)^4$ and those with 10 electrons whose configuration is $(\sigma)^2(\pi)^4(\delta)^2(\delta*)^2$. Theoretically, the first category has a 'true' triple bond since no antibonding level is filled but, in practice, it is very difficult to distinguish it from the second type in the absence of unequivocal geometrical arguments. We will study the two different classes separately.

Complexes involving Six Electrons

Complexes involving six electrons are generally ions having the isoelectronic structures Re_2^{4+}, Mo_2^{4+}, Ru_2^{6+}, Os_2^{6+}, W_2^{2+} or Te_2^{4+}. To illustrate the complexity of assigning such complexes by geometrical arguments, we give the structure below:

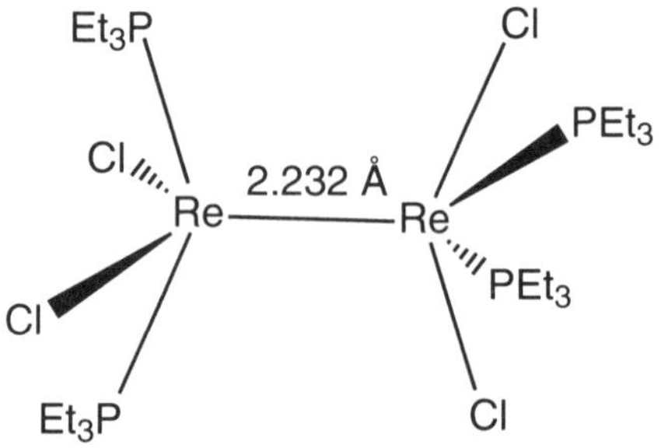

The coordination sphere of each metal is only slightly distorted from square planar and the Re—Re distance is 2.232 Å.[102] For comparison, the analogous quadruply bonded dimer, $Re_2Cl_6(PEt_3)_2$, has an Re—Re distance of 2.222 Å, which underlines nicely the difficulties involved in finding an unambiguous crystallographic bond order criterion for such species. We note that the electronic distribution $(\sigma)^2(\pi)^4$ in these triply bonded compounds leads to cylindrical symmetry about the M–M bond, whilst the filling of δ^2 and $\delta*^2$ means that there is no δ bond. The practical outcome of these effects is free rotation about the metal–metal axis.

Numerous bridged Os_2^{6+} and Ru_2^{6+} complexes are known, such as the following:[103]

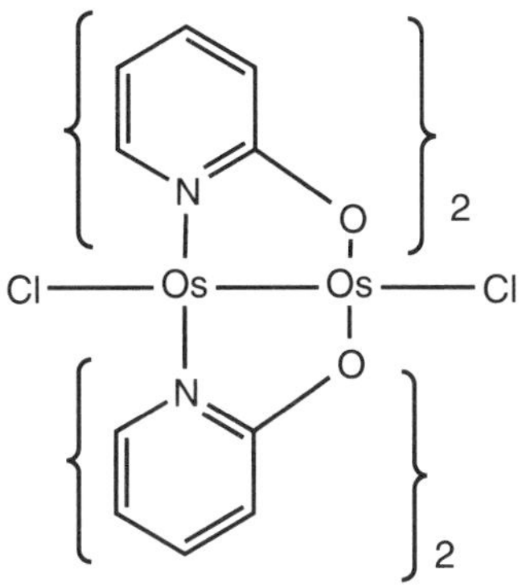

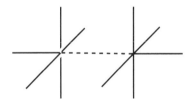

Only one of the two orthogonal planes is displayed.

Nearly all of the complexes containing these cores have bridging ligands to ensure the stability of the structure. They often result from the combination of two ML_5 fragments and are usually found in the eclipsed conformation (see the following figure).

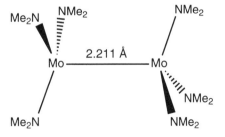

Complexes involving 10 d Electrons

Complexes involving 10 d electrons generally have the formula M_2X_6 where M = Mo, W; a typical example is given in the figure below.[104]

Me₂N NMe₂   NMe₂
  Mo —— 2.211 Å —— Mo
Me₂N   NMe₂
                       NMe₂

In this configuration, each metal coordination sphere adopts an essentially undistorted D_{3h} structure. Again, the electronic configuration $(\sigma)^2(\pi)^4$ has cylindrical symmetry and rotamers may only arise from non-bonding interactions between the supporting ligands.

Most of these dimers are obtained by the reaction of metal halides with organometallics or classical lithium or magnesium reagents:

$$WCl_6 + LiNMe_2 \rightarrow W(NMe_2)_6 + W_2(NMe_2)_6$$

In the dimer the W−W distance is 2.290 Å.

The enormous variety of complexes formally having triple bonds means that we have only been able to show a few representative examples of the most frequently encountered classes. To close this section, we highlight a number of complexes containing the lighter transition metals, particularly those with Cr^I_2 (and also Mo^I_2): these have bridging CO ligands, as shown in the figure below.[105]

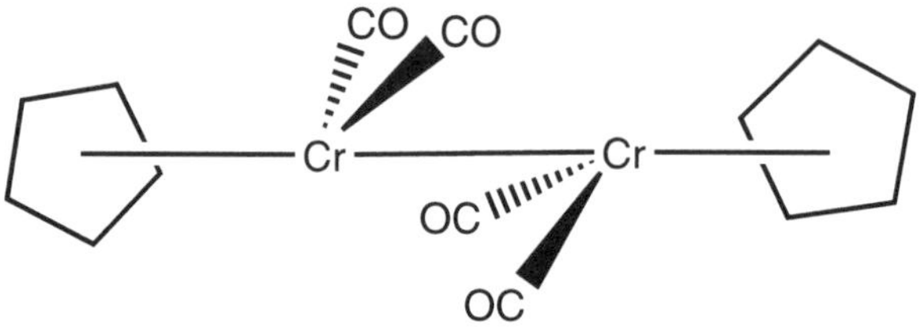

Quadruple M≡M bonds

This is a widely distributed motif, for which extensive experimental and structural data are available. It has been reviewed exhaustively by Cotton and Walton[81] whose excellent review is once again recommended for those wishing to explore the subject in detail. We will limit ourselves to a general outline of the salient points of this chemistry.

A brief consideration of the conventional MO ordering (σ, π, δ, δ^*, π^*, σ^*), reveals that, in the presence of ligands, a quadruple bond can only be generated by the electronic configuration: $(\sigma)^2(\pi)^4(\delta)^2$. It is thus necessary that the metals carry a total of eight electrons in their frontier MOs and it goes without saying that the simplest case, irrespective of whether neutral or ionic fragments are involved, is $d^4 + d^4$. Quadruply bonded compounds are characterized by M−M bonds which are either short or 'very short' by comparison with those which we have already discussed. This is a general rule which holds even in the absence of bridging ligands, as is shown by the ionic compound $Cs_2Re_2Br_8$ described in the following figure.[106]

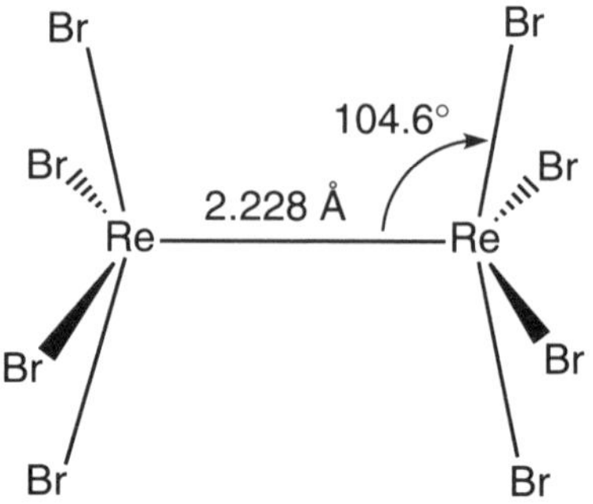

This compound is formed by the fusion of two essentially square planar ML_4 fragments and is therefore very typical of the general class. We'll recall here that two square-planar structures having an eclipsed geometry give D_{4h} symmetry whilst a staggered or twisted geometry transforms according to D_{4d}. We have already seen earlier in this chapter that the changes in the MOs between these two limiting forms are small; we expand this point a little, because it is frequently overlooked in many general MO treatments. In the eclipsed conformation (D_{4h}), the MOs δ and δ^* are separated by a weak but measurable energy gap, but they are entirely degenerate in the staggered (D_{4d}) conformation. It follows logically that the D_{4h} electronic configuration $(\sigma)^2(\pi)^4(\delta)^2$ is low spin (singlet, $S=0$), while the D_{4d} geometry gives a high spin (triplet, $S=1$) electronic configuration as in the following scheme (see also the beginning of this chapter).

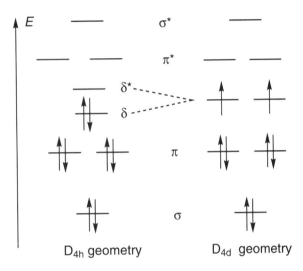

Consequently, the energetic differences between the singlet and triplet forms are very low, in spite of the fact that these quadruply bonded species are often, implicitly, assumed to be low spin. Once again, the eclipsed form is favoured by bridging ligands; in the table below, taken from reference 81, the reader may compare the geometries and internuclear bond lengths for a wide variety of bridged and unbridged structures. It is noteworthy that the M—M distances are relatively insensitive to their environment, varying only between 2.0 and 2.3 Å irrespective of the metals or their coordination spheres. The structures adopted by a number of the classical bridging ligands given in the table are appended below.

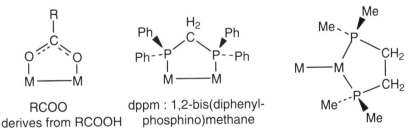

Compound	Structure	M−M (Å)	Bridging ligands
$Re_2Cl_8^{2-}$	D_{4h}	2.241	no
$Re_2Cl_6(PEt_3)_2$	C_{2h}	2.222	no
$Re_2(O_2CC_6H_5)_4Cl_2$	D_{4h}	2.235	yes
$Mo_2Cl_8^{4-}$	D_{4h}	2.139	no
$Mo_2Br_8^{4-}$	D_{4h}	2.135	no
$Mo_2Cl_4(PMe_3)_4$	D_{4d}	2.130	no
$Mo_2(O_2CCH_3)_4$	D_{4h}	2.093	yes
$Mo_2(O_2CC_6H_5)_4$	D_{4h}	2.096	yes
$Mo_2Cl_4(dppm)$	D_{2h}	2.138	yes
$W_2(O_2CCF_3)_4$	D_{4h}	2.211	yes
$W_2(CH_3)_8^{4-}$	D_{4h}	2.264	no
$W_2Cl_4(PMe_3)_4$	D_{2h}	2.262	no
$W_2Cl_4(dmpe)_2$		2.287	no
$Cr_2(O_2CCH_3)_4$	D_{4h}	2.288	yes
$CrMo(O_2CCH_3)_4$	pseudo D_{4h}	2.050	yes

(the counter-ion is not indicated for the ionic structures)

Certain Cr_2 derivatives are considered to have 'ultra short' bonds; examples of these include complex **1** where the Cr−Cr bond is 1.847 Å[107] and complex **2** having a Cr−Cr distance of 1.895 Å.[108] As usual when very short bonds are found, these complexes have four bridging ligands. The interesting exchange reaction of lithium dimethoxyphenylide (LiDMP) with chromium (II) acetate which leads to complex **1** is rather typical of the chemistry of these compounds:

$$Cr_2(O_2CCH_3)_4 + LiDMP \rightarrow Cr_2(DMP)_4 + 4\ CH_3 - COOLi$$

(in LiDMP, the Li is situated between the two methoxy groups)

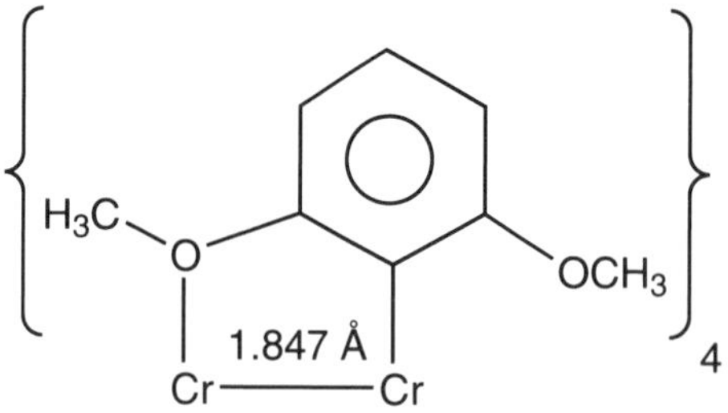

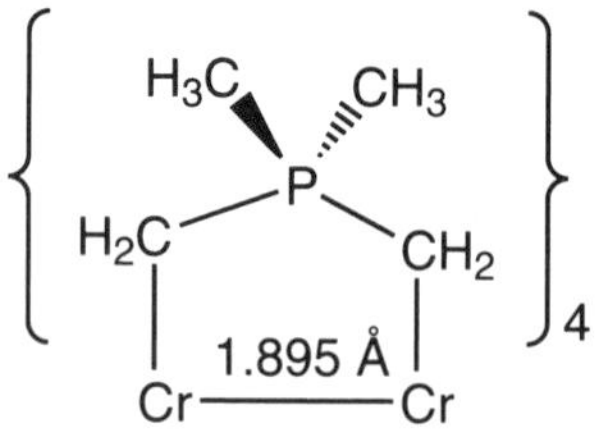

Conclusion

This section has demonstrated the great variety of bimetallic metal–metal bonded structures, and explained a little of their chemistry. It should be quite clear that an even higher level of complexity would be found were we to treat the case of larger polymetallic complexes, whose chemistry is under intensive investigation. Thus we will do no more than mention their existence here.

Throughout this section, we have shown that the notion of a bond can be generalized with the help of simple *orbital* concepts and schemes. However, we must be careful not to put the cart before the horse, and therefore underline that theory should be regarded as a workable model of reality rather than a substitute for experimentally verified data. The bond order concept provides a practical method of classification, but does not account for all of the electronic properties of a bimetallic complex.

3.8 Phosphorus Compounds

Heteroatomic compounds having a lone pair, such as ROH, R_2O, RSH, R_2S, R_3N, R_3P, etc., play a fundamental role as ligands in coordination chemistry. Amongst these, phosphorus ligands have aroused an ever-increasing interest since the pioneering studies of Chatt and Mann in 1938. This interest reflects two factors: the fine control which can be exerted over the sterical properties of organophosphorus compounds, and their ability to play the role of a σ donor and π acceptor simultaneously. This effect allows the ligand to modulate and adapt to the electronic characteristics of the metallic centre. The most significant and widely used phosphorus ligands are the phosphanes R_3P; amongst these, the phosphines (R = H, alkyl, aryl) and the phosphites (R = alkoxy or H) are the most important. They behave as terminal two-electron ligands: $R_3P \rightarrow M$.

Phosphane Synthesis

The nucleophilic attack of an organometallic reagent on a phosphorus halide probably constitutes the most common synthesis of phosphanes. When generalized to include non-carbon nucleophiles such as RO^-, RS^-, R_2N^-, ..., it gives phosphanes having P−OR, P−SR, P−NR$_2$, ... bonds:

$$PX_3 + 3\,RM \longrightarrow PR_3 \qquad M = Li, MgX, ZnX, SnR_3 \ldots$$

$$R_2PX + R'M \longrightarrow R_2PR' \qquad X = Cl, Br, OR \ldots \text{ solvent: } THF, Et_2O \ldots$$

An example of a nucleophilic attack is found in the fourth stage of the preparation of methyldiphenylphosphine from PCl_3:

$$\text{⬡} + PCl_3 + AlCl_3 \longrightarrow PhPCl_2, AlCl_3 \xrightarrow{L} L, AlCl_3 + PhPCl_2$$
$$L = \text{pyridine}$$

The first step, an electrophilic attack of $Cl_2P^+ AlCl_4^-$ on benzene, leads to the aromatic substitution of one hydrogen by phosphorus. In the second, the product $PhPCl_2$ is displaced from its Lewis base adduct by a more powerful donor.

$$2\,PhPCl_2 \xrightarrow[AlCl_3]{\Delta} Ph_2PCl + PCl_3$$

After a redistribution of aryl substituents catalysed by $AlCl_3$, a classical organometallic attack is employed:

$$Ph_2PCl + MeLi \longrightarrow Ph_2PMe + LiCl$$

The recent synthesis of tris(tetrathiafulvalenyl)phosphine[109] also uses a nucleophilic attack of this type:

$$3\ TTFLi + PBr_3 \longrightarrow P(TTF)_3$$

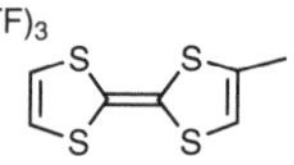

Such phosphines, capable of linking redox-active TTF substituents to other metal centres through the phosphorus atom, have great potential in coordination chemistry.

The inverse procedure exploits the reactivity of the phosphorus anions:

$$[R_2P]^{\ominus}\ M^{\oplus} + R'X \longrightarrow R_2PR'$$

M = Li, Na, K ; X = Cl, Br, I, OTs; solvent: in general THF

To illustrate this scheme, we outline the synthesis of DIOP developed by the French chemist Kagan.[110] This phosphine was a prototype for the optically active phosphines which are now routinely used to catalyse the asymmetric hydrogenation of olefins (see Section 5.2).

$$Ph_3P + 2\ Li \longrightarrow Ph_2PLi + PhLi$$

Alkali metals are capable of cleaving the P—Ph bond of triphenylphosphine in solvents such as THF. The mechanism involves a single electron transfer from the metal to the PPh_3 LUMO; this gives a radical anion which subsequently decomposes:

$$[Ph_3P]^{-\bullet} \longrightarrow Ph^{\bullet} + [Ph_2P]^{\ominus}$$

The co-reagent is a ditosylate derivative of D- or L-tartaric acid, one of the cheapest naturally occurring sources of optically active starting materials:

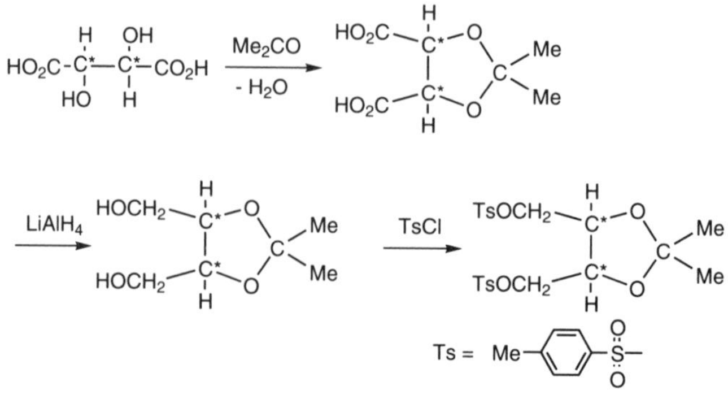

Reaction of the ditosylate with the phosphide gives DIOP.

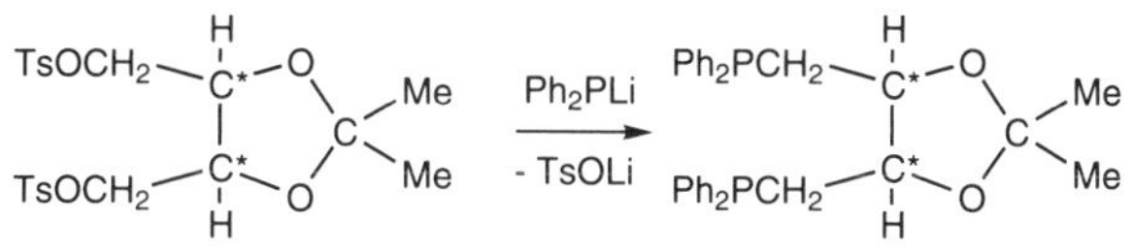

The reduction of pentavalent phosphorus derivatives provides another important route to phosphanes, for example:

$$R_3P{=}O\ (R_3P{=}S) \xrightarrow{\text{reduction}} R_3P$$

$$RP(O)X_2 \xrightarrow{\text{reduction}} RPH_2 \qquad (X = Cl, Br, OR'...)$$

$$R_2P(O)X \xrightarrow{\text{reduction}} R_2PH$$

The great strength of the P=O bond: D(P=O) ≈ 130–140 kcal/mol means that powerful (often non-selective) reducing agents such as $LiAlH_4$, $HSiCl_3$, $PhSiH_3$ or Si_2Cl_6 are usually necessary for the reduction of phosphine oxides. For the more easily reduced thiophosphorylated compounds, D(P=S) ≈ 90 kcal/mol, reagents such as $LiAlH_4$, Si_2Cl_6, Ni, Fe, and Bu_3P may be used. An elegant synthesis of a five-membered ring based upon the McCormack reaction, named after its discoverer at EI Dupont de Nemours, shows how such reductions can be used to allow straightforward entries into rather complex structures:

$$PCl_3 + AlCl_3 + MeCl \longrightarrow [Me\text{-}PCl_3]^{\oplus}[AlCl_4]^{\ominus} \xrightarrow{Al} MePCl_2$$

This is a simple example of the Kinnear-Perren reaction, applicable to a wide variety of alkyl halides. It involves the attack of the PCl_3 lone pair on the methyl carbocation $[Me]^+$, $[AlCl_4]^-$. The product is a pentavalent phosphonium salt, which may be reduced to the dichlorophosphine with aluminium powder.

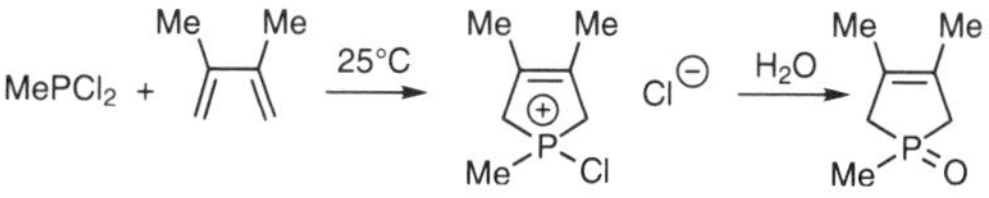

The reaction involves the lone pair (HOMO) and the low-lying LUMO in a cheletropic [4 + 1] cycloaddition. The methyldichlorophosphine can be replaced by RPX_2 or PX_3 (X = Cl or Br, R = alkyl or aryl). The last step is a classical reduction:

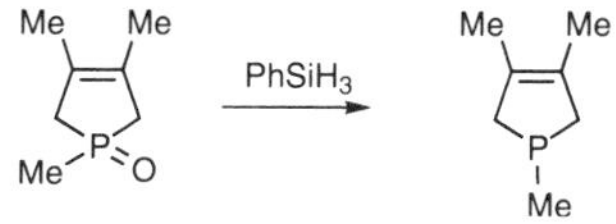

The direct reduction of $MePCl_2$ would give a primary phosphine:

$$MePCl_2 \xrightarrow{LiAlH_4} MePH_2$$

Another important and relatively sophisticated starting material may be prepared in a simple fashion:

$$Cl_3P=S \; + \; 3 \, MeLi \quad \longrightarrow \quad Me_3P=S$$

$$But: \qquad Cl_3P=S \quad \xrightarrow{\; MeMgX \;} \quad \underset{\underset{S \; \; S}{|| \; ||}}{Me_2P-PMe_2}$$

Here, unlike the organolithiums, the alkylmagnesium reagents can function as nucleophiles and reducing agents simultaneously, to give synthetically valuable diphosphines.

$$RMgX \quad \longrightarrow \quad R^{\bullet} \; + {}^{\bullet}MgX \quad \xrightarrow{\; R_2P(S)Cl \;} \quad ClMgX \; + \; R_2P(S)^{\bullet}$$

$$\downarrow \qquad\qquad\qquad\qquad\qquad\quad \downarrow$$

$$R\text{-}R + R\text{-}H + R(\text{-}H) \qquad\qquad\qquad [R_2P(S)]_2$$

Electrochemical studies have shown that the reducing power of Grignard reagents decreases with decreasing basicity: $R = Me_3C > Me_2CH > Et > Me > Ph$, which explains why the arylmagnesiums do not undergo this type of chemistry. The difference between lithium and magnesium reflects the greater ionization of the $C^{\delta-}-Li^{\delta+}$ bond. Lithium is substantially more electropositive than magnesium as the following data shows:

- electronegativity: Li 0.74; Mg 1.56
- ionization potentials: $Li \rightarrow Li^+$ 5.377 V; $Mg \rightarrow Mg^+ \rightarrow Mg^{2+}$ 7.646 and 15.035 V.

$$\underset{\underset{S \; \; S}{|| \; ||}}{Me_2P-PMe_2} \quad \xrightarrow[- \, Bu_3P=S]{\; Bu_3P \;} \quad Me_2P-PMe_2$$

The diphosphine is spontaneously inflammable in air, but is less reducing than tributylphosphine which, therefore, may be used to remove the sulfur.

The quarternization of phosphanes by alkyl halides is one of their most typical reactions:

$$R_3P \; + \; R'X \quad \longrightarrow \quad [R_3PR']^{\oplus}X^{\ominus}$$

Essentially, it involves the nucleophilic attack of the phosphorus lone pair on the electrophilic carbon of the C−X bond. If the phosphane has an alkoxy substituent, the phosphonium salt evolves further:

$$R_2P\text{-}OR' \quad \xrightarrow{\; R''X \;} \quad \underset{R''}{R_2\overset{\oplus}{P}\text{-}O\text{-}R'} \; + \; X^{\ominus} \quad \xrightarrow{\; \Delta \;} \quad \underset{R''}{R_2P=O} \; + \; R'X$$

This transformation, known as the Arbuzov reaction, is driven by the energy gain during conversion of the P−O single bond into a double bond: $D(P=O)-D(P-O) \approx 50$ kcal/mol.

The oxidation state 5 is thermodynamically favoured, but it is possible to return to the trivalent state through 'dequarternization'. Thus, the hydrolysis of a phosphonium salt can lead to either a phosphine (attack on R) or to a phosphine oxide (attack on P):

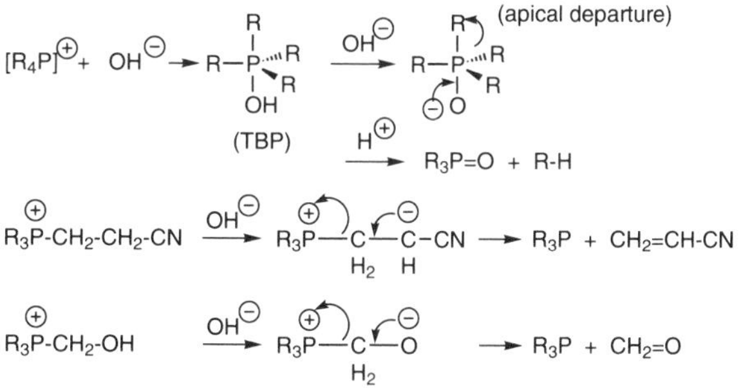

The phosphonium salt can also be converted into a phosphine electrochemically, or by chemical reduction using a hydride.

$$R_4P^{\oplus} + H^{\ominus} \longrightarrow [R_4P\text{-}H] \xrightarrow{\text{reductive elimination}} R_3P + R\text{-}H$$

$$R_4P^{\oplus} + e^{-} \longrightarrow R_4P^{\bullet} \longrightarrow R_3P + R^{\bullet}$$

For the reaction to proceed cleanly, either the R substituents should be the same, or one should be a significantly better leaving group than the others. Benzyl is often the preferred leaving group because both the $PhCH_2^-$ anion and the $PhCH_2^{\bullet}$ radical are relatively stable delocalized species. The following example is representative:

$$(PhCH_2)_3P + EtBr \longrightarrow [(PhCH_2)_3P\text{-}Et]^{\oplus} \xrightarrow{H^{\ominus}} (PhCH_2)_2P\text{-}Et + PhCH_3$$

One of the many important differences between phosphorus and nitrogen chemistry is the relative strength of their bonds to hydrogen. Thus, the N–H bond is quite strong: D(N–H) ≈ 93 kcal/mol, but the P–H bond is relatively weak: D(P–H) ≈ 77 kcal/mol. Consequently, the P–H functionality can be added across a wide variety of unsaturated molecules (alkenes, alkynes, carbonyls). The reaction scheme with olefins is as follows:

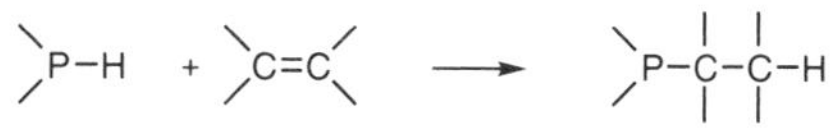

The reaction can be catalysed by bases, Lewis acids or free radicals. In the last case, careful choice of the radical is necessary to limit competing olefin polymerization. Base catalysis is particularly efficient for activated (electron-poor) olefins, where the catalytic cycle can be represented as follows:

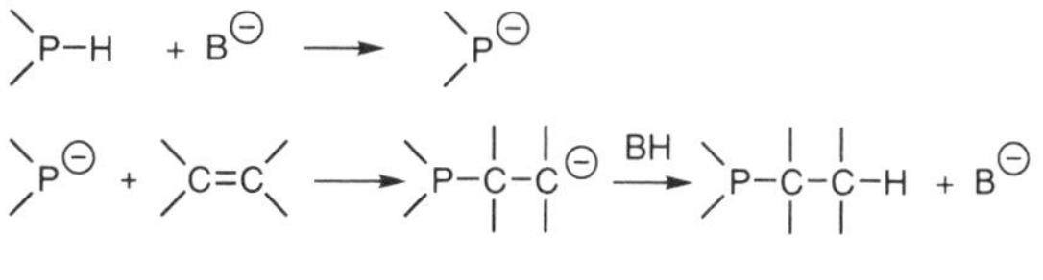

Suitable bases include potassium hydroxide, butyllithium and potassium *tert*-butoxide. The reaction of P–H bonds with acrylonitrile provides a good example:

$$PH_3 + 3\ CH_2{=}CH{-}CN \xrightarrow[\text{H}_2\text{O, CH}_3\text{CN}]{\text{KOH}} P(CH_2CH_2CN)_3$$

A recent application of this scheme permitted the first synthesis of phosphanes having a fullerene-type substituent:[111]

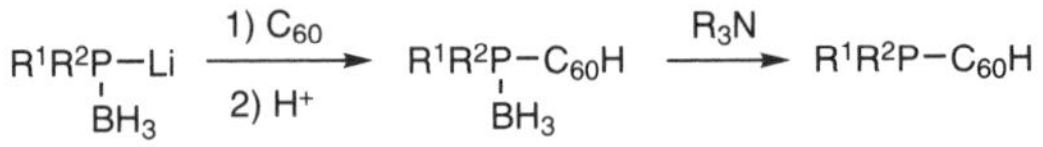

$$R^1R^2P{-}\underset{\overset{|}{BH_3}}{Li} \xrightarrow[\text{2) H}^+]{\text{1) C}_{60}} R^1R^2P{-}\underset{\overset{|}{BH_3}}{C_{60}H} \xrightarrow{R_3N} R^1R^2P{-}C_{60}H$$

The addition of the phosphide ion occurs at a double bond between two carbon hexagons.

Acidic catalysis, by sulfonic acids, phosphoric acids, boron trifluoride, or transition metal salts (NiCl$_2$, PtCl$_4^{2-}$, etc.) is preferred with electron-rich olefins. In this case, it is the olefin rather than the phosphine that is activated by the catalyst:

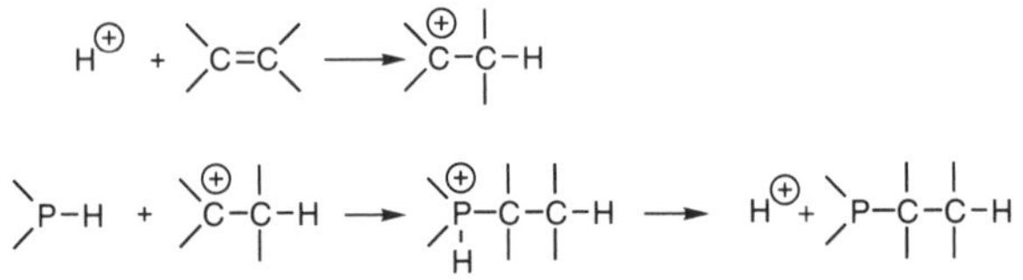

Nonetheless, the most general procedure is a radical-induced catalysis using organic peroxides, azobisisobutyronitrile, or UV irradiation:

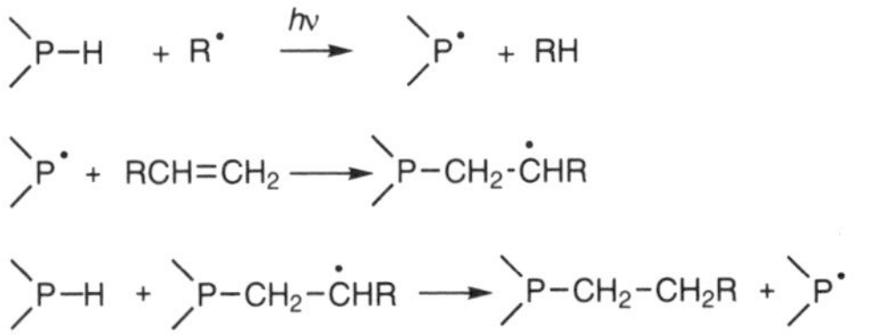

For example, it is used in the industrial preparation of tributylphosphine from 1-butene:

$$PH_3 + 3\ EtCH{=}CH_2 \longrightarrow PBu_3$$

and may also be used to prepare cyclic phosphines:

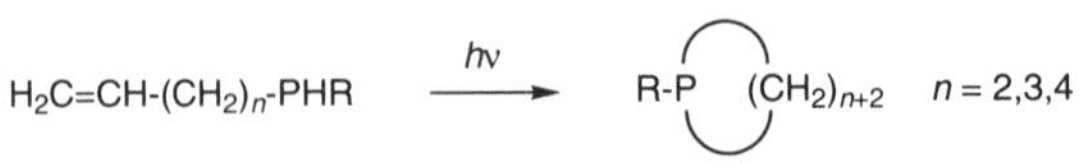

In the case of alkynes, the general scheme is very similar:

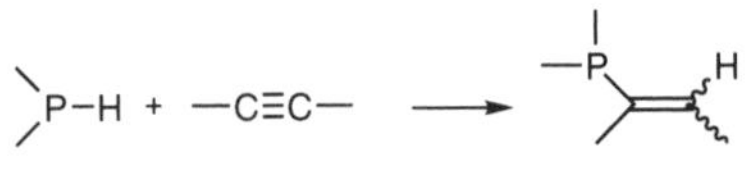

but the stereochemistry of the addition at the double bond is not clearly defined. Additionally, a second reaction can occur to the first-formed vinylphosphine, which means that the potential of the reaction is rather limited. Nonetheless, it is useful for the production of heterocycles:

$$R-C\equiv C-C\equiv C-R \ + \ PhPH_2 \ \xrightarrow{\ BuLi\ } \ R-\!\!\overbrace{}^{}\!\!-R \quad \text{(phosphole)}$$

Finally, carbonyl compounds also insert into the P—H bond without difficulty:

$$\text{\textbackslash}P-H \ + \ \text{\textbackslash}C{=}O \ \longrightarrow \ \text{\textbackslash}P-\overset{|}{\underset{|}{C}}-OH$$

The industrial synthesis of tris(hydroxymethyl)phosphine is a good example of this reaction:

$$PH_3 \ + \ 3\,H_2C{=}O \ \longrightarrow \ P(CH_2OH)_3$$

However, in cases where the phosphorus atom is appreciably nucleophilic, a migration of the oxygen function is observed. This obviously limits the applications of the method:

$$\text{\textbackslash}P-C\underset{OH}{\overset{/}{\text{\textbackslash}}} \ \longrightarrow \ \left[\ \overset{H}{\underset{O}{P{-}C}}\ \right] \ \longrightarrow \ \text{\textbackslash}P\overset{||}{\underset{O}{-}}C-H$$

The Stereoelectronic Properties of Phosphanes and their Implications in Coordination Chemistry

Traditionally, the metal–phosphorus bond is described analogously to the M—CO interaction, with phosphorus simultaneously playing the role of a σ donor (through its lone pair) and a π acceptor. Initially, it was thought that the vacant 3d orbitals were responsible for the π-acceptor properties, but the unoccupied σ* of the P—R bonds are now considered to be the principal back-acceptors.[112] Thus PF_3 is a good π acceptor because the empty σ*s of the P—F bonds are low in energy.

The strength of the M—P bond reflects both its σ-donor and π-acceptor components. To a first approximation, the σ-donor interaction can be estimated through the strength of the interaction between the phosphane and a naked proton: $R_3P + H^+ \rightarrow [R_3P{-}H]^+$. However, for phosphanes of low basicity, it is much easier to measure the lone pair ionization energy: $R_3P \rightarrow R_3P^+ + e^-$. The values in Table 3.6 show that this σ-donor power varies considerably according to the nature of the substituents at phosphorus.

The π-acceptor ability can be measured by the electron affinity: $R_3P + e^- \rightarrow [R_3P]^-$ of the phosphorus centre. The results of such measurements, and related theoretical calculations of LUMO energies, show that some phosphanes are excellent acceptors. The classification of the ligands in declining order of π-acceptor ability is the following:

NO > CO ≈ RNC ≈ PF_3 > PCl_3 > $PCl_2(OR)$ > PCl_2R >
PBr_2R > $PCl(OR)_2$ > $PClR_2$ > $P(OR)_3$ > PR_3 ≈ SR_2 > RCN > RNH_2 ≈ OR_2.

Table 3.6 Basicity of phosphines

Phosphine	pK_a	Phosphine	pK_a
$^tBu_3P$	11.4	Ph_2PMe	4.6
Et_3P	8.7	Ph_3P	2.73
Me_3P	8.65	Ph_2PEt	2.62
$^nBu_3P$	8.43	$P(4-ClC_6H_4)_3$	1.0
Me_2PPh	6.5		

A few gas phase ionization energies complete this table:

$$PPh_3 \ 7.80; \ PH_3 \ 10.58; \ PCl_3 \ 10.59; \ PF_3 \ 12.23 \ eV.$$

An analysis of the infrared spectrum of phosphane-$Ni(CO)_3$ complexes provides the most common and simple way to estimate the combined donor–acceptor character of a phosphane. The C_{3v} symmetry of the $Ni(CO)_3$ fragment means that the IR spectrum comprises only two CO bands, corresponding to the A_1 and E modes (see Section 3.2). The position of the higher frequency A_1 mode is used to classify the overall donor–acceptor properties (Table 3.7).

Table 3.7 Frequency of A_1 mode in $R_3P \rightarrow Ni(CO)_3$

Phosphane	$v \ (cm^{-1})$	Phosphane	$v \ (cm^{-1})$	
$^tBu_3P$	2056.1	Ph_3P	2068.9	stronger
$^nBu_3P$	2060.3	$Ph_2P(OMe)$	2072	π acceptor
Et_3P	2061.7	$(MeO)_3P$	2079.8	$\downarrow$
Et_2PPh	2063.7	F_3P	2111	
Me_3P	2064.1			

Phosphanes are pyramidal compounds, whose inversion barrier lies in the range 30–35 kcal/mol. This means that phosphorus does not invert at room temperature (unlike nitrogen whose barrier is much lower at less than 10 kcal/mol). This non-inversion permits the preparation of phosphines having a chiral centre localized at phosphorus, which are useful in asymmetric catalysis (see Section 5.2). This pyramidality has further consequences, particularly because the ability of phosphorus to coordinate to a metal centre depends partially upon how closely it can approach the metal. The M—P distance will be increased (and the M—P bond weakened) as the steric hindrance of the phosphorus ligand increases. This hindrance is measured by the 'Tolman cone angle'[113] (see scheme) which reflects the pyramidal nature of phosphorus and the intrinsic steric bulk of each substituent.

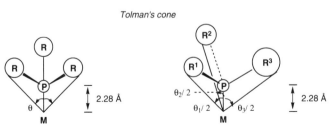

Tolman's cone

symmetrical phosphine PR_3 non-symmetrical phosphine $PR^1R^2R^3$

By convention, the metal is placed at 2.28 Å from the phosphorus atom (a typical Ni−P distance). A cone is then drawn which just touches the outermost atoms of the groups bound to phosphorus and comes to an apex at the metal. The Tolman angle measured at this apex then gives an indication of the steric demand of the phosphine. When the phosphorus is not symmetrical, the average value of the angles for each of the different groups attached to phosphorus is taken:

$$\theta = 2/3 \sum_{i=1}^{3} \theta_i/2$$

Table 3.8 gives some θ values. It can be seen that the phosphane bulk can be modified to a considerable degree by variation of its substituents.

Table 3.8 Cone angles of some phosphanes

Phosphane	θ (°)	Phosphane	θ (°)
PH^3	87	$Ph_2P-\hexagon$	153
$PhPH_2$	101	$(Me_2N)_3P$	157
$MePH_2$	103	$^iPr_3P$	160
PF_3	104	$(PhCH_2)_3P$	165
$(MeO)_3P$	107	$\left(\hexagon-\right)_3 P$	170
Me_3P	118	$(^tBuO)_3P$	172
PCl_3	124	$^tBu_3P$	182
Ph_2PH	128	$(C_6F_5)_3P$	184
$(PhO)^3P$	128	$\left(\hexagon-\right)_3 P$ (Me)	194
PBr_3	131		
Et_3P	132	$\left(Me-\hexagon-\right)_3 P$ (Me, Me)	212
Ph_3P	145		

Finally, it is clear that the type of metal influences the strength of the M−P interaction. The phosphanes are polarizable ligands with a marked π-acceptor character (or 'soft' ligands in the sense of Pearson's HSAB theory). They will therefore preferentially coordinate to 'soft' metal fragments; this differentiates them from the less polarizable amines and ethers, for example, which prefer to coordinate with 'hard' metal centres. As a general rule, the polarizability (softness) of a metal centre increases with the number of filled subshells (screening effect) and with the number of valence electrons. The 'soft' metal centres, which coordinate preferentially to phosphines, are thus electron-rich (therefore usually in low oxidation states) and are associated with transition metals situated to the bottom and the right of the periodic table.

Other Types of Phosphorus Ligands

In addition to phosphanes, there are many other kinds of phosphorus ligands. The majority, which involve one- or two-coordinate phosphorus, have been the subject of well documented studies.[114] We will briefly review the most important cases.

The Phosphinidenes [RP]

These are carbene-like species, which have monovalent phosphorus centres and triplet ground states. They are not stable in the free form, but have recently been prepared and characterized as follows in a low temperature matrix-isolation study:[115]

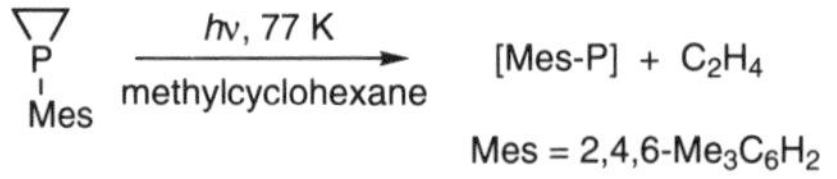

$$\underset{\substack{\text{P}\\|\\\text{Mes}}}{\triangledown} \xrightarrow[\text{methylcyclohexane}]{h\nu,\ 77\ \text{K}} \text{[Mes-P]} + C_2H_4$$

Mes = 2,4,6-Me$_3$C$_6$H$_2$

The organic chemistry of phosphinidenes is still underdeveloped, but their coordination chemistry is very rich. Six types of phosphinidene complexes can be distinguished:

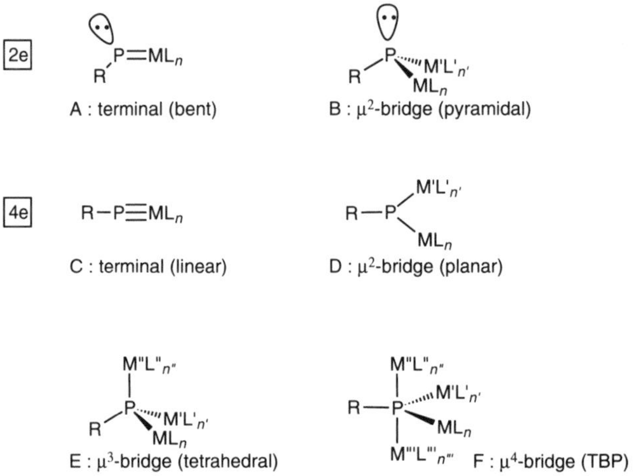

2e

A : terminal (bent)

B : μ^2-bridge (pyramidal)

4e   R−P≡ML$_n$

C : terminal (linear)

D : μ^2-bridge (planar)

E : μ^3-bridge (tetrahedral)

F : μ^4-bridge (TBP)

The terminal two-electron type (A) is the phosphorus analogue of the metal–carbene complexes that we have already discussed. As in the carbon case, the electrophilic $RP^{\delta+} = ML_n$ and nucleophilic complexes $RP^{\delta-} = ML_n$ have distinctly different behaviour. The electrophilic complexes, which are only known as unstable transient species, have a very rich chemistry. Reactions with carbon–carbon double and triple bonds give three-membered phosphorus-containing rings:

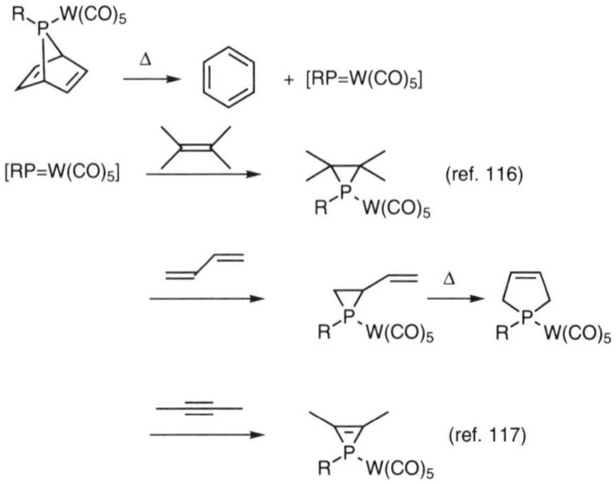

Phosphines give phospha-ylides and carbene complexes give phosphaalkenes:

$$[RP=W(CO)_5] \quad + R^1{}_3P \quad \longrightarrow \quad \overset{\ominus}{R}P\overset{\oplus}{-}PR^1{}_3 \quad (ref.\ 118)$$
$$\underset{W(CO)_5}{|}$$

$$[RP=W(CO)_5] \quad + R^1{}_2C=Cr(CO)_5 \quad \longrightarrow \quad \left[\begin{array}{c} R^1 \\ R^1 \end{array}\!\!>\!\!\begin{array}{c} C-Cr(CO)_5 \\ \diagup \\ P \end{array}\!\!<\!\!\begin{array}{c} \\ W(CO)_5 \end{array} \right]$$

$$\xrightarrow[-\ Cr(CO)_5]{\Delta} \quad \underset{(OC)_5W}{\overset{R}{\diagdown}}P=C\overset{R^1}{\underset{R^1}{\diagup}} \quad (ref.\ 119)$$

The nucleophilic phosphinidene complexes appear to have greater stability, which has allowed their isolation in a number of cases.[120]

$$Cp_2M\overset{.H}{\underset{\diagdown Li}{\diagup}} \quad + ArPCl_2 \quad \longrightarrow \quad Cp_2M=PAr$$

$$Ar = 2,4,6\text{-}^tBu_3C_6H_2$$

$$M = Mo,\ W$$

Their chemistry, which is of the 'phospha-Wittig' type, strongly resembles that of Schrock's carbenes.[121]

$$\underset{PMe_3}{Cp_2Zr=PAr} + Ph(R)C=O \quad \longrightarrow \quad ArP=C(R)Ph + \underset{PMe_3}{[Cp_2Zr=O]}$$

The μ^2 two-electron phosphinidene complexes (type B) may be considered as phosphanes having organometallic substituents. They are well known but their chemistry has not been subjected to systematic study because it is a combination of the classical reactivity of the lone pair and the metal phosphorus σ bond.

The first terminal four-electron phosphinidene complexes (type C) were characterized recently.[122] Type D was discovered by Huttner in 1975:[123]

$$L_nM\overset{\overset{Ph}{|}}{\underset{\underset{H}{|}}{-}P-H} \xrightarrow{BuLi} L_nM\overset{\overset{Ph}{|}}{\underset{\underset{Li}{|}}{-}P-H} \xrightarrow{RNCl_2} \overset{Ph}{\diagdown}\overset{\overset{Ph}{\overset{|}{P}-ML_n}}{\underset{\underset{\underset{L_nM}{}}{P}\ \overset{|}{\underset{Ph}{P}-ML_n}}{}} \xrightarrow{\Delta} L_nM\overset{\overset{Ph}{\overset{|}{P}}}{\diagdown}ML_n$$

$$ML_n = CpMn(CO)_2$$

The C and D complexes, along with those in class A, are characterized by their extreme ^{31}P NMR chemical shifts. Classical phosphorus derivatives generally lie in a range from -100 to $+100$ ppm with respect to H_3PO_4. Here, chemical shifts of $+800$ ppm are relatively commonplace. The possible interconversion of type A with the type B should be noted:

$$L_nM\overset{\overset{R}{\overset{|}{P}}}{\diagup\diagdown}ML_n \quad \rightleftharpoons \quad L_nM\overset{\overset{\ddot{}}{\underset{}{P}}{\diagdown}^R}{--}ML_n$$

$$(4e,\ planar) \qquad\qquad (2e,\ pyramidal)$$

The longest-known complexes, types E and F, are the normal products formed in reactions between dihalophosphines and organometallic anions. They may also be prepared by the thermal decomposition of primary phosphine complexes $L_nM \leftarrow P(R)H_2$. The class of μ^3 and μ^4 phosphinidenes are often used to build polymetallic clusters.[124] As a general rule, the presence of bulky substituents R on the phosphorus stabilizes those phosphinidene complexes whose phosphorus is relatively weakly coordinated (A–D).

The Phosphides $[R_2P]^-$

The phosphide anions R_2P^- are easily made by metallation of secondary phosphines:

$$R_2PH \xrightarrow{\text{BuLi}} R_2P^{\ominus} \; Li^{\oplus}$$

Their complexes may be prepared from R_2PH, R_2P^-, R_2PX (X = Cl, Br) or R_2P-PR_2. The three following types can be defined:

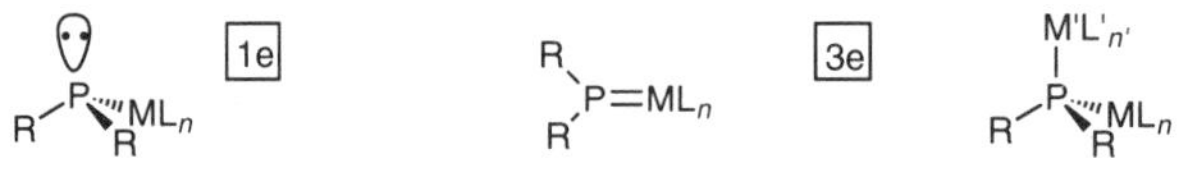

A : terminal (pyramidal)   B : terminal (planar)   C : μ^2-bridge (tetrahedral)

The first and third types are classical. The second, discovered much more recently (1978), contains a metal phosphorus double bond whose chemistry has been studied in depth.

The phosphorus has nucleophilic character at its lone pair in the type A complexes:

On the contrary, phosphorus is electrophilic in the B types which, consequently, are best considered as derivatives of the phosphenium cation R_2P^+ (a species isoelectronic with the silylenes R_2Si and carbenes). The following example is characteristic:

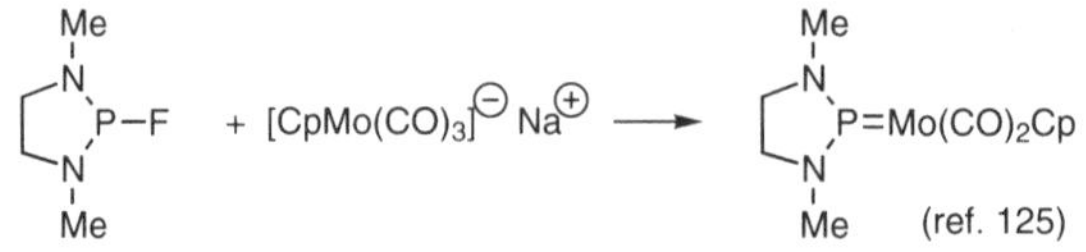

The substantial shortening of the P=Mo distance 2.213 Å with respect to single P–Mo bonds 2.40–2.57 Å indicates the presence of the double bond. When the metal has a spin, the $^1J(PM)$ coupling constants become very large in the type B complexes: $^1J(P=W) = 550$–850 Hz versus 200–400 Hz for $^1J(P-W)$. The P=W double bond undergoes 1,2 addition and [2 + 1] cycloaddition reactions, for example:[126]

$$CpW=P\begin{smallmatrix}{}^tBu\\{}^tBu\end{smallmatrix} \quad \xrightarrow{\text{HCl}} \quad CpW-P\begin{smallmatrix}{}^tBu\\Cl\\{}^tBu\end{smallmatrix}$$

(CO ligands)

$$\xrightarrow{[X]} \quad CpW\underset{X}{-}P\begin{smallmatrix}{}^tBu\\{}^tBu\end{smallmatrix}$$

X = S, Se, CH$_2$ (from CH$_2$N$_2$), MeP (from (MeP)$_5$),
Fe(CO)$_4$ (from Fe$_2$(CO)$_9$)

Finally, the action of a ligand L can convert a B type complex into an A type:

$$R_2P=ML_n \quad + \ L \quad \longrightarrow \quad R_2P-ML_{n+1}$$

The Phosphoranides [R$_4$P]$^-$

A few terminal one-electron complexes of the type R$_4$P$-$M have been discovered recently, for example:[127]

$$\left[\text{Ph}-P\underset{O}{\overset{(CO)_3}{\diagdown}}Mo\underset{N}{-}H \right]^{\oplus} \quad \xrightarrow[60°C]{\text{MeLi}} \quad Cp(OC)_2Mo-P-N$$

Methyllithium eliminates the proton at nitrogen, provoking P$-$N bond formation. The molybdenum then migrates from the P$-$N bond to one of the P$-$O bonds.

The Phosphaalkenes [RP=CR$_2$] and π Phosphorus Complexes

The first stable phosphaalkene was discovered by Becker in 1976.[128] In this compound, the phosphorus atom is sp^2 hybridized and the lone pair, P$-$R single bond and P=C double bond are coplanar, making angles of 120° with each other. The double bond length of 1.60–1.70 Å may be compared with 1.84 Å for a P$-$C single bond. The π system is weakly polarized in the sense P$^{\delta+}-$C$^{\delta-}$. The frontier orbitals are the lone pair and π-bonding combinations, which lie very close in energy. The main synthetic routes to phosphaalkenes involve either a silatropic rearrangement or a 1,2-elimination reaction:

$$RP(SiMe_3)_2 + R^1C(O)Cl \longrightarrow RP\begin{smallmatrix}C(O)R^1\\SiMe_3\end{smallmatrix} \xrightarrow{\text{silatropy}} RP=C\begin{smallmatrix}OSiMe_3\\R^1\end{smallmatrix}$$
(ref. 128)

$$MesPCl_2 \xrightarrow[\text{2) R}_3\text{N}]{\text{1) Ph}_2\text{CHMgX}} MesP=CPh_2 \quad \text{(ref. 129)}$$

Mes = 2,4,6-Me$_3$C$_6$H$_2$

A transposition of the Wittig reaction gives an easy access to phosphaalkenes coordinated at the lone pair:[130]

$$RP-P(O)(OEt)_2 \ + \ R^1R^2C=O \ \longrightarrow \ RP=CR^1R^2 \ + \ (EtO)_2PO_2^{\ominus}$$
$$\underset{W(CO)_5}{\quad} \qquad\qquad\qquad\qquad \underset{W(CO)_5}{\quad}$$

In addition to reactions at the lone pair, numerous 1,2 additions and [2+1], [2+3] and [2+4] cycloadditions are observed; these underline the fundamental analogy between P=C and C=C double bonds.[131] The phosphaalkenes also form three types of metal complexes, implicating both the lone pair and the π system in M−L bonds:

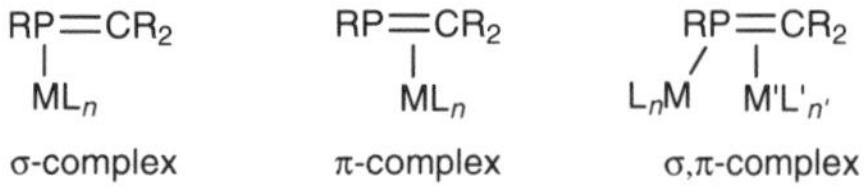

$$\underset{ML_n}{RP=CR_2} \qquad \underset{ML_n}{RP=CR_2} \qquad \underset{L_nM \quad M'L'_{n'}}{RP=CR_2}$$

σ-complex   π-complex   σ,π-complex

The low energy separation of the π and n orbitals may lead to interconversions between the σ- and π-type coordination modes. Such changes can easily be detected by ^{31}P NMR measurements:[132]

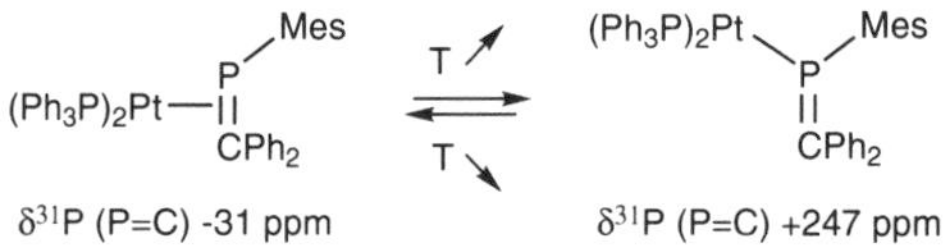

$$\delta^{31}P \ (P=C) \ {-}31 \ ppm \qquad\qquad \delta^{31}P \ (P=C) \ {+}247 \ ppm$$

The existence of P=C and P=P double bonds has spawned a whole series of phosphorus analogues of classical carbon π complexes.[133] The major classes known at present are given in Table 3.9. Except for four π-electron antiaromatic rings, all the corresponding phosphorus species are stable in the free state. Phosphaalkynes RC≡P are also well known and have a very rich and varied coordination chemistry.[134]

Table 3.9 Phosphorus analogues of π complexes

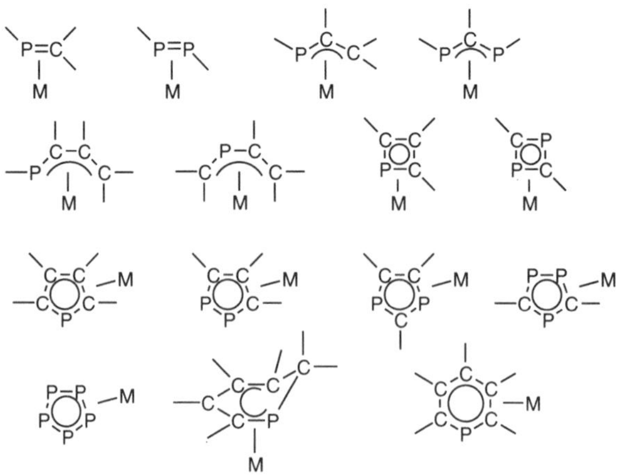

Detection of Phosphorus Derivatives

The only stable natural isotope, phosphorus-31 (^{31}P), has a nuclear spin of 1/2. Its convenience and high sensitivity mean that ^{31}P NMR is the most important technique for the analysis and detection of phosphorus compounds. However, there is no simple overall correlation between the ^{31}P chemical shift and the coordination number or stereoelectronic properties at phosphorus. Table 3.10 gives examples of NMR parameters for some typical phosphorus-containing compounds.

Table 3.10 ^{31}P chemical shifts of some representative compounds*

Compound	δ^{31}P	Coordination number of P
$^{t}Bu-C{\equiv}P$	-69	1
Na^{+}, Ph_2P^{-}	-22	
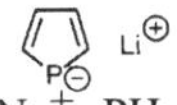 $Li^{\oplus}$	$+75$	
Na^{+}, PH_2^{-}	-303	
$(Me_3Si)_3C-P{=}P-C(SiMe_3)_3$	$+600$	2
$^{t}Bu-P{=}N-^{t}Bu$	$+472$	
$PhP{=}CH-NMe_2$	$+70$	
$Me_3Si-P{=}CPh_2$	$+286$	
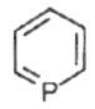	$+211$	
PH_3	-241	
PCl_3	$+219$	
PBr_3	$+227$	
PMe_3	-62	
PPh_3	-8	
$MePH_2$	-163	
$MePCl_2$	$+191$	
$PhPH_2$	-122	3
$PhPCl_2$	$+165$	
Me_2PH	-98	
Me_2PCl	$+94$	
Ph_2PH	-41	
Ph_2PCl	$+81$	
$(MeO)_3P$	$+141$	
$(PhO)_3P$	$+127$	
$O{=}PCl_3$	$+3$	
$O{=}PMe_3$	$+36$	
$O{=}PPh_3$	$+27$	
$O{=}P(OMe)_3$	$+2$	
$O{=}P(OPh)_3$	-18	4
$S{=}PCl_3$	$+28$	
$S{=}P(OMe)_3$	$+74$	
$S{=}PPh_3$	$+42$	
Me_4P^{+}, I^{-}	$+25$	
$(EtO)_5P$	-71	5
Cl_6P^{-}	-296	6

* 85% H_3PO_4 as external reference; δ positive for downfield shifts (ppm).

In general, complexation of a tricoordinate phosphorus compound induces a small change in chemical shift ($\Delta\delta$), which may be either positive or negative according to the metal. However, complexation of the π bond of a doubly coordinated phosphorus provokes impressive shifts to high field ($\Delta\delta$ $-$ve). Where the metal has a detectable nuclear spin, $^1J(P-M)$ is invariably much higher for σ complexes than for π complexes.

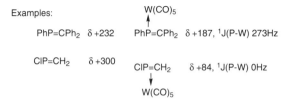

3.9 References

1. W. Hieber and F. Leutert, *Naturwissenschaften*, 1931, **19**, 360.

2. W. Hieber and E. Fack, *Z. Anorg. Allg. Chem.*, 1938, **236**, 83.

3. See: E. A. McNeill and F. R. Scholer, *J. Am. Chem. Soc.*, 1977, **99**, 6243.

4. R. D. Wilson and R. Bau, *J. Am. Chem. Soc.*, 1976, **98**, 4687.

5. T. A. Budzichowski, M. H. Chisholm, J. C. Huffman and O. Eisenstein, *Angew. Chem., Int. Ed. Engl.*, 1994, **33**, 191.

6. S. Martinengo, G. Ciani, A. Sironi and P. Chini, *J. Am. Chem. Soc.*, 1978, **100**, 7096.

7. G. J. Kubas, *Acc. Chem. Res.*, 1988, **21**, 120.

8. P. J. Fagan, J. M. Manriquez, E. A. Maata, A. M. Seyam and T. J. Marks, *J. Am. Chem. Soc.*, 1981, **103**, 6650.

9. M. J. Bunker and M. L. H. Green, *J. Chem. Soc., Dalton Trans.*, 1981, 85.

10. W. Tam, W. K. Wong and J. A. Gladysz, *J. Am. Chem. Soc.*, 1979, **101**, 1589.

11. C. R. Eady, B. F. G. Johnson and J. Lewis, *J. Chem. Soc., Dalton Trans.*, 1975, 2606.

12. J. K. Burdett, *Coord. Chem. Rev.*, 1978, **27**, 1.

13. For a short review, see: L. Weber, *Angew. Chem., Int. Ed. Engl.*, 1994, **33**, 1077.

14. R. Colton and C. J. Commons, *Aust. J. Chem.*, 1975, **28**, 1673.

15. W. A. Herrmann, H. Biersack, M. L. Ziegler, K. Weidenhammer, R. Siegel and D. Rehder, *J. Am. Chem. Soc.*, 1981, **103**, 1692.

16. H. Willner, J. Schaebs, G. Hwang, F. Mistry, R. Jones, J. Trotter and F. Aubke, *J. Am. Chem. Soc.*, 1992, **114**, 8972.

17. G. Hwang, C. Wang, F. Aubke, H. Willner and M. Bodenbinder, *Can. J. Chem.*, 1993, **71**, 1532.

18. F. A. Cotton and C. S. Kraihanzel, *J. Am. Chem. Soc.*, 1962, **84**, 4432.

19. L. H. Jones and B. I. Swanson, *Acc. Chem. Res.*, 1976, **9**, 128.

20. L. J. Todd and J. R. Wilkinson, *J. Organomet. Chem.*, 1974, **77**, 1.

21. F. Calderazzo and G. Pampaloni, *J. Organomet. Chem.*, 1983, **250**, C33.

22. See for example: C. G. Dewey, J. E. Ellis, K. L. Fjare, K. M. Pfahl and G. F. P. Warnock, *Organometallics*, 1983, **2**, 388.

23. B. E. Hanson, *J. Am. Chem. Soc.*, 1989, **111**, 6442.

24. R. S. Berry, *J. Chem. Phys.*, 1960, **32** 933; see also: E. L. Mutterties, *Acc. Chem. Res.*, 1970, **3**, 266.

25. D. Seyferth and R. S. Henderson, *J. Am. Chem. Soc.*, 1979, **101**, 508.

26. H. Schumann, W. Genthe and N. Bruncks, *Angew. Chem., Int. Ed. Engl.*, 1981, **20**, 119.

27. M. F. Lappert, C. L. Raston, B. W. Skelton and A. H. White, *J. Chem. Soc., Chem. Commun.*, 1981, 485.

28. D. L. Reger and P. J. McElligott, *J. Am. Chem. Soc.*, 1980, **102**, 5923.

29. For a review on this topic, see: R. H. Crabtree, *Chem Rev.*, 1985, **85**, 245. Very recently, promising results in the activation of C–H bonds have been obtained with a cationic Ir(III) complex: B. A. Arndtsen and R. G. Bergman, *Science*, 1995, **270**, 1970.

30. J. W. Bruno, G. M. Smith, T. J. Marks, C. K. Fair, A. J. Schultz and J. M. Williams, *J. Am. Chem. Soc.*, 1986, **108**, 40; C. M. Fendrick and T. J. Marks, *J. Am. Chem. Soc.*, 1986, **108**, 425.

31. J. A. Connor, M. T. Zafarani-Moattar, J. Bickerton, N. I. El Saied, S. Suradi, R. Carson, G. Al Takhin and H. A. Skinner, *Organometallics*, 1982, **1**, 1166; J .L. Goodman, K. S. Peters and V. Vaida, *Organometallics*, 1986, **5**, 815.

32. R A. Fischer and S. L. Buchwald, *Organometallics*, 1990, **9**, 871.

33. S. L. Buchwald, B. T. Watson and J. C. Huffman, *J. Am. Chem. Soc.*, 1986, **108**, 7411.

34. S. L. Buchwald, R. J. Lum and J .C. Dewan, *J. Am. Chem. Soc.*, 1986, **108**, 7441.

35. M. Brookhart and M. L. H. Green, *J. Organomet. Chem.*, 1983, **250**, 395.

36. R. Di Cosimo and G. M. Whitesides, *J. Am. Chem. Soc.*, 1982, **104**, 3601.

37. H. W. Turner, R. R. Schrock, J. D. Fellmann and S. J. Holmes, *J. Am. Chem. Soc.*, 1983, **105**, 4942.

38. Review on CO insertion: E. J. Kuhlmann and J. Alexander, *Coord. Chem. Rev.*, 1980, **33**, 195.

39. T. C. Flood, J. E. Jensen and J. A. Statler, *J. Am. Chem. Soc.*, 1981, **103**, 4410.

40. W. J. Evans, A. L. Wayda, W. E. Hunter and J. L. Atwood, *J. Chem. Soc., Chem. Commun.*, 1981, 706.

41. P. Braunstein, D. Matt and D. Nobel, *Chem. Rev.*, 1988, **88**, 747.

42. S. L. Buchwald and R. B. Nielsen, *Chem. Rev.*, 1988, **88**, 1047.

43. S. B. Butts, S. H. Strauss, E. M. Holt, R. E. Stimson, N. W. Alcock and D. F. Shriver, *J. Am. Chem. Soc.*, 1980, **102**, 5093.

44. J. R. Blickensderfer and H. D. Kaesz, *J. Am. Chem. Soc.*, 1975, **97**, 2681.

45. R. R. Schrock, *J. Am. Chem. Soc.*, 1974, **96**, 6796.

46. M. F. Semmelhack and R. Tamura, *J. Am. Chem. Soc.*, 1983, **105**, 4099.

47. E. O. Fischer, R. Reitmeier and K. Ackermann, *Z. Naturforsch.*, 1984, **39b**, 668.

48. R. A. Pickering, R. A. Jacobson and R. J. Angelici, *J. Am. Chem. Soc.*, 1981, **103**, 817.

49. N. A. Bailey, P. L. Chell, C. P. Manuel, A. Mukhopadhyay, D. Rogers, H. E. Tabbron and M. J. Winter, *J. Chem. Soc., Dalton Trans.*, 1983, 2397.

50. W. E. Buhro, A. Wong, J. H. Merrifield, G.-Y. Lin, A. C. Constable and J. A. Gladysz, *Organometallics*, 1983 **2**, 1852.

51. C. P. Casey, L. D. Albin and T. J. Burkhardt, *J. Am. Chem. Soc.*, 1977, **99**, 2533.

52. C. P. Casey, S. W. Polichnowski, A. J. Shusterman and C. R. Jones, *J. Am. Chem. Soc.*, 1979, **101**, 7282.

53. G. J. Baird, S. G. Davies, R. H. Jones, K. Prout and P. Warner, *J. Chem. Soc., Chem. Commun.*, 1984, 745.

54. T. Bodner and A. R. Cutler, *J. Organomet. Chem.*, 1981, **213**, C31.

55. M. A. Gallop and W. R. Roper, *Adv. Organomet. Chem.*, 1986, **25**, 121.

56. R. M. Vargas, R. D. Theys and M. M. Hossain, *J. Am. Chem. Soc.*, 1992, **114**, 777.

57. W. A. Herrmann, J. L. Hubbard, I. Bernal, J. D. Korp, B. L. Haymore and G. L. Hillhouse, *Inorg. Chem.*, 1984, **23**, 2978.

58. A. J. Arduengo III, S. F. Gamper, J. C. Calabrese and F. Davidson, *J. Am. Chem. Soc.*, 1994, **116**, 4391.

59. H. Schumann, M. Glanz, J. Winterfeld, H. Hemling, N. Kuhn and T. Kratz, *Angew. Chem., Int. Ed. Engl.*, 1994, **33**, 1733.

60. P. B. Hitchcock, M. F. Lappert and P. L. Pye, *J. Chem. Soc., Dalton Trans.*, 1978, 826; M. J. Doyle, M. F. Lappert, P. L. Pye and P. Terreros, *J. Chem. Soc., Dalton Trans.*, 1984, 2355.

61. R. P. Beatty, J. M. Maher and N. J. Cooper, *J. Am. Chem. Soc.*, 1981, **103**, 238.

62. R. R. Schrock, *J. Am. Chem. Soc.*, 1979, **101**, 3210.

63. A. J. Arduengo III, R. L. Harlow and M. Kline, *J. Am. Chem. Soc.*, 1991, **113**, 361.

64. T. E. Taylor and M. B. Hall, *J. Am. Chem. Soc.*, 1984, **106**, 1576.

65. H. Nakatsuji, J. Ushio, S. Han and T. Yonezawa, *J. Am. Chem. Soc.*, 1983, **105**, 426.

66. This 'backside' attack is currently considered as the best hypothesis to rationalize the formation of cyclopropanes from carbene complexes and olefins, see: H. Fischer, E. Mauz, M. Jaeger and R. Fischer, *J. Organomet. Chem.*, 1992, **427**, 63. The former proposal involving the formation of cyclopropanes by reductive elimination from metallacyclobutanes is less favoured.

67. M. Brookhart, M. B. Humphrey, H. J. Kratzer and G. O. Nelson, *J. Am. Chem. Soc.*, 1980, **102**, 7802.

68. K. H. Dötz, J. Muehlemeier, U. Schubert and O. Orama, *J. Organomet. Chem.*, 1983, **247**, 187.

69. V. Dragisich, C. K. Murray, B. P. Warner, W. D. Wulff and D. C. Yang, *J. Am. Chem. Soc.*, 1990, **112**, 1251.

70. For a short review on recent synthetic applications of carbene chemistry, see: H.-G. Schmalz, *Angew. Chem., Int. Ed. Engl.*, 1994, **33**, 303.

71. F. N. Tebbe, G. W. Parshall and G. S. Reddy, *J. Am. Chem. Soc.*, 1978, **100**, 3611.

72. C. P. Casey, T. J. Burkhardt, C. A. Bunnell and J. C. Calabrese, *J. Am. Chem. Soc.*, 1977, **99**, 2127.

73. R. R. Schrock, *J. Am. Chem. Soc.*, 1976, **98**, 5399; R. R. Schrock and J. D. Fellmann, *J. Am. Chem. Soc.*, 1978, **100**, 3359.

74. E. O. Fischer, G. Kreis, C. G. Kreiter, J. Muller, G. Huttner and H. Lorenz, *Angew. Chem., Int. Ed. Engl.*, 1973, **12**, 564.

75. J. H. Wengrovius, J. Sancho and R. R. Schrock, *J. Am. Chem. Soc.*, 1981, **103**, 3932.

76. R. A. Andersen, M. H. Chisholm, J. F. Gibson, W. W. Reichert, I. P. Rothwell and G. Wilkinson, *Inorg. Chem.*, 1981, **20**, 3934.

77. R. R. Schrock, M. L. Listemann and L. G. Sturgeoff, *J. Am. Chem. Soc.*, 1982, **104**, 4291.

78. J. Manna, R. J. Kuk, R. F. Dallinger and M. D. Hopkins, *J. Am. Chem. Soc.*, 1994, **116**, 9793.

79. M. A. Gallop and W. R. Roper, *Adv. Organomet. Chem.*, 1986, **25**, 121.

80. A. C. Filippou, W. Grünleitner, C. Völkl and P. Kiprof, *Angew. Chem., Int. Ed. Engl.*, 1991, **30**, 1167.

81. F. A Cotton and R. A. Walton, *Multiple Bonds between Metal Atoms*, Wiley, New York, 1982.

82. EHT calculation, the MO basis is taken from D. A Brown and A. Owens, *Inorg. Chim. Acta*, 1971, **5**, 675.

83. P. Chaquin, A. Sevin and H. T .Yu, *J. Mol. Struct. (Theochem.)*, 1985, **21**, 121.

84. Same MO basis as in reference 82. Mo$-$Cl$=2.45$ Å, in the dimer, Mo$-$Mo$=2.30$ Å.

85. J. V. Brencic and F. A. Cotton, *Inorg. Chem.*, 1969, **8**, 7.

86. 1 eV $=23.06$ kcal/mol$=96.5$ kJ/mol.

87. M. Bennett and R. Mason, *Nature*, 1965, **205**, 760.

88. R A. Belmonte, R. R. Schrock and C. S. Day, *J. Am. Chem. Soc.*, 1982, **104**, 3082.

89. B. K. Teo, M. B. Hall, R. F. Fenske and L. F. Dahl, *Inorg. Chem.*, 1975, **14**, 3103.

90. L. Arndt, T. Delord and M .Y. Darensbourg, *J. Am. Chem. Soc.*, 1984, **106**, 456.

91. R. F. Bryan, P. T. Greene, M. J. Newlands and D. S. Field, *J. Chem. Soc. (A)*, 1970, 3068.

92. P. O. Nabel and T. L. Brown, *J. Am. Chem. Soc.*, 1984, **106**, 644.

93. A. L. Balch, L. S. Benner and M. Olmstead, *Inorg. Chem.*, 1979, **18**, 2996.

94. F. A. Cotton, B. G. De Boer, M. D. La Prade, J. R. Ripal and D. A. Ucko, *J. Am. Chem. Soc.*, 1970, **92**, 2926.

95. W. Geilmann and F. W. Wrigge, *Z. Anorg. Allg. Chem.*, 1935, **223**, 144.

96. A. Broll, H. G. von Schnering and H. Schäfer, *J. Less Common Met.*, 1970, **22**, 243.

97. (a) J. L. Templeton, W. C. Dorman, J. C. Clardy and R. E. McCarley, *Inorg. Chem.*, 1978, **17**, 1263; (b) F. A. Cotton and R. C. Najjar, *Inorg. Chem.*, 1981, **20**, 2716.

98. L. B. Anderson, F. A. Cotton, D. DeMarco, A. Fang, W. H. Ilsley, B. W. Kolthammer and R. A. Walton, *J. Am. Chem. Soc.*, 1981, **103**, 5078.

99. (a) F. A. Cotton, B. A. Frenz and L. Kruczynski, *J. Am. Chem. Soc.*, 1973, **95**, 951; (b) L. N. Lewis and K. G. Caulton, *Inorg. Chem.*, 1980, **19**, 1840.

100. K. Nicholas, L. S. Bray, R. E. Davis and R. Pettit, *J. Chem. Soc., Chem. Commun.*, 1971, 608.

101. (a) J. P. Collman and H. J. Arnold, *Acc. Chem. Res.*, 1993, **26**, 586; (b) J. P. Collman, Y. Ha, P. S. Wagenknecht, M. A. Lopez and R. Guilard, *J. Am. Chem. Soc.*, 1993, **115**, 9080; (c) J. P. Collman, H. J. Arnold, J. P. Fitzgerald and K. J. Weissman, *J. Am. Chem. Soc.*, 1993, **115**, 9309.

102. F. A. Cotton, B. A. Frenz, J. R. Ebner and R. A. Walton, *Inorg. Chem.*, 1976, **15**, 1630.

103. F. A. Cotton and J. L. Thompson, *Inorg. Chim. Acta*, 1980, **44**, L247.

104. M. H. Chisholm, F. A. Cotton, B. A. Frenz, W. W. Reichert, L. W. Shive and B. R. Stults, *J. Am. Chem. Soc.*, 1976, **98**, 4469.

105. J. Podenza, P. Giordano, D. Mastropaolo and A. Efraty, *Inorg. Chem.*, 1974, **13**, 2540.

106. F. A. Cotton, B. G. DeBoer and M. Jeremic, *Inorg. Chem.*, 1970, **9**, 2143.

107. F. A. Cotton, S. A. Koch and A. Millar, *Inorg. Chem.*, 1978, **17**, 2087.

108. (a) E. Kurras, U. Rosenthal, M. Mennenga, G. Oehme and G. Engelhardt, *Z. Chem.*, 1974, **14**, 160; (b) F. A. Cotton, B. E. Hanson, W. H. Ilsley and G. W. Rice, *Inorg. Chem.*, 1979, **18**, 2713.

109. M. Fourmigué and P. Batail, *J. Chem. Soc., Chem. Commun.*, 1991, 1370.

110. H. B. Kagan and T. P. Dang, *J. Am. Chem. Soc.*, 1972, **94**, 6429.

111. S. Yamago, M. Yanagawa and E. Nakamura, *J. Chem. Soc., Chem. Commun.*, 1994, 2093.

112. For a recent discussion of this problem, see: G. Pacchioni and P. S. Bagus, *Inorg. Chem.*, 1992, **31**, 4391.

113. C. A. Tolman, *Chem. Rev.*, 1977, **77**, 313.

114. M. Regitz and O. J. Scherer, *Multiple Bonds and Low Coordination in Phosphorus Chemistry*, Thieme, Stuttgart, 1990.

115. X. Li, S. I. Weissman, T.-S. Lin, P. P. Gaspar, A. H. Cowley and A. I. Smirnov, *J. Am. Chem. Soc.*, 1994, **116**, 7899.

116. A Marinetti and F. Mathey, *Organometallics*, 1984, **3**, 456.

117. A Marinetti, F. Mathey, J. Fischer and A. Mitschler, *J. Am. Chem. Soc.*, 1982, **104**, 4484.

118. P. Le Floch, A. Marinetti, L. Ricard and F. Mathey. *J. Am. Chem. Soc.*, 1990, **112**, 2407.

119. N. H. Tran Huy, L. Ricard and F. Mathey, *Organometallics*, 1988, **7**, 1791.

120. P. B. Hitchcock, M. F. Lappert and W.-P. Leung, *J. Chem. Soc., Chem. Commun.*, 1987, 1282.

121. T. L. Breen and D. W. Stephan, *J. Am. Chem. Soc.*, 1995, **117**, 11914.

122. A. H. Cowley, B. Pellerin, J. L. Atwood and S. G. Bott, *J. Am. Chem. Soc.*, 1990, **112**, 6734; C. C. Cummins, R. R. Schrock and W. M. Davis, *Angew. Chem., Int. Ed. Engl.*, 1993, **32**, 756.

123. G. Huttner and K. Evertz, *Acc. Chem. Res.*, 1986, **19**, 406.

124. G. Huttner and K. Knoll, *Angew. Chem., Int. Ed. Engl.*, 1987, **26**, 743.

125. L. D. Hutchins, R. T. Paine and C. F. Campana, *J. Am. Chem. Soc.*, 1980, **102**, 4521.

126. W. Malisch, K. Jörg, U. Hofmockel, M. Schmeusser, R. Schemm and W. S. Sheldrick, *Phosphorus Sulfur*, 1987, **30**, 205.

127. J. Wachter. B. F. Mentzen and J. G. Riess, *Angew. Chem., Int. Ed. Engl.*, 1981, **20**, 284.

128. G. Becker, *Z. Anorg. Allg. Chem.*, 1976, **423**, 242.

129. Th. C. Klebach, R. Lourens and F. Bickelhaupt, *J. Am. Chem. Soc.*, 1978, **100**, 4886.

130. A. Marinetti and F. Mathey, *Angew. Chem., Int. Ed. Engl.*, 1988, **27**, 1382.

131. F. Mathey, *Acc. Chem. Res.*, 1992, **25**, 90.

132. Th. A. Van der Knaap, F. Bickelhaupt, J. G. Kraaykamp, G. van Koten, J. P. C. Bernards, H. T. Edzes, W. S. Veeman, E. de Boer and E. J. Baerends, *Organometallics*, 1984, **3**, 1804.

133. See: F. Mathey, *New J. Chem.*, 1987, **11**, 585; J. F. Nixon, *Chem. Rev.*, 1988, **88**, 1327; F. Mathey, *Coord. Chem. Rev.*, 1994, **137**, 1.

134. P. Binger, Chapter 2.5 in ref. 114.

Applications

In the previous chapters, we have reviewed the fundamental laws of organometallic chemistry and looked at the nature of the bonds between metals and their ligands. This means that we are in a position to understand the applications of this rather distinctive chemistry.

The transition metals are crucial to many processes. In nature,[1] iron is the active centre in haemoglobin, which allows us to fix oxygen. Nitrogenase, the enzyme which allows the bacterial fixation of nitrogen, includes an Fe_4S_4 cluster and a second molybdenum–iron–sulfur cluster whose structure is not yet fully understood. Vitamin B_{12} contains cobalt. As regards medicine, one of the most potent treatments for cancer is 'cis-platin' (cis-$PtCl_2\{NH_3\}_2$), which blocks DNA replication. Transition metals are also well represented in the field of molecular materials: take, for example molecular ferromagnets,[2] molecular diodes, or the copper-sulfur complexes which become superconducting below 12.8 K.[3] At present, a whole new domain of molecular electronics based upon the transition metals is under development.[4]

Because the materials above incorporate highly sophisticated structures which are too complex to treat here, we will concentrate upon one of the more straightforward applications of these metals: their use in the areas of organic synthesis and catalysis. Even here, the area is undergoing such rapid and impressive development that we will only touch upon the most important aspects.

References

1. M. N. Hughes; in *Comprehensive Coordination Chemistry*, eds. G. Wilkinson, R. D. Gillard and J. A. McCleverty, Pergamon, Oxford, 1987, vol 6, pp. 545–754; H. E. Howard-Lock and C. J. L. Lock, *ibid.*, pp. 755–778.
2. J. S. Miller, A. J. Epstein and W. M. Reiff, *Acc. Chem. Res.*, 1988, **21**, 114.
3. J. M. Williams, A. M. Kini, H. H. Wang, K. D. Carlson, U. Geiser, L. K. Montgomery, G. J. Pyrka, D. M. Watkins, J. M. Kommers, S. J. Boryschuk, A. V. Strieby Crouch, W. K. Kwok, J. E. Schirber, D. L. Overmyer, D. Jung and M.-H. Whangbo, *Inorg. Chem.*, 1990, **29**, 3272.
4. O. Kahn, *C. R. Acad. Sci. Paris, Ser. La Vie des Sciences*, 1988, **5**, 187.

4 Some Applications in Organic Synthesis

4.1 The Hydrozirconation of Alkenes and Alkynes

The addition of a metallic hydride to an alkene (or an alkyne) provides a versatile route for the preparation of functionalized organic molecules. During this process, the double bond becomes saturated and a useful functional group is introduced. The scheme is as follows:

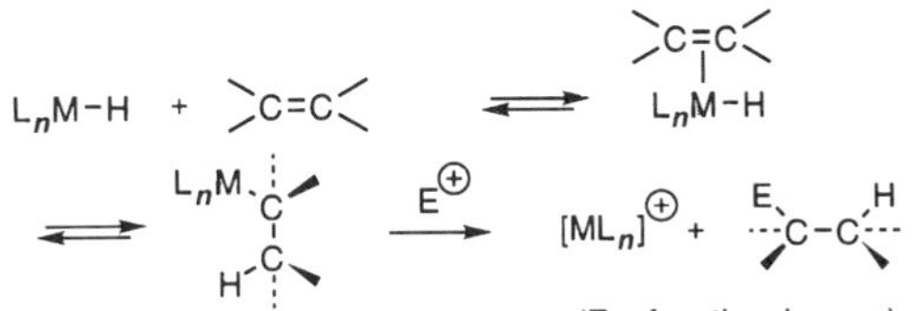

(E = functional group)

For this sequence to operate efficiently, the equilibria have to be strongly displaced towards the σ-organometallic. This requirement is satisfied in the case of zirconium (IV), where Schwartz[1] has developed numerous applications in organic synthesis.

The Synthesis of the Schwartz Reagent $Cp_2Zr(H)Cl$

The first synthesis of the Schwartz reagent, which involved the partial reduction of Cp_2ZrCl_2 by a 'mild' aluminium hydride, gave a very impure product.

$$Cp_2ZrCl_2 \xrightarrow[NaAlH_2(OR)_2,\ THF]{LiAlH(O^tBu)_3\ or} Cp_2Zr(H)Cl$$

A considerably improved preparation relies on reaction of the dihydride Cp_2ZrH_2 with CH_2Cl_2.[2] This allows the use of the 'harsher' lithium aluminium hydride as the reducing agent:

$$Cp_2ZrCl_2 \xrightarrow{LiAlH_4} Cp_2Zr(H)Cl + Cp_2ZrH_2 \xrightarrow{CH_2Cl_2} Cp_2Zr(H)Cl$$

Here, the purity of the product is about 95%.

The Hydrozirconation Reaction

The Schwartz reagent combines with a wide variety of activated or non-activated alkenes under mild conditions (25°C). With terminal olefins, the zirconium favours the terminal position for both electronic and steric reasons:

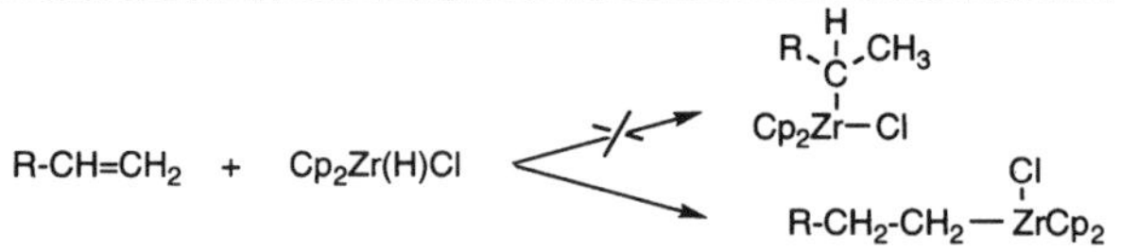

R-CH=CH₂ + Cp₂Zr(H)Cl

$$R-CH=CH_2 + Cp_2Zr(H)Cl$$

The alkylzirconium products have quite acceptable thermal stability. This is because the preferred decomposition pathway for most metal–alkyl bonds is β-elimination, which is favoured by an electronic transfer from the metal d orbitals into the antibonding σ^* orbital of the β C–H bond (see Section 3.3, properties of the M–C bond). The empty d shell found in the Zr(IV) configuration thus inhibits the decomposition process and gives relatively stable products.

Addition of Cp₂Zr(H)Cl to internal olefins is more difficult, as the following hydrozirconation reactivity series shows:

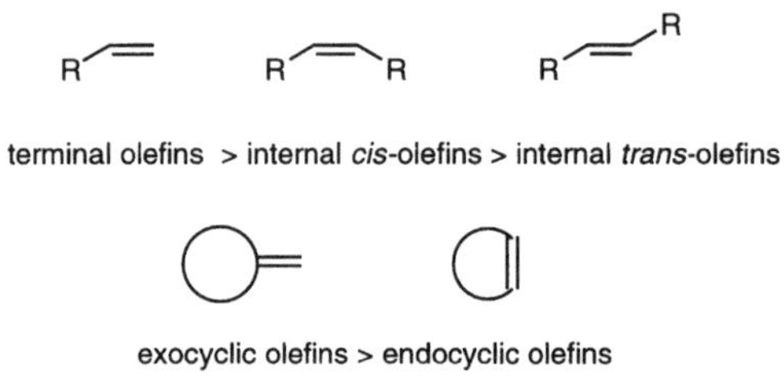

terminal olefins > internal *cis*-olefins > internal *trans*-olefins

exocyclic olefins > endocyclic olefins

Tri- and tetra-substituted olefins do not react. It is clear that steric factors must play a dominant role because the zirconium invariably adds to the least hindered carbon. The hydrozirconation of non-terminal olefins is instructive:

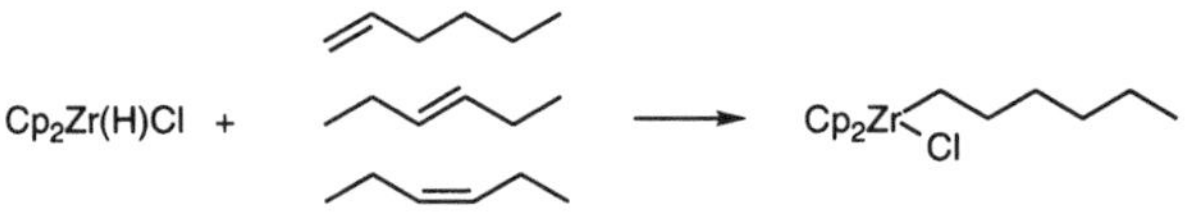

Cp₂Zr(H)Cl + ⟶ Cp₂Zr–Cl

Although the σ/π equilibria are strongly displaced towards the σ alkyls, the existence of transient π complexes allows the zirconium to migrate along the carbon chain, to give the thermodynamically most stable compound:

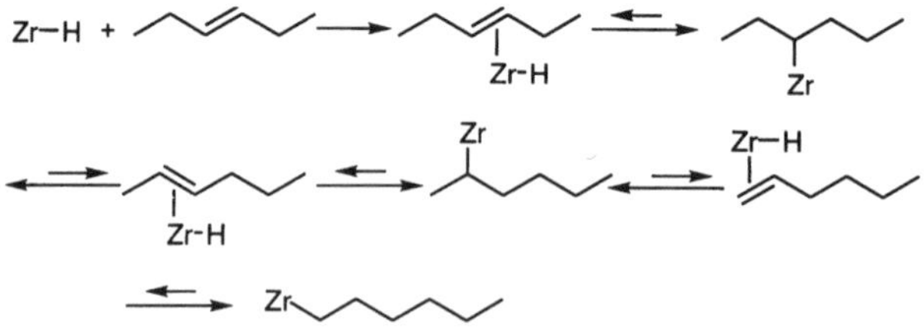

The hydrozirconation of 1,3-dienes occurs at the more accessible double bond; it leads to γ,δ-unsaturated metal derivatives:

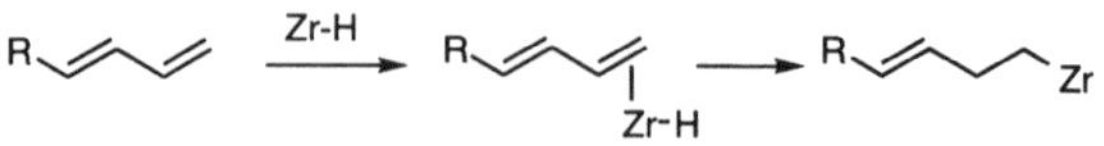

With alkynes, the less hindered *cis*-alkenylzirconium is formed:

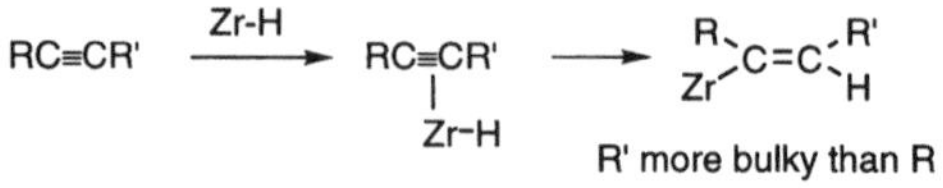

R' more bulky than R

When there is little difference in the steric influence of the two groups, (e.g. R = Me, R' = Et), a mixture of the two possible products is usually obtained.

Cleavage of the Zirconium–Carbon Bond

Electrophiles cleave the σ Zr–C bond easily. The reaction scheme is given below:

$$Cp_2Zr\overset{Cl}{\underset{R}{<}} \quad + \quad E^{\oplus} \quad \longrightarrow \quad [ZrClCp_2]^{\oplus} \quad + \quad R\text{–}E$$

The following reagents are used frequently:

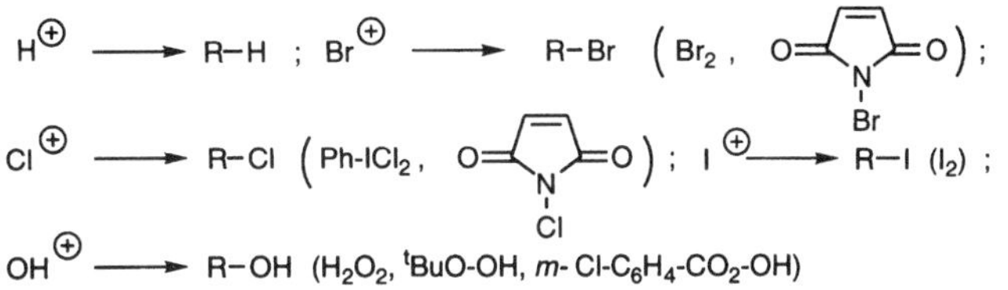

Generally, electrophilic cleavage of the Zr–C bond proceeds through a concerted mechanism, with retention of configuration at the carbon atom. For example with Br_2:

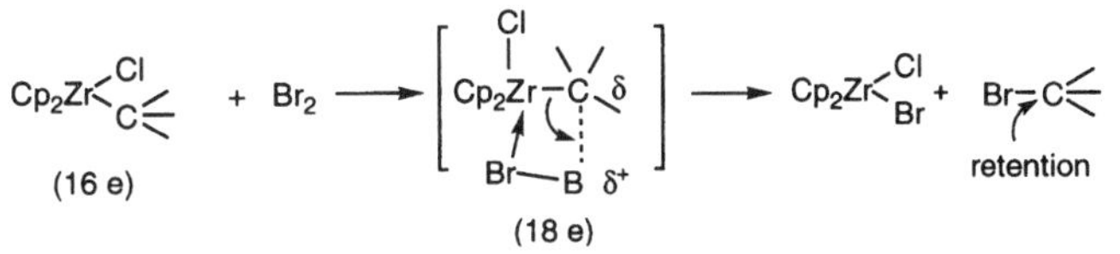

Carbon–zirconium bonds are inert towards alkyl halides R–X and react very slowly with acyl halides R'C(O)X [$E^+ = R^+$ and $R'C(O)^+$ respectively]. Fortunately, the facile carbonylation of the C–Zr bond provides a straightforward preparation of carbonyl complexes:

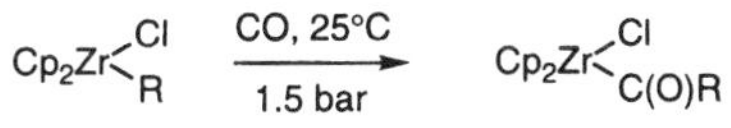

$$Cp_2Zr\overset{Cl}{\underset{R}{<}} \quad \xrightarrow[\text{1.5 bar}]{\text{CO, 25°C}} \quad Cp_2Zr\overset{Cl}{\underset{C(O)R}{<}}$$

These acylzirconium products are also attacked by a wide range of electrophiles:

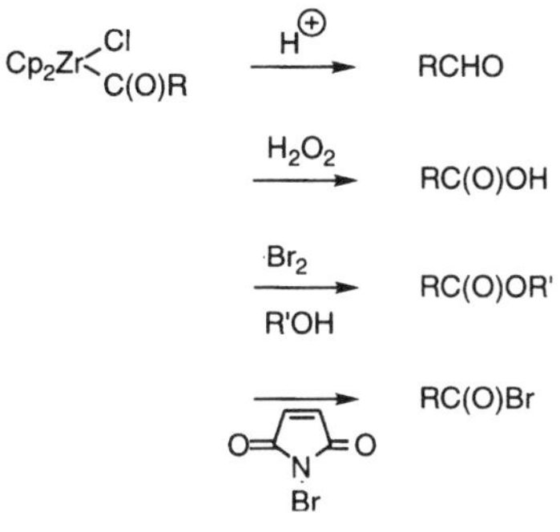

The C–Zr σ bond may also be activated by transmetallation:

$$Cp_2Zr\overset{Cl}{\underset{R}{<}} \quad + \quad M^{\oplus} \quad \longrightarrow \quad [ZrClCp_2]^{\oplus} \quad + \quad R-M$$

The reaction is particularly easy with $AlCl_3$:

$$Cp_2Zr\overset{Cl}{\underset{R}{<}} \quad + \quad AlCl_3 \quad \longrightarrow \quad Cp_2ZrCl_2 \quad + \quad R-AlCl_2$$

Reaction of an acyl halide with the alkylaluminium product allows the preparation of ketones, which would not be available directly from the organozirconium:[3]

$$R-AlCl_2 \quad + \quad R'C(O)X \quad \longrightarrow \quad XAlCl_2 \quad + \quad R-C(O)R'$$

Additionally, nickel acetylacetonate gives a nickel alkyl which may be used to effect 1,4 additions to α,β-unsaturated ketones:[4]

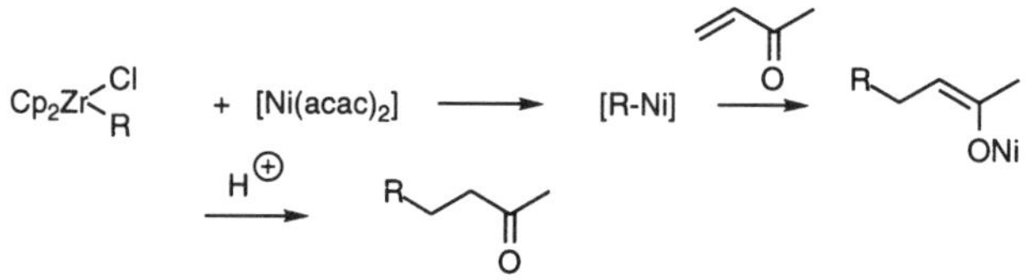

Cyanocuprates behave analogously:[5]

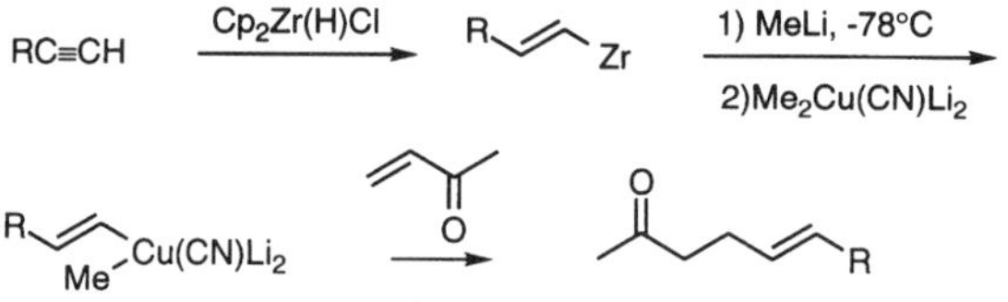

They have recently been used in alkylation reactions: in order to avoid competing alkylations, a thienyl copper reagent is used in place of the more frequently encountered methyl copper systems:[6]

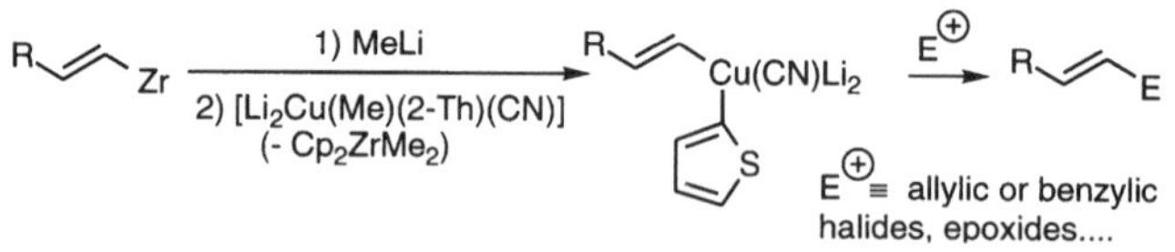

Transmetallation processes are widely employed to modulate the reactivity of the products of the hydrozirconation reaction. The process can be applied to any metal which is less electropositive than zirconium; further important examples include palladium[7] and mercury.[8] Boron has also been employed;[9] it offers a versatile method for the creation of C–C bonds.

To illustrate the potential of organozirconium chemistry in synthesis, we finish by giving a number of important applications from the literature:

(a) The olefin metathesis reaction gives internally functionalized olefins which have little practical interest. Hydrozirconation allows their transformation into much more useful products having a terminal functional group; these are used for the manufacture of surfactants:

$$C_{14}H_{29}\text{-CH}=\text{CH-}C_{14}H_{29} \xrightarrow[\text{2) }^{t}\text{BuO-OH}]{\text{1) }Cp_2Zr(H)Cl} CH_3(CH_2)_{29}OH \quad \text{(ref. 10)}$$

The 1-triacontanol product stimulates plant growth.

(b) γ,δ-unsaturated aldehydes can be easily prepared from dienes:

$$R\diagup\diagdown \xrightarrow{Cp_2Zr(H)Cl} R\diagup\diagdown\text{-Zr(Cl)Cp}_2$$

$$\xrightarrow[\text{2) }H^{\oplus}]{\text{1) CO}} R\diagup\diagdown\text{-CHO} \quad \text{(ref. 11)}$$

(c) α,β-unsaturated ketones are accessible through transalumination:

$$\rangle\text{-C}\equiv\text{C-CH}_3 \xrightarrow{Cp_2Zr(H)Cl} \overset{CH_3}{\underset{H}{\rangle C=C\diagup_{Zr(Cl)Cp_2}}} \quad \text{(ref. 3)}$$

$$\xrightarrow{AlCl_3} \overset{CH_3}{\underset{H}{\rangle C=C\diagup_{AlCl_2}}} \xrightarrow{CH_3C(O)Cl} \overset{CH_3}{\underset{H}{\rangle C=C\diagup_{C(O)CH_3}}}$$

A good illustration of the complementary reactivity of zirconium and aluminium is given by the fact that direct hydroalumination of this alkyne is impossible.

4.2 Carbonylation by Collman's Reagent

In spite of its industrial potential, the use of carbon monoxide in organic synthesis has been relatively limited. In theory, it should serve as a valuable source of carbonyl compounds:

$$[Nu]^{\ominus} + CO \longrightarrow \left[\underset{O}{Nu-C}\right]^{\ominus} \xrightarrow{E^{\oplus}} \underset{O}{Nu-C-E}$$

However, reality is less attractive. Nucleophilic attack on CO is very difficult to control and organolithiums, for instance, give a complex mixture of products. A detailed study of the reaction has confirmed that acyl anions are formed initially; they may even be trapped by a chlorosilane.[12]

$$RLi + CO \xrightarrow{-110^{\circ}C} [R\text{-}C(O)]^{\ominus}\ Li^{\oplus} \xrightarrow{Me_3SiCl} \underset{O}{R-C-SiMe_3}$$

However, they are very unstable and evolve easily:[13]

$$\left[\underset{O}{Ph-C^{\ominus}} \longleftrightarrow \overset{Ph}{\underset{\ominus O}{\cdots C}}\right] \longrightarrow \overset{Ph}{\underset{\ominus O}{C}}=\overset{Ph}{\underset{O\ominus}{C}} \xrightarrow{H^{\oplus}} \underset{O}{Ph-C-CH(OH)Ph}$$

Consequently organic chemists have had to develop some rather sophisticated chemistry to obtain stable equivalents of acyl anions.[14] The situation is completely different when we look at transition metals. We have seen in Section 3.3 that carbon monoxide can be easily inserted into carbon–transition metal σ bonds; this gives metal–acyl complexes which have significant practical applications:

$$R-ML_n \xrightarrow{CO} \left[\overset{CO}{\underset{|}{R-ML_n}}\right] \xrightarrow{CO} \overset{CO}{\underset{|}{R-\overset{||}{\underset{O}{C}}-ML_n}}$$

(16 e)   (18 e)   (18 e)

For this chemistry to be viable on a stoichiometric basis, we need a reagent containing a cheap, abundant transition metal. This obviously means iron.

The Synthesis of Collman's Reagent $Na_2Fe(CO)_4$

The essence of this synthesis involves the simple reduction of iron pentacarbonyl by an electron source. Sodium is the preferred metal:

$$Fe(CO)_5 + 2e^- \longrightarrow [Fe(CO)_4]^{2-} + CO$$

If good contact between the metal surface and the iron carbonyl solution is not maintained, polymetallic byproducts are produced:[15]

$$[Fe(CO)_4]^{2-} + Fe(CO)_5 \longrightarrow [Fe_2(CO)_8]^{2-} + CO$$

To ensure clean generation of the reagent, sodium amalgam (Collman's original synthesis[16]), the sodium-benzophenone radical anion in dioxane ($Ph_2CO + Na \rightarrow Ph_2C^\cdot-O^-$, Na^+)[17] or ultrasonic activation[18] are usually employed.

The Synthetic Applications of Collman's Reagent[16]

Collman's reagent is a powerful nucleophile, which attacks halides and organic tosylates with the formation of iron–carbon bonds:

$$Na_2Fe(CO)_4 + R-X \longrightarrow [R-Fe(CO)_4]^{\ominus}$$

The mechanism involves the classical SN_2 substitution which is familiar from conventional organic chemistry. The kinetics are second order ($v = k[RX][[Fe(CO)_4]^{2-}]$) and the reaction is favoured if X is a good leaving group. Aryl and vinyl halides do not react:

R-I > R-Br > R-OTs > R-Cl

primary R > secondary R > tertiary R

We observe the usual inversion of configuration at the carbon bound to X which typifies the SN_2 mechanism.

$$Fe^{\ominus} \overset{a}{\underset{b}{\underset{c}{\diagup}}}C-X \longrightarrow Fe-\overset{a}{\underset{c}{\underset{b}{C}}}$$

The alkyliron and acyliron forms are in an equilibrium which may be displaced to the right by the addition of a two-electron ligand L (CO or PPh_3):

$$[R-Fe(CO)_4]^{\ominus} \rightleftharpoons [R(CO)-Fe(CO)_3]^{\ominus} \xrightarrow{L} [R(CO)-Fe(CO)_3(L)]^{\ominus}$$

(See Sections 2.3 and 3.3 for the mechanistic details of these transformations.)

The alkyliron can be protonated, alkylated or oxidized by cleavage of the metal–carbon bond:

$$[\text{R-Fe(CO)}_4]^{\ominus} \xrightarrow{\text{H}^{\oplus}} \text{RH}$$

$$\xrightarrow{\text{R'X}} \text{RC(O)R'}$$

$$\xrightarrow{\text{O}_2} \text{RC(O)OH} \xrightarrow{\text{H}_2\text{O}} \text{RC(O)OH}$$

$$\xrightarrow{\text{X}_2} \text{RC(O)X} \xrightarrow{\text{R'OH}} \text{RC(O)OR'}$$

$$\xrightarrow{\text{R'}_2\text{NH}} \text{RC(O)NR'}_2$$

This provides an easy access to hydrocarbons, ketones and carboxylic acid derivatives. The acyliron (which can also be obtained from acyl halides and Collman's reagent) is cleaved by the same reagents:

$$[\text{R-Fe(CO)}_4]^{\ominus} \xrightarrow{\text{L (CO, PPh}_3)} [\text{R(CO)-Fe(CO)}_3(\text{L})]^{\ominus} \xrightarrow{\text{H}^{\oplus}} \text{RCHO}$$

$$\text{RC(O)X} + [\text{Fe(CO)}_4]^{2-} \xrightarrow{\text{(L = CO)}} \qquad \xrightarrow{\text{R'X}} \text{RC(O)R'}$$

$$\xrightarrow{\text{O}_2} \text{RC(O)OH}$$

$$\xrightarrow{\text{X'}_2} \text{RC(O)X'}$$

Thus aldehydes and ketones may be prepared from alkyl or acyl halides. The Collman reagent reacts very specifically with the C−X bond in the R−X or R−C(O)X compound, so the reaction tolerates a wide variety of functional groups (ester, ketone or nitrile ...) in the R fragment.

An interesting variation on these reactions involves an additional insertion of ethylene.[19]

$$[\text{R(CO)-Fe(CO)}_3(\text{L})]^{\ominus} \xrightarrow{\text{CH}_2\text{=CH}_2} [\text{RC(O)-CH}_2\text{-CH}_2\text{-Fe(CO)}_3(\text{L})]^{\ominus}$$

$$\xrightarrow{\text{H}^{\oplus}} \text{RC(O)C}_2\text{H}_5$$

An intramolecular olefin insertion allows the synthesis of cyclic ketones:[20]

Finally, it is sometimes possible to omit the tedious isolation of the Collman reagent. For certain applications, it may be prepared and used *in situ* from Fe(CO)$_5$ and aqueous hydroxide in H$_2$O/CH$_2$Cl$_2$ under phase transfer catalysis by Bu$_4$N$^+$, X$^-$. Methylketones can then be prepared very easily:[21]

$$\frac{\text{CH}_2\text{Cl}_2 \text{ layer:} \quad \text{RX} + \text{Fe(CO)}_5 + \text{CO}}{\text{H}_2\text{O layer:} \quad \text{NaOH} + [\text{NBu}_4]^{\oplus} \text{X}^{\ominus}} \longrightarrow [\text{RC(O)-Fe(CO)}_4]^{\ominus} [\text{NBu}_4]^{\oplus}$$

$$\downarrow \text{IMe}$$

$$\text{RC(O)Me}$$

4.3   The Cyclopropanation of Alkenes by Carbene Complexes

One of the most important routes to cyclopropanes in classical organic chemistry involves the cycloaddition of a carbene CR_2 or carbenoid $[CR_2X]^-$ with an alkene:

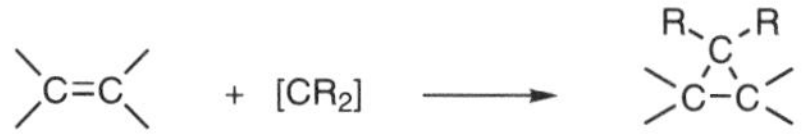

The relative stability and high reactivity of transition metal carbene complexes make them attractive candidates as carbene sources; their application as reagents for the cyclopropanation of olefins therefore comes to mind.[22] We have already seen (Section 3.4) that carbene complexes react with olefins by two processes, cyclopropanation and metathesis:

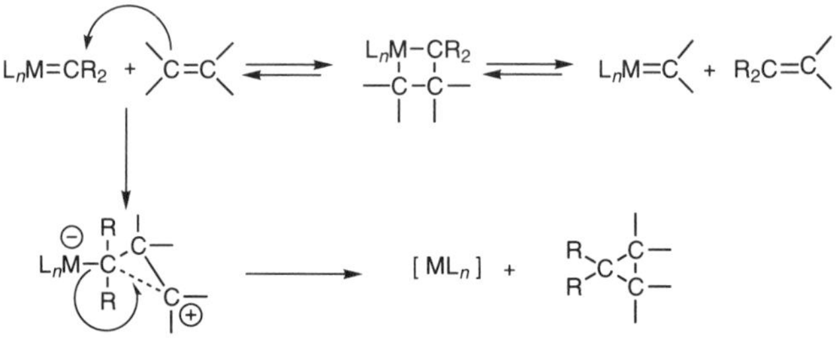

According to recent results,[23] cyclopropane formation probably results from a 'backside' attack of the nucleophilic olefin on the electrophilic carbenic carbon. Thus, the objective is to favour this attack over the competing metathesis. Steric hindrance at the metal, a high oxidation state and a positive charge all tend to promote cyclopropanation (see Section 3.4). Again, for obvious economic reasons, iron is the best metal for such a synthetic stoichiometric application. Taken together, these criteria explain why cationic iron(II) carbene complexes $[Fe(Cp)(L)_2=CR_2]^+$ are currently the reagents of choice for olefin cyclopropanation by transition metals.

Synthesis and Use of Cationic Iron(II) Carbene Complexes $[Fe(Cp)(L)_2=CR_2]^+$

In general, these cationic complexes are unstable and must be generated and used *in situ*. The simplest member of the series, $[Fe=CH_2]^+$, is obtained by thermal decomposition of the corresponding (dimethylsulfonium)methyl complex:[24]

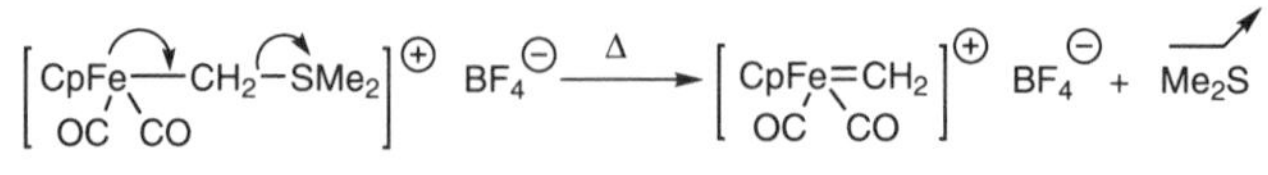

Once formed, it effects cyclopropanation with retention of the stereochemistry of the parent olefin. The yields are excellent.

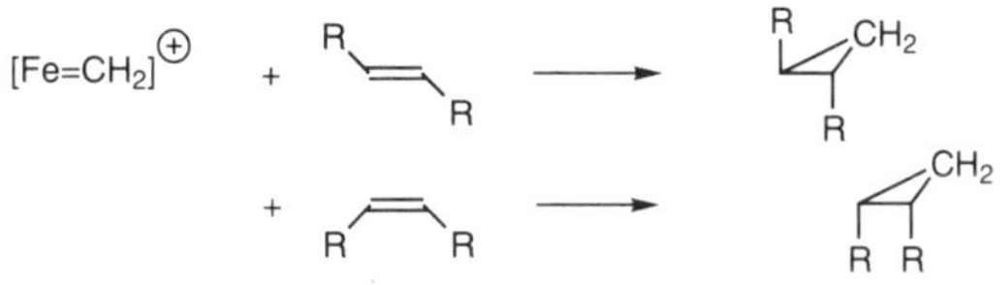

By a similar technique,[25] the ethylidene complex [Fe=CHMe]$^+$ may be prepared:

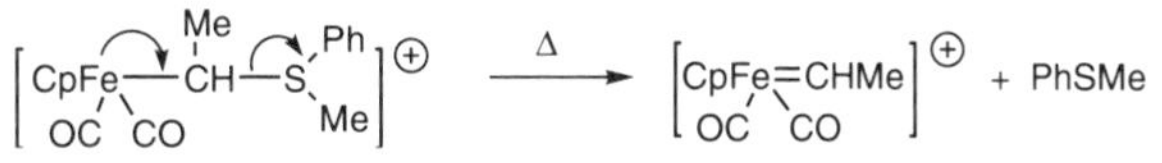

Its reaction with an olefin leads to a mixture of two cyclopropanes, which differ according to the relative orientation of the 'carbenic' fragment and the double bond in the transition state:

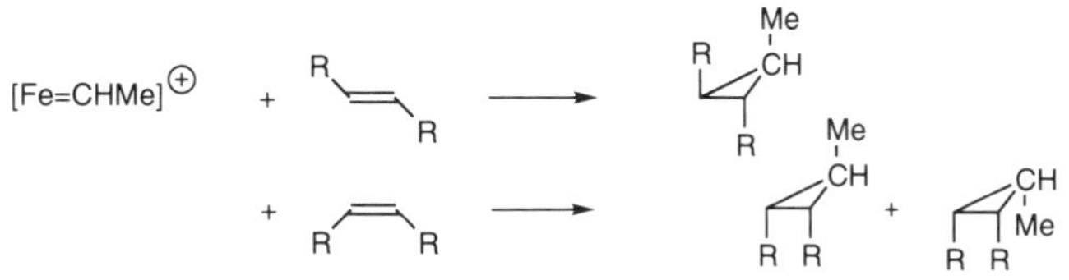

An alternative synthetic approach has been developed by Brookhart:[26]

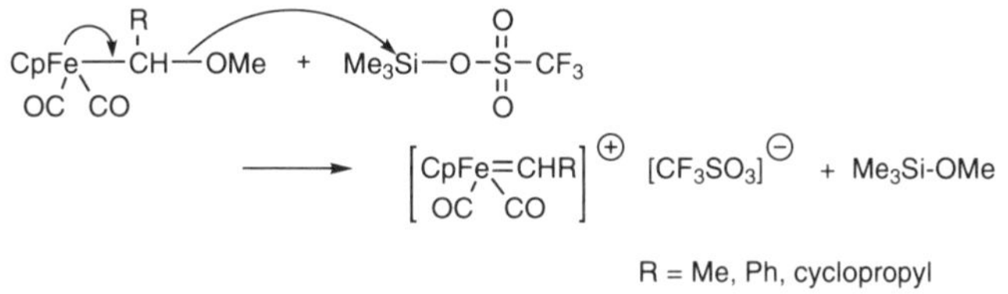

R = Me, Ph, cyclopropyl

Finally, one of the carbonyl ligands may be replaced by an optically active phosphine, to give a chiral carbene complex which allows the synthesis of optically active cyclopropanes from monosubstituted olefins.[27] The carbene complex is obtained as a mixture of two optically active diastereomers which can be separated. The two isomers react to give cyclopropanes which are the mirror image of each other:

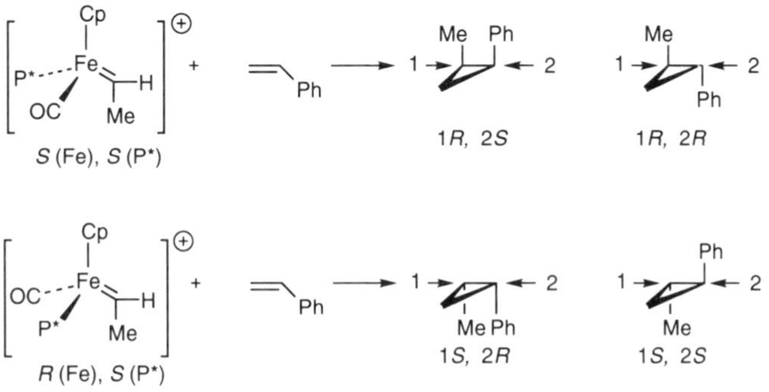

The asymmetric induction results from the chirality at iron rather than at the phosphine.

4.4 η^4-Diene-Iron-Tricarbonyls in Organic Synthesis

Conjugated dienes are amongst the most important building blocks in organic chemistry. They undergo a vast series of additions and cycloadditions, including the well known Diels–Alder reaction with olefins:

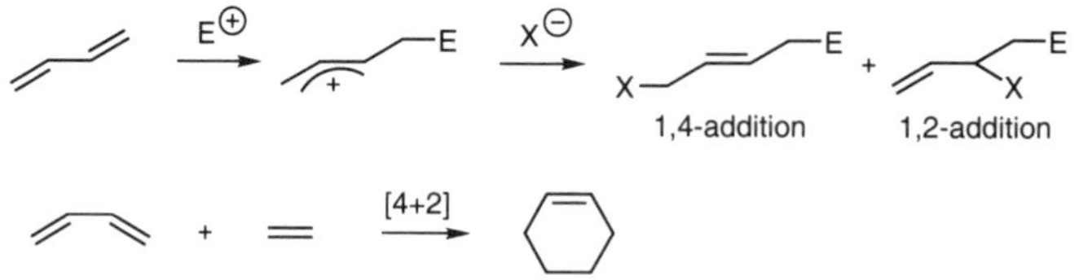

However, the chemistry of functionally substituted dienes is poorly developed, because they are difficult to prepare. First of all, grafting a functional group on to a diene is a delicate operation, because electrophilic substitution suffers from a competing electrophilic addition:

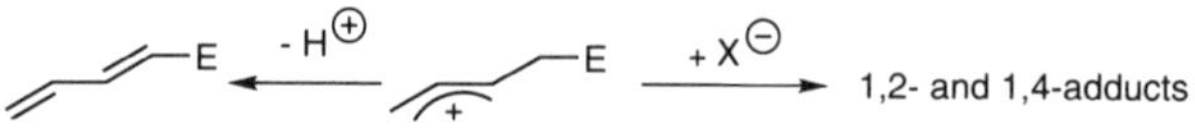

Furthermore, once the functional group is installed, it is difficult to carry out a subsequent reaction with electrophiles or nucleophiles without destroying the diene. Thus to allow this area to be developed, it is essential to develop a system that masks or unmasks a diene at will. Up to now, the best method consists of complexing the diene to iron tricarbonyl.

Synthesis of η^4-Diene-Iron-Tricarbonyls

In the majority of cases, it suffices to heat an diene with an iron carbonyl (Fe(CO)$_5$, Fe$_2$(CO)$_9$ or Fe$_3$(CO)$_{12}$) to obtain the desired η^4 complex. Iron pentacarbonyl, the commonest and cheapest compound, is also the least reactive; it requires high reaction temperatures which may cause isomerization:

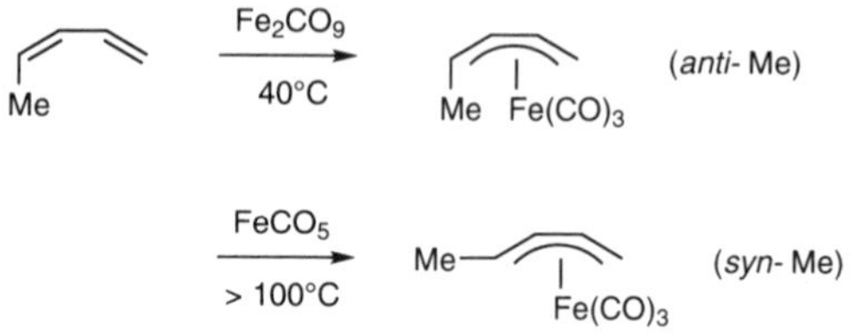

Of course, the *syn* isomer is thermodynamically more stable than the *anti* isomer owing to its lesser steric hindrance. At high temperatures, it is also possible to convert a non-conjugated diene into a η^4 complex:

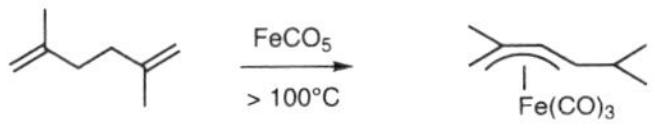

The migration of the double bonds probably involves a η^3-allyl-iron hydride, as below:

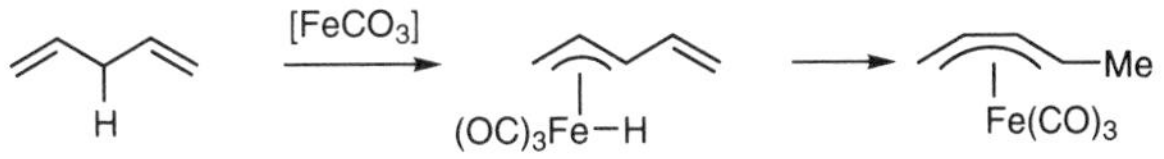

Diene complexes may also be prepared by an exchange of η^4 ligands. Labile η^4 complexes are easily obtained by the reaction of iron carbonyls with α,β-unsaturated ketones, which may be subsequently displaced by other dienes:

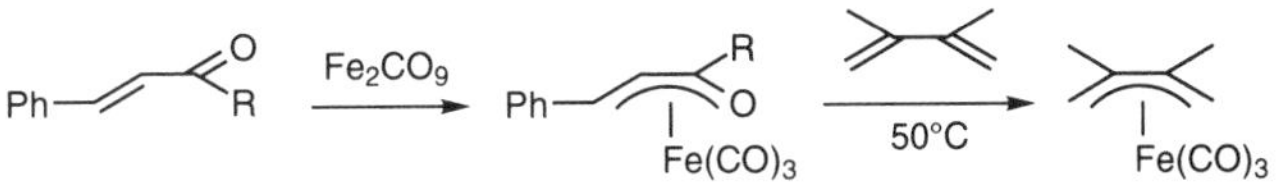

Another widely employed route to η^4 diene complexes consists of treating Collman's reagent with an unsaturated dihalide, for example:[28]

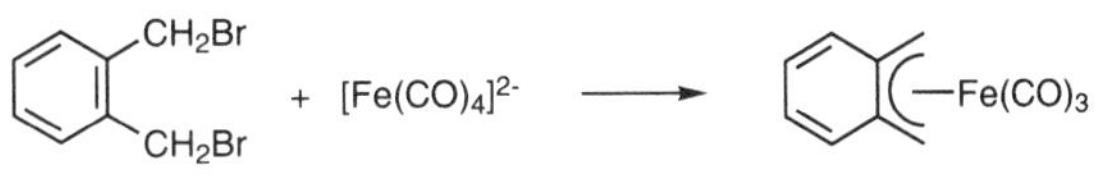

This route is particularly valuable in the case above, because the free diene doesn't exist. It is possible to replace Collman's reagent by a neutral iron carbonyl, but in this case half the metal is lost in the form of ferrous halide. The most celebrated example of this second type of approach is the synthesis of cyclobutadiene iron tricarbonyl:[29]

Some Structural and Spectral Characteristics of η^4-Diene-Iron-Tricarbonyls

The structure of a $(\eta^4$-diene)Fe(CO)$_3$ complex is represented pictorially in Figure 4.1. The iron shows tetragonal pyramidal geometry and the four carbons of the diene are co-planar, with the substituents slightly bent out of plane away from the metal. Their displacement enhances the π backdonation from the occupied metal d orbitals to the LUMO of the diene.

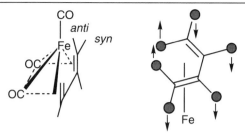

Figure 4.1   The geometry of a $(\eta^4\text{-diene})Fe(CO)_3$ complex

The distance between the iron and the diene plane is around 1.64 Å and the diene C=C bond lengths are approximately equal, at 1.40–1.42 Å. The proton and ^{13}C NMR resonances generally shift to high fields upon coordination; the values in Figure 4.2 are typical.

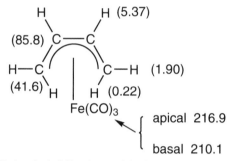

Figure 4.2 ^1H and ^{13}C chemical shifts observed in the NKR spectra of $[Fe(\eta^4\text{-}C_4H_6(CO)_3]$ (δ in ppm, Me_4Si as internal reference)

The medium energy bonding interaction between the iron and the diene $[\Delta H = 47.8$ kcal/mol for $[Fe(\eta^4\text{-}C_4H_6) (CO)_3]$, ref. 30], is relatively easy to break; it is thus well adapted to the liberation of the diene from the complex.

Synthetically useful Properties of $[Fe(\eta^4\text{-diene})(CO)_3]$ Complexes

Many electrophiles ($E^+ = H^+$, $[MeC(O)]^+$, SO_2, BF_3, etc.) add to iron η^4-diene complexes to give η^3-allyl-iron cations, for example:

In certain cases (such as CF_3CO_2D), the addition is reversible and the deuterium appears in the *syn* position. Note that the electrophilic addition of E^+ gives an unstable 16e complex, which generally captures a 2e ligand before it can be intercepted (cf. Cl^- in the scheme above). With non-coordinating counter-ions (e.g. $AlCl_4^-$, BF_4^-, ...), the iron completes its 18 valence electrons by re-coordination of the carbon which underwent the original electrophilic attack, giving a substituted η^4-diene. The two schemes are summarized below:

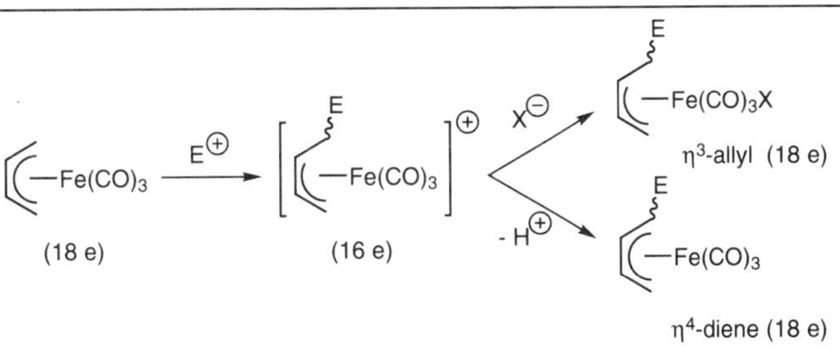

The second offers an interesting path to complexes containing functionally substituted dienes. A more precise treatment of acetylation is given below:

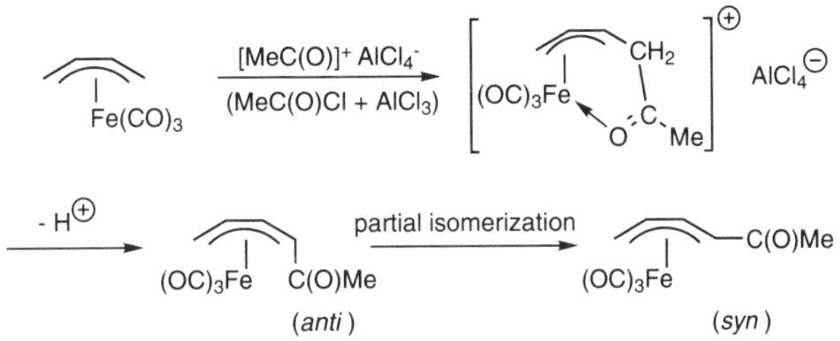

It should be noted that butadiene iron tricarbonyl is acetylated 3850 times faster than benzene (but less rapidly than ferrocene [Fe(η^5-C$_5$H$_5$)$_2$], which is acetylated 3×10^6 times faster than benzene). This high reactivity means that the more difficult Vilsmeier formylation of η^4-diene complexes, by a mixture of P(O)Cl$_3$ + HC(O)NR$_2$ (an equivalent of [HC(O)]$^+$), is also possible.

Three important synthetic applications, the isomerization of dienes, protection of dienes, and addition of a functional group to dienes, are all suggested by the chemistry above. In practice, all are feasible. A further important nuance results from the non-equivalence of the two diene faces after the complexation to iron. In the scheme below, we show pictorially how a substituted carbon centre (C—R) becomes chiral upon coordination.

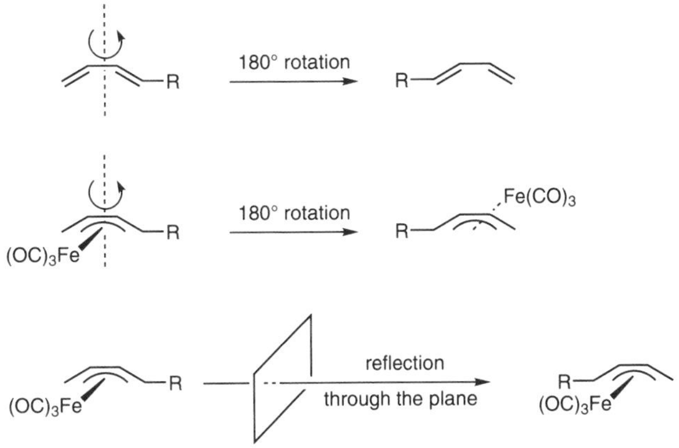

Thus, it becomes possible to prepare functional, optically active η^4-diene complexes, the two enantiomers below being good examples:[32]

$$\text{Me} \overset{|}{\underset{\text{Fe(CO)}_3}{\diagup\diagdown}}\text{CHO} \qquad \text{OCH} \overset{|}{\underset{\text{Fe(CO)}_3}{\diagup\diagdown}}\text{Me}$$

In turn, these complexes open up the possibility of asymmetric synthesis. Their racemization involves a ΔG of 31 kcal/mol, which means that they are useful up to 90–100°C.[32] The following steps are involved:

$$\text{Me} \overset{(+)}{\underset{\text{Fe(CO)}_3}{\diagup\diagdown}}\text{CHO} \longrightarrow \text{Me} \overset{(+)}{\underset{\text{Fe(CO)}_3}{\diagup\diagdown}}\text{R} \; (+\,-)$$

(mixture of two diastereomers)

separation

decomplexation

$$\text{Me} \diagup\diagdown\text{R}_{(+)} \longleftarrow \text{Me} \overset{(+)}{\underset{\text{Fe(CO)}_3}{\diagup\diagdown}}\text{R}_{(+)}$$

(pure enantiomer)

The liberation of the diene is carried out by destroying the complex through mild oxidation of the iron. The most frequently used reagents are cerium (IV) in MeOH, basic H_2O_2, *meta*-chloroperbenzoic acid or Me_3NO, although the last compound seems to give indifferent results.[32] A whole series of optically active dienes has been prepared in this manner, for example:

$$\text{Me} \diagdown\diagup\diagdown \overset{\displaystyle \text{C*-Me}}{\underset{\text{HO Ph}}{|}}$$

Another important reaction of η^4-diene complexes is their conversion into cationic η^5-pentadienyl complexes. Two general methods exist:

(a) The abstraction of H^- by Ph_3C^+, BF_4^-:

$$\left[\underset{(OC)_3Fe}{\diagup} \right] \xrightarrow{[Ph_3C^{\oplus}]} \left[\underset{(OC)_3Fe}{\diagup} \right]^{\oplus} \quad (+ Ph_3CH)$$

(b) The protonation of an alcohol functionality α to the complexed diene by a strong acid:

$$\text{R} \overset{|}{\underset{\text{Fe(CO)}_3}{\diagup\diagdown}}\text{CH}_2\text{OH} \xrightarrow{H^{\oplus}} \left[\text{R} \overset{|}{\underset{\text{Fe(CO)}_3}{\diagup}} \right]^{\oplus} \quad (+ H_2O)$$

These complexed η^5-pentadienyl cations react smoothly with nucleophiles, which attack the face of the diene which is remote from the iron, for steric reasons (see Section 2.3):

$$\left[\underset{(OC)_3Fe}{\diagup} \right]^{\oplus} + Nu^{\ominus} \longrightarrow \underset{(OC)_3Fe}{\diagup}\!\!\underset{Nu}{|}$$

$Nu = RO$, R_2N, H ($NaBH_4$), R (R_2Cd, R_2Zn, R_2CuLi), etc...

Electrophilic aromatic substitutions also work well:

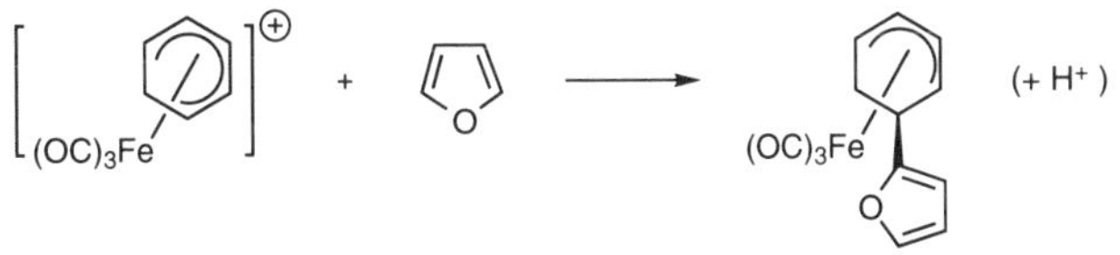

The functionalized products may again be liberated by the mild oxidations described above.

4.5 η^6-Arene-Chromium-Tricarbonyls in Organic Synthesis

To an organic chemist, an arene is an 'electron-rich' molecule which usually reacts with electrophiles. One of the most characteristic reactions of such a system is electrophilic aromatic substitution:

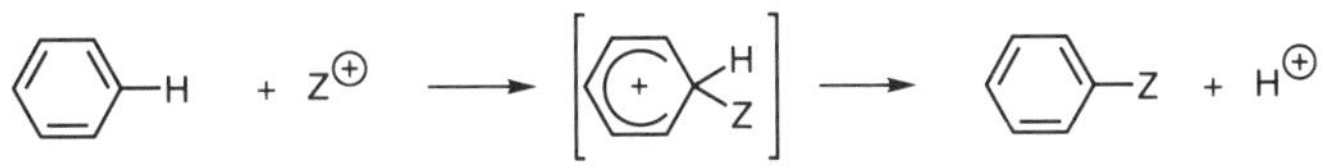

The final elimination of H^+ restores the aromaticity of the system. On the other hand, nucleophilic substitution is very difficult:

Often, the situation changes completely if the arene is converted into a η^6-arene complex. The complexing group usually acts as an 'electron sink' and causes the arene to become 'electron-poor'. In such cases, nucleophilic substitution becomes possible, as is observed for arene-chromium-tricarbonyls. These compounds are finding increasing use in organic synthesis because they are particularly easy to prepare, transform and demetallate.

Synthesis and Demetallation of η^6-Arene-Chromium-Tricarbonyls

In the majority of cases it suffices to heat the free arene with $Cr(CO)_6$:

The reaction is enhanced by donor substituents (R, R_2N, ...) and inhibited by electron attractors such as Cl and CO_2Me. The addition of donor solvents (THF, Bu_2O,

MeO$-$CH$_2$$-CH_2$$-O-CH_2$$-CH_2$$-$OMe) promotes the reaction; nonetheless it remains inefficient or fails completely for aromatics bearing very strong attractors (CN, NO$_2$, CHO, CO$_2$H). The use of UV is also sometimes beneficial. In difficult cases, it is better to replace Cr(CO)$_6$ by Cr(CO)$_3$L$_3$ where L is chosen to be easily displaced by the arene.

$$\text{(benzene)} \quad + \quad [\text{Cr(CO)}_3\text{L}_3] \longrightarrow \text{(benzene·Cr(CO)}_3) \quad + \quad 3\,\text{L}$$

L = CH$_3$CN, (pyridine)

It is possible to prepare an η^6-styrene complex without competing polymerization by this methodology:[33]

$$\text{(C}_6\text{H}_5)-\text{CH}=\text{CH}_2 \quad + \quad \left[\text{Cr(CO)}_3\left(\text{pyridine}\right)_3\right]$$

$$\longrightarrow \quad \text{[(}\eta^6\text{-styrene)Cr(CO)}_3\text{]}-\text{CH}=\text{CH}_2 \quad + 3 \quad \text{(pyridine)}$$

The polystyrene complex can equally be prepared directly:[34]

$$\left(-\text{CH}-\text{CH}_2-\right)_n \quad + \quad [\text{Cr(CO)}_3(\text{CH}_3\text{CN})_3] \longrightarrow \left(-\text{CH}-\text{CH}_2-\right)_n$$

Another synthetic option involves arene exchange, which is merely a question of displacing an equilibrium. The stability order of η^6 complexes at 142°C in THF is the following:[35]

Me$_6$C$_6$ > Me$_4$C$_6$H$_2$ > Me$_3$C$_6$H$_3$ > Me$_2$N$-$C$_6$H$_5$ > Me$_2$C$_6$H$_4$ > MeC$_6$H$_5$ $\approx$ C$_6$H$_6$ > MeC(O)$-$C$_6$H$_5$ $\approx$ MeO$-$C$_6$H$_5$ > MeO$_2$C$-$C$_6$H$_5$ > ClC$_6$H$_5$ $\approx$ FC$_6$H$_5$ > C$_{10}$H$_8$.

Thus, the best arene for displacement is naphthalene:

$$\text{(naphthalene·Cr(CO)}_3) \quad + \quad \text{arene} \longrightarrow [\text{Cr(CO)}_3(\text{arene})] \quad + \quad \text{(naphthalene)}$$

Finally, it is sometimes possible to employ a direct synthesis of the arene within the chromium coordination sphere, through the interaction of a chromium carbene complex with an alkyne (see Section 3.4).

The decomplexation of an arene from its chromium complex is easily achieved by treatment with a mild oxidant, such as Ce^{4+}, I$_2$, or air. Sunlight promotes the process.

Physiochemical Properties of η^6-Arene-Chromium-Tricarbonyls

IR spectroscopy is particularly useful for the detection of an arene-chromium-tricarbonyl because the local C_{3v} symmetry gives two intense bands A_1 and E (see Section 3.2). With substituted arenes, the perturbation of local symmetry sometimes provokes a splitting of the E band.

Examples:

$(\eta^6\text{-}C_6H_6)Cr(CO)_3$	ν (CO)	1987, 1917 cm^{-1}
$(\eta^6\text{-}MeO_2C{-}C_6H_5)Cr(CO)_3$		1997, 1927 cm^{-1}
$(\eta^6\text{-}Me_2N{-}C_6H_5)Cr(CO)_3$		1969, 1895, 1889 cm^{-1}

The carbonyl frequencies are influenced by the electronic character of the arene: the stronger its potential as a donor, the more the COs are polarized and the lower the force constants and stretching frequencies.

In structure, these complexes resemble a piano stool. The arene is planar and lies 1.73 Å from the chromium, orthogonal to the threefold symmetry axis. In the solid state, the COs are staggered with respect to the carbons of the arene but the barrier to arene rotation about the $Cr(CO)_3$ axis is very weak. The chromium–arene bond is of medium energy:

$$D(Cr-C_6H_6) = 43 \text{ kcal/mol}$$
$$D(Cr-C_6Me_6) = 49 \text{ kcal/mol}$$
$$D(Cr-\text{naphthalene}) = 36 \text{ kcal/mol}$$

The $Cr(CO)_3$ group behaves as a strongly electron-attracting arene substituent. Thus, it reinforces the acidity of benzoic acid:

$$C_6H_5\text{-}CO_2H \ : \ pK_a \ 5.48 \qquad \underset{Cr(CO)_3}{\bigcirc}\!\!-CO_2H \ : \ pK_a \ 4.52$$

Synthetic Applications of Arene-Chromium-Tricarbonyls[36]

The scheme below summarizes the influence of the complexing group. Each effect has synthetic applications.

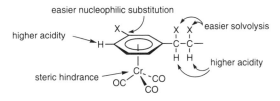

Modification of arene properties upon complexation

Addition of Nucleophiles[37]

Carbanions attack the non-complexed face of the arene, giving η^5-cyclohexadienyl complexes:

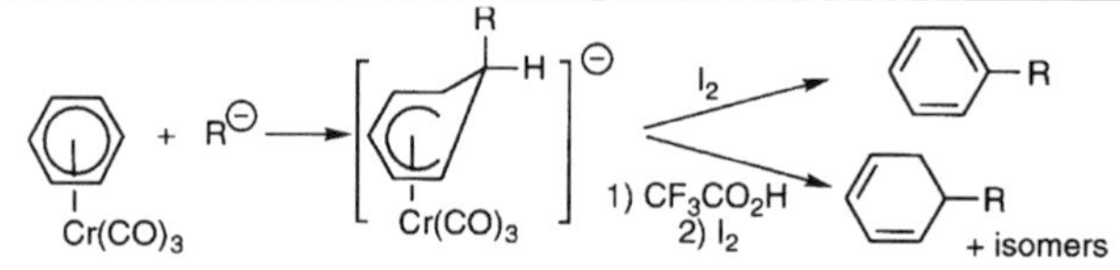

Protonation by a strong acid (CF_3CO_2H) and subsequent oxidation with iodine leads to a substituted cyclohexadiene; on the other hand, direct treatment with iodine produces the corresponding arene. Numerous applications of this synthetic scheme have been developed in the literature:

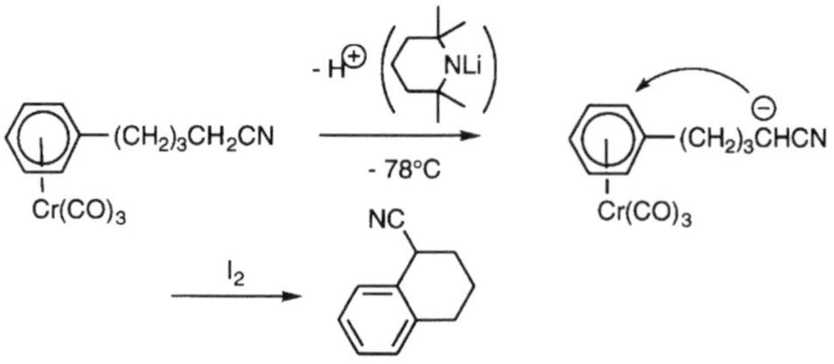

Lithium tetramethylpiperidide behaves as a mild, very hindered base towards an arene-chromium-tricarbonyl.

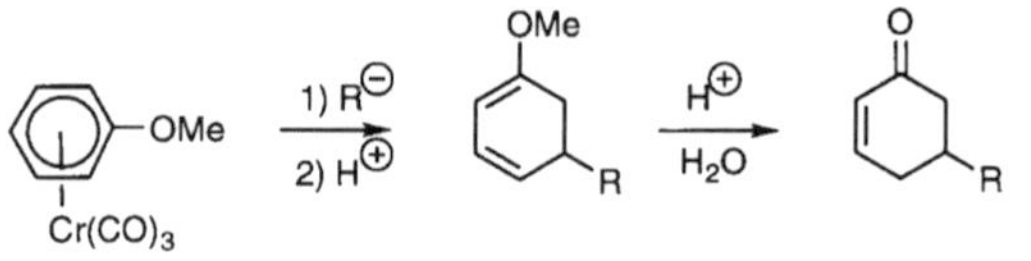

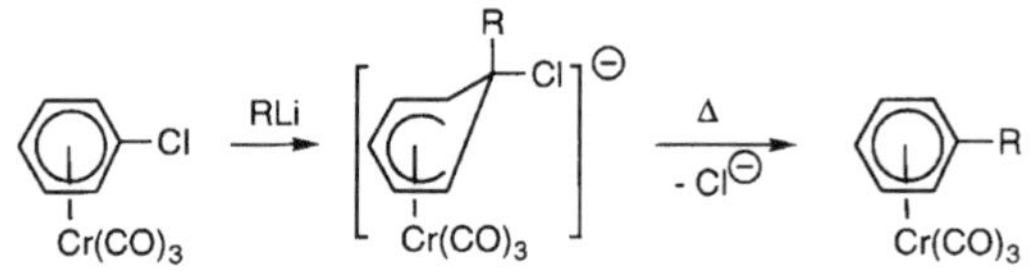

Lithiation of the Arene

A very powerful nucleophile such as butyllithium is able to abstract the acidic protons from the arene. This process is favoured over nucleophilic addition. In the absence of specific directing effects, lithiation takes place at all the possible positions; by using a directing effect (for example chelation of the lithium ion by a functional group on the arene), the metallation becomes regioselective and synthetically useful.

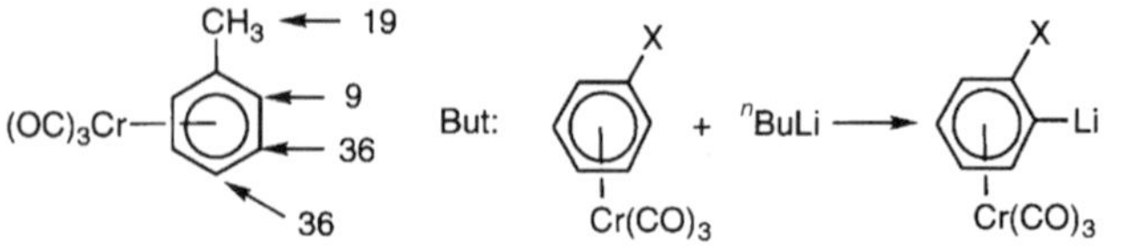

% lithiation at the various positions of
(η^6-toluene)chromium-tricarbonyl

X = F, Cl, OMe

The following example combines regioselective lithiation with nucleophilic addition:

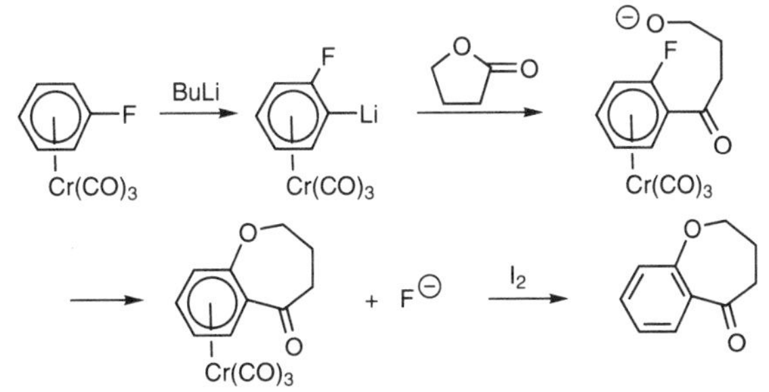

Activation of the Benzylic Positions[38]

The protons in the α position of toluene-chromium-tricarbonyl and its relatives show a marked acidity. Even double alkylations are possible, as shown in the example below:

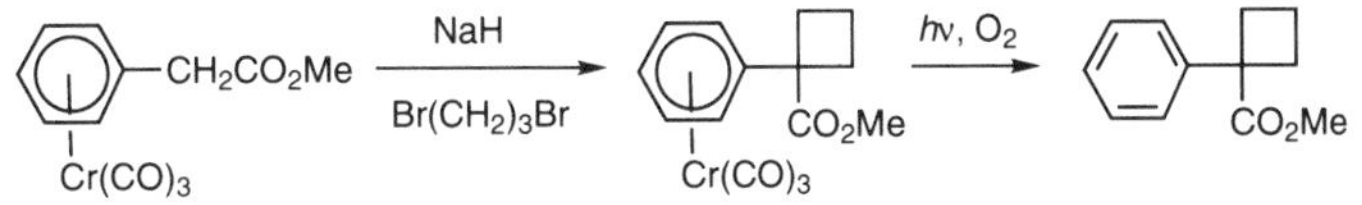

Steric and Asymmetric Effects

The steric influence of the $Cr(CO)_3$ fragment may be used to control the stereochemistry of a reaction:

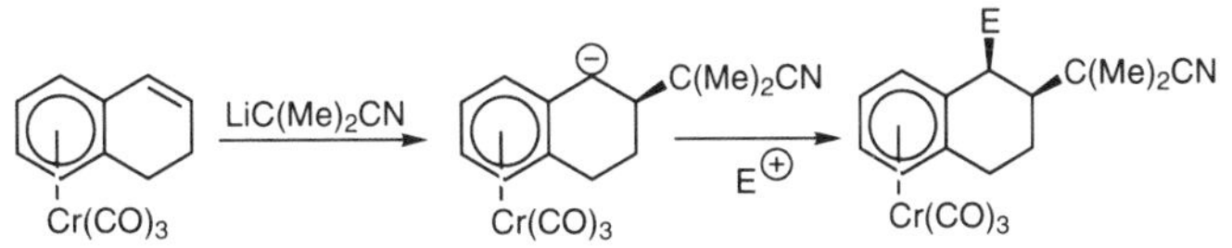

From another standpoint, an arene having two different substituents becomes chiral upon complexation because the plane of symmetry is eliminated:

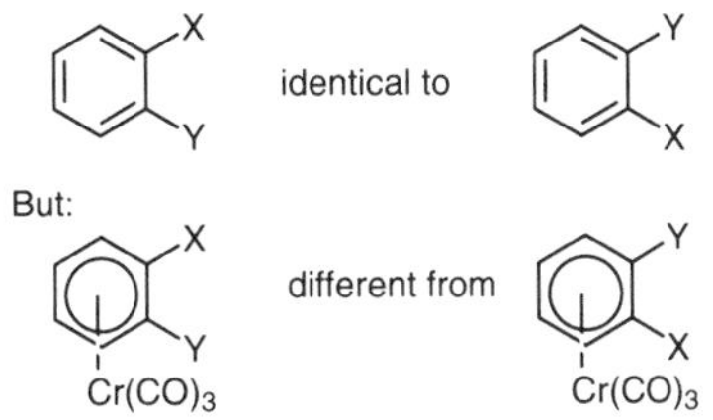

This allows asymmetric syntheses. For example, starting with $(+)$–(S) amphetamine:[39]

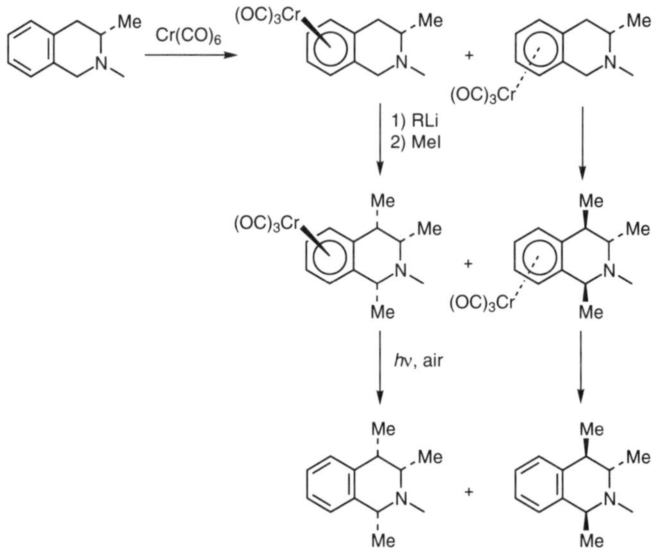

4.6   References

1. J. Schwartz and J. A. Labinger, *Angew. Chem., Int. Ed. Engl.*, 1976, **15**, 333; see also: E. Negishi and T. Takahashi, *Synthesis*, 1988, 1.
2. S. L. Buchwald, S. J. La Maire, R. B. Nielsen, B. T. Watson and S. M. King, *Tetrahedron Lett.* 1987, **28**, 3895.
3. D. B. Carr and J. Schwartz, *J. Am. Chem. Soc.*, 1977, **99**, 638.
4. M. J. Loots and J. Schwartz, *J. Am. Chem. Soc.*, 1977, **99**, 8045.
5. K. A. Babiak, J. R. Behling, J. H. Dygos, K. T. McLaughlin, J. S. Ng, V. J. Kalish, S. W. Kramer and R. L. Shone, *J. Am. Chem. Soc.*, 1990, **112**, 7441.
6. B. H. Lipshutz and K. Kato, *Tetrahedron Lett.*, 1991, **32**, 5647.
7. M. Yoshifuji, M. J. Loots and J. Schwartz, *Tetrahedron Lett.*, 1977, 1303.
8. R. A. Budnik and J. K. Kochi, *J. Organomet. Chem.*, 1976, **116**, C3.
9. T. E. Cole, S. Rodewald and C. L. Watson, *Tetrahedron Lett.*, 1992, **33**, 5295.
10. T. Gibson, *Tetrahedron Lett.*, 1982, **23**, 157.
11. C. A. Bertelo and J. Schwartz, *J. Am. Chem. Soc.*, 1976, **98**, 262.
12. D. Seyferth and R. M. Weinstein, *J. Am. Chem. Soc.*, 1982, **104**, 5534.
13. N. S. Nudelman and A. A. Vitale, *J. Org. Chem.*, 1981, **46**, 4625.
14. See, for example: O. W. Lever Jr, *Tetrahedron*, 1976, **32**, 1943.
15. J. P. Collman, R. G. Finke, P. L. Matlock, R. Wahren, R. G. Komoto and J. I. Brauman, *J. Am. Chem. Soc.*, 1978, **100**, 1119.
16. J. P. Collman, *Acc. Chem. Res.*, 1975, **8**, 342.
17. R. G. Finke and T. N. Sorrell, *Org. Synth.*, 1980, **59**, 102.
18. K. S. Suslick and R. E. Johnson, *J. Am. Chem. Soc.*, 1984, **106**, 6856.
19. M. P. Cooke and R. M. Parlman, *J. Am. Chem. Soc.*, 1975, **97**, 6863.
20. J. Y. Mérour, J. L. Roustan, C. Charrier, J. Collin and J. Benaïm, *J. Organomet. Chem.*, 1973, **51**, C24.
21. P. Laurent, G. Tanguy and H. des Abbayes, *J. Chem. Soc., Chem. Commun.*, 1986, 1754.

22. This topic has been reviewed: M. Brookhart and W. B. Studbaker, *Chem. Rev.*, 1987, **87**, 411.

23. See, for example: H. Fischer, E. Mauz, M. Jaeger and R. Fischer, *J. Organomet. Chem.*, 1992, **427**, 63.

24. E. J. O'Connor and P. Helquist, *J. Am. Chem. Soc.*, 1982, **104**, 1869.

25. K. A. M. Kremer, P. Helquist and R. C. Kerber, *J. Am. Chem. Soc.*, 1981, **103**, 1862.

26. M. Brookhart *et al.*, *J. Am. Chem. Soc.*, 1977, **99**, 6099; 1980, **102**, 7802; 1981, **103**, 979; 1983, **105**, 258; *Organometallics*, 1985, **4**, 943.

27. M. Brookhart, D. Timmers, J. R. Tucker, G. D. Williams, G. R. Husk, H. Brunner and B. Hammer, *J. Am. Chem. Soc.*, 1983, **105**, 6721.

28. B. F. G. Johnson, J. Lewis and D. J. Thompson, *Tetrahedron Lett.*, 1974, 3789.

29. G. F. Emerson, L. Watts and R. Pettit, *J. Am. Chem. Soc.*, 1965, **87**, 131.

30. J. A. Connor, C. P. Demain, H. A. Skinner and M .T. Zafarani-Moattar, *J. Organomet. Chem.*, 1979, **170**, 117.

31. T. H. Whitesides and R. W. Arhart, *J. Am. Chem. Soc.*, 1971, **93**, 5296.

32. R. Grée, *Synthesis*, 1989, 341.

33. G. A. Moser and M. D. Rausch, *Synth. React. Inorg. Metal-Org. Chem.*, 1974, **4**, 37.

34. A. Munoz-Escalona and G. Di Filippo, *Makromol. Chem.*, 1977, **178**, 659.

35. C. A. L. Mahaffy and P. L. Pauson, *J. Chem. Res.* (M), 1979, 1752.

36. For a comprehensive review, see: V. N. Kalinin, *Russ. Chem. Rev. (Engl. Transl.)*, 1987, **56**, 1190.

37. M. F. Semmelhack, G. R. Clark, J. L. Garcia, J. J. Harrison, Y. Thebtaranonth, W. Wulff and A. Yamashita, *Tetrahedron*, 1981, **37**, 3957.

38. G. Jaouen, A. Meyer and G. Simmoneaux, *J. Chem. Soc., Chem. Commun.*, 1975, 813.

39. S. G. Davies, *Chem. Ind. (London)*, 1986, 506.

5 Some Applications in Homogeneous Catalysis

First of all, let's recall some basic principles. A catalyst changes the rate of a chemical reaction without being incorporated into the final reaction products. (It generally raises it: when a catalyst slows a reaction it is called an inhibitor. Inhibitors of oxidation and corrosion, anti-knock petrol additives, etc., are of high industrial importance). Thus, it intervenes as a reaction partner during the intermediate steps of the catalytic cycle but is recovered chemically unchanged at the end (see scheme).

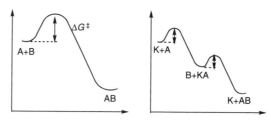

Catalysis of the reaction A + B → AB

The catalyst acts by reducing the energetic barriers to the reaction in question. This may be achieved by bringing the reagents (here A and B) into contact within the coordination sphere of a metal, which explains the importance of coordination chemistry in catalysis.

Reaction A+B uncatalysed   Reaction A+B catalysed by K

Catalysis is a common phenomenon, which arises to a considerable extent in nature (enzymes, etc.). As far as chemists are concerned, it exists in three categories: homogeneous, heterogeneous and hybrid catalysis. Homogeneous catalysis, where the molecular catalyst is dissolved in the same solvent as the reagents, is easy to follow and study by traditional spectroscopic methods, such as IR and NMR. Heterogeneous catalysis involves reactions which take place at the surface of a solid material (metal, oxide, sulfide, etc.). In these cases phenomena are more complex and only partly understood (the physical and chemical properties of the catalyst surface play an important role). In hybrid catalysis, a molecular catalyst is grafted onto a solid support. This facilitates the separation of the reaction products from the catalyst, whilst largely retaining the mild operating conditions and selectivity associated with homogeneous catalysis.

In the following Chapter, we will discuss only the first type.

5.1 Alkene Hydrogenation and Related Reactions

The hydrogenation of alkenes follows the pathway:

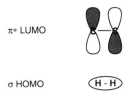

In theory, a concerted addition of H_2 to C=C is forbidden for reasons of symmetry, so to get around this problem we use a catalyst. The catalysed process generally operates by homolysis of the H−H bond.

$\pi*$ LUMO

σ HOMO   (H - H)

The mechanistic principles underlying catalytic hydrogenation at a metallic centre are described in Figure 5.1. The complex must meet three criteria to act as an effective homogeneous hydrogenation catalyst:

1. hydrogen must not reduce it to the metal;

2. it must form a labile complex with the alkene (if the complex with the alkene is too stable, the catalytic cycle will be blocked);

3. it should be soluble in the organic solvents used as reaction media.

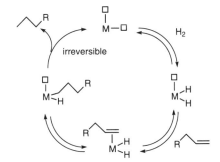

Figure 5.1 Idealized mechanism for the hydrogenation of an alkene by oxidative addition of H_2 to a metallic centre (the square corresponds to a vacant site or a weakly coordinated solvent)

The prototype for these systems is Wilkinson's catalyst, [RhCl(PPh$_3$)$_3$] (16 electrons),[1] which normally operates at 25°C under hydrogen at atmospheric pressure. It is prepared by the reaction of excess ethanolic triphenylphosphine with RhCl$_3$:

$$RhCl_3 \xrightarrow[\text{EtOH}]{\text{4 Ph}_3\text{P}} [RhCl(PPh_3)_3] \quad (+ Ph_3PO)$$

The triphenylphosphine serves both as a ligand and as a reducing agent. The complex, which has a very slight tetrahedral distortion from the ideal square planar geometry, effects the specific *cis* addition of hydrogen to alkenes:

The ease of reaction decreases with increasing steric hindrance at the olefin:

The precise mechanism in the case of Wilkinson's catalyst is more complex than the idealized scheme outlined above. Nevertheless, some useful conclusions can be drawn from the general mechanism. First of all, it is necessary to coordinate the two reagents within the coordination sphere of the metal, which implies the intermediacy of a 14-electron fragment derived from the rhodium precursor somewhere within the cycle. Thus it can be deduced that one of the triphenylphosphine ligands ought to be labile, as is the case:

$$RhClL_3 \ \rightleftharpoons \ L + RhClL_2 \ (L = PPh_3)$$

This 14-electron complex tends to dimerize, to give a stable chlorine-bridged dimer which is inactive as a catalyst:

Thus, the reaction is usually accomplished in a weak two-electron donor solvent (S) EtOH, THF, etc. ... which stabilizes the reactive 14-electron species by reversible coordination.

$$RhClL_2 + S \ \longrightarrow \ [RhCl(S)L_2]$$

By the same logic, the replacement of triphenylphosphine by a more powerful ligand such as triethylphosphine (a better and less hindered donor, see Section 3.8) should block catalysis through stabilizing the 16e form; this is again observed. Thus the 14-electron species, more or less solvated, is the true hydrogenation catalyst, [RhClL_3] being merely its precursor. [RhClL_2] is evidently too reactive and unstable to be detected and the fact that it adds hydrogen 10^4 times faster than does [RhClL_3] demonstrates its fantastic reactivity. For most olefins, the true catalytic cycle strongly resembles the general scheme above, but it should be noted that reduction of olefins having strong coordination properties ($CH_2=CH_2$, $PhCH=CH_2$, $CH_2=CH-CH=CH_2$) follows a different pathway. Wilkinson's catalyst constitutes the prototype for a whole series of very active hydrogenation catalysts; two other extremely efficient systems are shown below:[2,3]

The cationic rhodium complex[2] is representative of a class of compound which is frequently employed in asymmetric catalysis (see Section 5.2).

$$\left[(cod)Ir \begin{array}{c} Py \\ PCy_3 \end{array} \right]^{\oplus} (16e)$$

Py = [pyridine structure]

Cy = [cyclohexyl structure]

To conclude, it should be noted that some alkene hydrogenation catalysts show a potential for the reduction of functional groups such as C=O, −N=N−, −CH=N−, and −NO$_2$. As an up-to-date example, we cite the hydrogenation of supercritical CO$_2$ (T > 31°C, p > 73 atm) in formic acid:[4]

$$\underset{\text{(120 atm)}}{CO_2} + \underset{\text{(85 atm)}}{H_2} \xrightarrow[50°C]{\text{catalyst}} H\text{-}CO_2H$$

catalyst: *cis*- RuH$_2$(PMe$_3$)$_4$ + NEt$_3$

Here, 1400 moles of formic acid per mole of catalyst per hour may be produced without using inflammable or toxic solvents.

Two reactions related to alkene hydrogenation are of particular importance; they are hydrosilylation and hydrocyanation of carbon–carbon double bonds.

The hydrosilylation of alkenes follows the general scheme:

$$\underset{}{>\!C\!=\!C\!<} \; + \; R_3Si\text{-}H \; \longrightarrow \; \underset{}{R_3Si \diagdown \; \diagup H \atop C\text{-}C}$$

The addition is *cis*, as in hydrogenation. The most frequently employed catalyst precursor for this reaction is H$_2$PtCl$_6$.[5] A few research workers have suggested that, in this case, the true catalyst is colloidal platinum suspended in an organic medium. Whatever the truth, authentically homogeneous catalysts such as Co$_2$(CO)$_8$, Ni(cod)$_2$, [NiCl$_2$(PPh$_3$)$_2$] and [RhCl(PPh$_3$)$_3$] have also proved to be effective. The catalytic mechanism is represented in Figure 5.2.

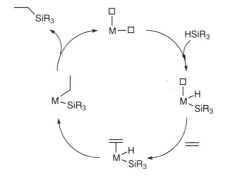

Figure 5.2 Simplified mechanism of the hydrosilylation of alkenes

With terminal olefins, the reaction gives predominantly linear silanes with only low concentrations of their branched isomers:

$$RCH=CH_2 + H\text{-}SiR'_3 \longrightarrow RCH_2\text{-}CH_2\text{-}SiR'_3 + RCH(SiR'_3)\text{-}CH_3$$

The use of alkyne substrates gives *trans*-vinylsilanes, as one would expect for a *cis*-addition mechanism:

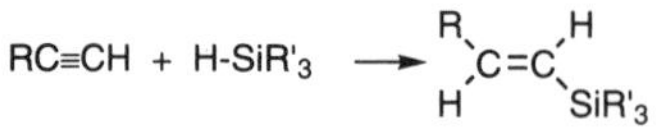

$$RC\equiv CH + H\text{-}SiR'_3 \longrightarrow \underset{H}{\overset{R}{}}C=C\underset{SiR'_3}{\overset{H}{}}$$

One can also hydrosilylate dienes (giving mainly 1,4-addition products) and carbonyl compounds (alkoxysilanes are obtained). Each of these reactions has industrial applications in the preparation of monomeric precursors for silicon based rubbers.[6]

The hydrocyanation of alkenes is described in the following equation:

$$RCH=CH_2 + H\text{-}CN \longrightarrow RCH_2\text{-}CH_2\text{-}CN$$

The most frequently employed catalyst is NiL_4 (L = phosphine or phosphite) and the active species is $L_2Ni(H)CN$. The mechanism is shown in Figure 5.3.

The main application of this reaction class is in the Dupont adiponitrile synthesis, which involves the double addition of HCN to butadiene:

$$H_2C=CH\text{-}CH=CH_2 + 2\,HCN \xrightarrow{\;[Ni(P(OAr)_3)_4]\;} NC\text{-}(CH_2)_4\text{-}CN$$

The adiponitrile is used to prepare the precursors of nylon 6,6:

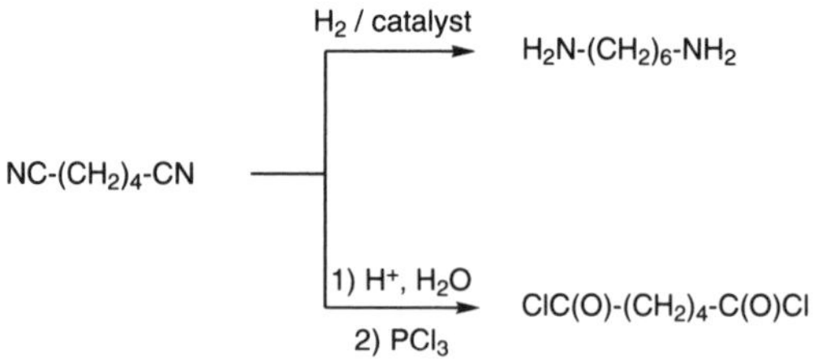

NC-(CH$_2$)$_4$-CN

$\xrightarrow{\quad H_2 / catalyst \quad}$ H$_2$N-(CH$_2$)$_6$-NH$_2$

$\xrightarrow[\text{2) PCl}_3]{\text{1) H}^+\text{, H}_2\text{O}}$ ClC(O)-(CH$_2$)$_4$-C(O)Cl

Condensation of the acid chloride with the diamine then furnishes nylon.[7]

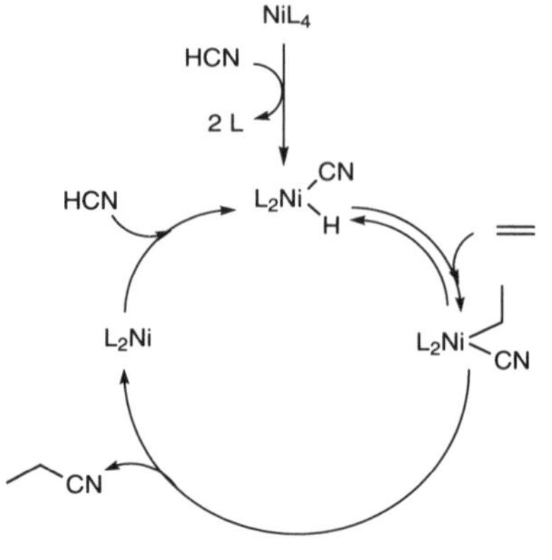

Figure 5.3 Simplified mechanism of the hydrocyanation of alkenes

5.2 Asymmetric Hydrogenation

An alkene having two different substituents on one of the olefinic carbons is prochiral because a process which saturates the double bond will generate a chiral carbon atom. Hydrogenation is one of the most important methods for the creation of this chirality.

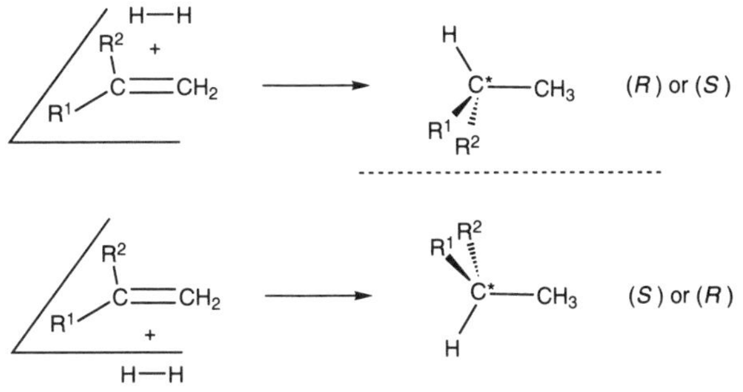

The nature of the enantiomer obtained depends upon which face of the olefin is hydrogenated. The π complex of a prochiral alkene is also chiral; here, the chirality is dictated by which face of the alkene is complexed. This is obvious when we realize that the initially prochiral olefin carbon becomes tetracoordinated through complexation.

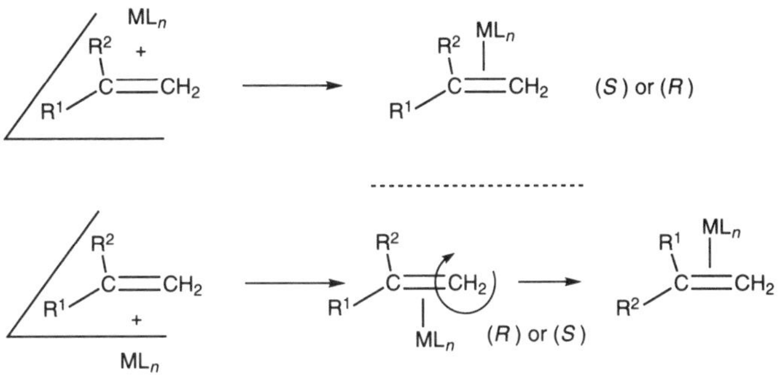

It can be verified that 180° rotation around the olefin axis does not convert the π complex into its mirror image. In the case where the metal or one of its ligands L* is chiral, a prochiral olefin will give two separable diastereomeric complexes $(R(C),R(L^*) + S(C),S(L^*)$ and $R(C),S(L^*) + S(C),R(L^*))$. Their physical and chemical properties will not be identical and their hydrogenation rates will differ. If, in a further refinement, we carry out complexation to a metal having an optically pure ligand L* (for example S), each diastereomer will include a single enantiomer of each of the pairs above. Each of these enantiomers will be hydrogenated at a different rate [because they are no longer mirror images of each other; in a mirror (S)–L* becomes (R)–L*].

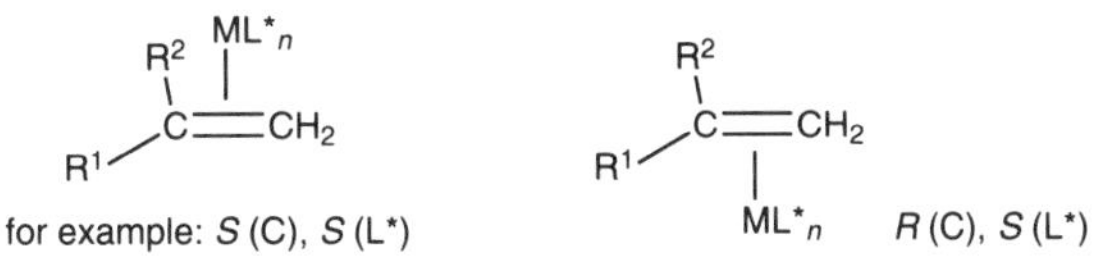

for example: S (C), S (L*) R (C), S (L*)

If, arbitrarily, we assume that the (S,S) enantiomer is more rapidly hydrogenated, we will obtain more R than S alkane (unfortunately, the convention which defines the carbon configuration changes when M is replaced by H; the chirality itself is retained rather than inverted). This is because the reversible chemical equilibrium relating the two enantiomers will be displaced towards the more rapidly consumed (S,S) enantiomer.

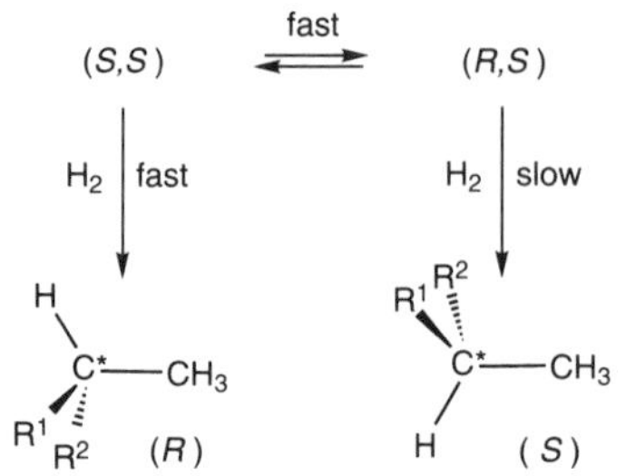

With the correct catalyst, the (R) alkane is formed almost exclusively. Under such conditions, the level of optical activity in the system (a small trace in the catalyst at the outset, a large quantity in the product) is multiplied by a large factor. This is the principle used by enzymes to create optical activity in nature. The performance of an asymmetry-inducing catalyst is measured by the enantiomeric excess of one product over the other:

$$ee = \left| \frac{q(R) - q(S)}{q(R) + q(S)} \right|$$

where $q(R)$ and $q(S)$ designate the quantities of each enantiomer obtained. A good asymmetric hydrogenation catalyst easily gives an enantiomeric excess greater than 95% if the prochiral alkene has a functional substituent capable of coordinating the metal.

One of the first truly efficient asymmetric hydrogenation catalysts was developed by Kagan.[8] It is derived from the Schrock and Osborn catalyst $[L_2RhS_2]^+$ (L = phosphine, S = solvent, see Section 5.1) by replacement of the usual phosphines by the chiral ligand DIOP. The synthesis of DIOP from tartaric acid was described in Section 3.8; the catalyst itself is obtained by its reaction with $[Rh(cod)_2]^+$, $[PF_6]^-$.

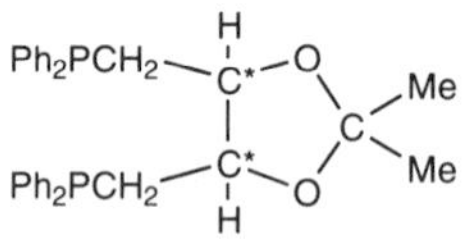

It should be noted that the asymmetric centres of DIOP are far removed from the phosphorus atoms and the catalytic centre of the complex. Numerous other phosphines have been proposed as auxiliaries for asymmetric hydrogenation. Amongst these, we highlight two particularly important examples:

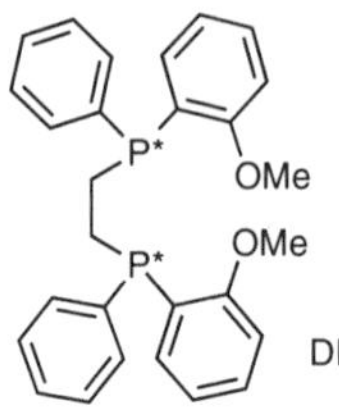

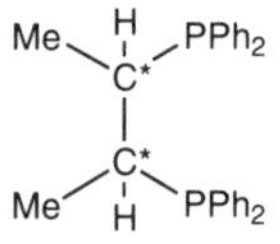

In DIPAMP, the phosphorus atoms themselves are optically active. The inversion barrier for a tricoordinated pyramidal phosphorus is above 30 kcal/mol (see Section 3.8), which means that phosphines whose chirality is localized at phosphorus are configurationally stable up to about 100°C. Recently, Noyori and his group[11] have introduced BINAP, whose chirality results from a steric barrier to rotation around the C−C bridge of the binaphthyl (atropisomerism). This phosphine is remarkably effective for both the asymmetric rhodium-catalysed hydrogenation of alkenes and the ruthenium(II)-catalysed hydrogenation of functional ketones.

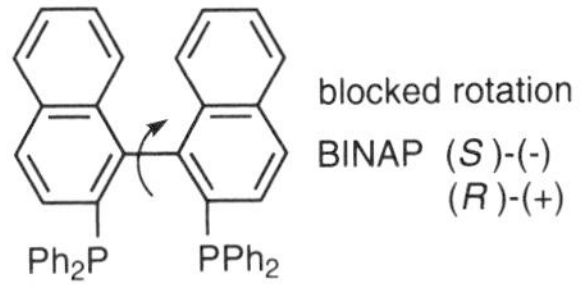

blocked rotation

BINAP (S)-(-)
(R)-(+)

Ph₂P   PPh₂

The following examples[12] illustrate the potential for the reduction of carbonyls perfectly:

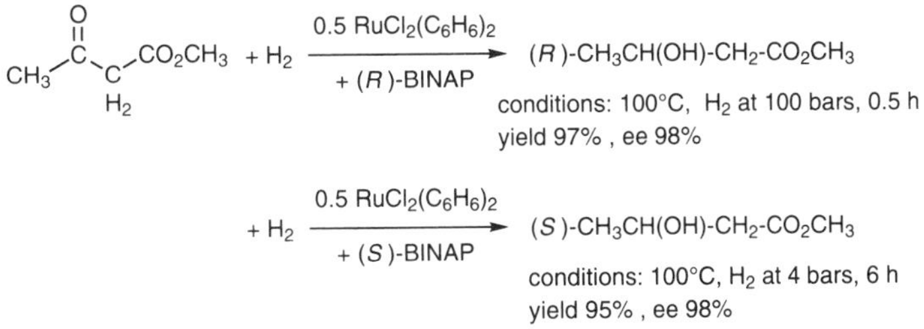

$$CH_3-\overset{\overset{\displaystyle O}{\|}}{C}-\underset{H_2}{C}-CO_2CH_3 + H_2 \xrightarrow[+ (R)\text{-BINAP}]{0.5\ RuCl_2(C_6H_6)_2} (R)\text{-}CH_3CH(OH)\text{-}CH_2\text{-}CO_2CH_3$$

conditions: 100°C, H₂ at 100 bars, 0.5 h
yield 97% , ee 98%

$$+ H_2 \xrightarrow[+ (S)\text{-BINAP}]{0.5\ RuCl_2(C_6H_6)_2} (S)\text{-}CH_3CH(OH)\text{-}CH_2\text{-}CO_2CH_3$$

conditions: 100°C, H₂ at 4 bars, 6 h
yield 95% , ee 98%

Other applications of BINAP have been developed, most notably in asymmetric olefin isomerization. The industrial implications of this research are nicely illustrated by two examples. The first is the Knowles Monsanto synthesis of L-DOPA,[13] a compound used in the treatment of Parkinson's disease.

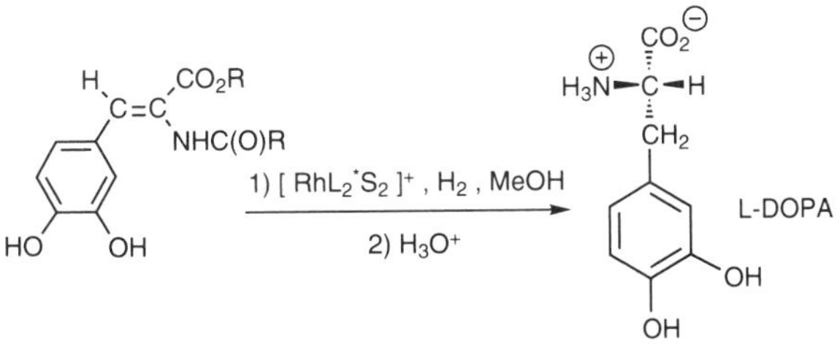

1) [RhL₂·S₂]⁺ , H₂ , MeOH
2) H₃O⁺

L-DOPA

The second is the Japanese BINAP-based industrial synthesis of L-menthol.

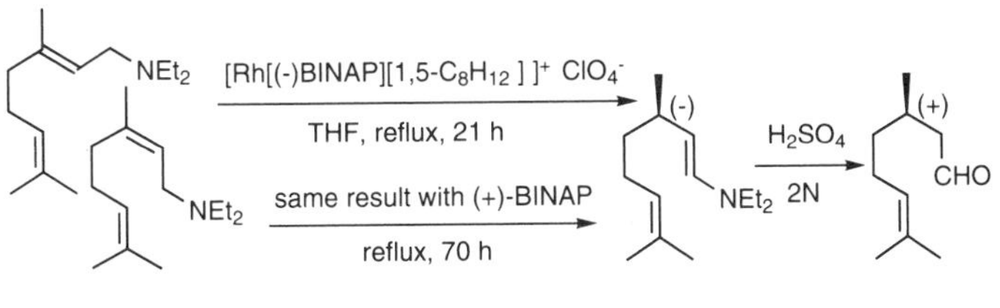

[Rh[(-)BINAP][1,5-C₈H₁₂]]⁺ ClO₄⁻

THF, reflux, 21 h

same result with (+)-BINAP

reflux, 70 h

H₂SO₄
2N

Currently, research is aimed at developing catalysts for the enantioselective hydrogenation of simple alkenes which cannot chelate to the transition metal. Encouraging progress has been made, for example:[14]

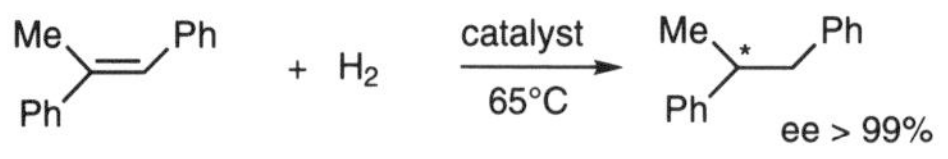

The catalyst, an optically active titanocene having C_2 symmetry, is stabilized by the addition of $PhSiH_3$.

In more general terms, there is intense interest in the development of asymmetric variants of any homogeneously catalysed process. This is particularly true in the area of pharmaceuticals, where highly undesirable biological effects may be associated with the presence of unwanted enantiomers; here, enantiomeric purity is an essential condition for the commercialization of new drugs.

5.3 Alkene Hydroformylation

The alkene hydroformylation reaction (also called the 'oxo' reaction, synthesis or process) is outlined in the following scheme:

$$R\text{-}CH{=}CH_2 + H_2 + CO \longrightarrow R\text{-}CH_2\text{-}CH_2\text{-}CHO + R\text{-}CH(CH_3)\text{-}CHO$$

Although, from a *purely formal* viewpoint, it consists of adding the elements of formaldehyde to an olefin, formaldehyde itself is never a reaction intermediate. Today, this oxo synthesis, discovered by Roelen in 1938, is one of the most important catalytic industrial reactions, producing about 6 million tons of aldehydes and alcohol derivatives world wide each year. The main application of 'the oxo' is the manufacture of linear *n*-butyraldehyde from propene. The suppression of side reactions leading to the branched *iso*-butyraldehyde which is systematically found with the main product, is still a major research area:

$$H_3C\text{-}CH{=}CH_2 + H_2 + CO \longrightarrow H_3C\text{-}CH_2\text{-}CH_2\text{-}CHO + H_3C\text{-}CH(CH_3)\text{-}CHO$$

(propene or
propylene)   (*n*- butyraldehyde)   (*iso*- butyraldehyde)

The *n*-butyraldehyde may be converted to *n*-butanol, 2-ethylhexanol or 2-ethylhexanoic acid:

$$H_3C\text{-}CH_2\text{-}CH_2\text{-}CHO \xrightarrow{H_2} H_3C\text{-}CH_2\text{-}CH_2\text{-}CH_2OH$$

(*n*- butanol)

$$H_3C\text{-}CH_2\text{-}CH_2\text{-}CH|O| \;+\; H_3C\text{-}CH_2\text{-}C|H_2|\text{-}CHO$$

$$\xrightarrow{-H_2O} \; H_3C\text{-}CH_2\text{-}CH_2\text{-}CH{=}C\text{-}CHO$$
$$\qquad\qquad\qquad\qquad\qquad\quad | $$
$$\qquad\qquad\qquad\qquad\qquad CH_2\text{-}CH_3$$

$$\xrightarrow{H_2} \; H_3C\text{-}CH_2\text{-}CH_2\text{-}CH_2\text{-}\underset{|}{C}H\text{-}CH_2OH$$
$$\qquad\qquad (2\text{-ethylhexanol}) \quad CH_2\text{-}CH_3$$

$$\xrightarrow{[O]} \; H_3C\text{-}CH_2\text{-}CH_2\text{-}CH_2\text{-}\underset{|}{C}H\text{-}CO_2H$$
$$\qquad\qquad\qquad\qquad\qquad\qquad CH_2\text{-}CH_3$$

(2-ethylhexanoic acid))

Commerical interest in these products reflects their use in the manufacture of plasticizers for polyvinyl chloride:

$$R = H_3C\text{-}(CH_2)_3\text{-}\underset{|}{C}H\text{-}CH_2\text{-}$$
$$\qquad\qquad\qquad C_2H_5$$

(2-ethylhexyl phthalate)

Another major application is the manufacture of 1-nonanol from 1-octene. This alcohol is used in detergent production.

The 'Oxo' Processes and their Mechanisms

Original Roelen Process

The original Roelen process is still widely used today. It operates between 120 and 170°C under 200–300 atmospheres of a mixture of CO and H_2. The precatalyst is cobalt carbonyl $Co_2(CO)_8$ and the true catalyst is the unsaturated 16-electron species [$HCo(CO)_3$]:

$$Co_2(CO)_8 + H_2 \longrightarrow 2\,H\text{-}Co(CO)_4 \underset{+CO}{\overset{-CO}{\rightleftharpoons}} 2\,[H\text{-}Co(CO)_3]$$

The first stage of the catalytic cycle consists of the addition of the olefin to the 16-electron catalyst. The first-formed product undergoes a migration-insertion reaction which leads to two isomeric cobalt-alkyls:

$$[H\text{-}Co(CO)_3] \underset{-RCH=CH_2}{\overset{\overset{\delta+\;\;\delta-}{+RCH=CH_2}}{\rightleftharpoons}}$$

R-CH$_2$-CH$_2$-Co(CO)$_3$ R-CH(CH$_3$)-Co(CO)$_3$

(formal addition of H$^-$)   (formal addition of H$^+$)

The formation of the linear derivative is obviously favoured by steric bulk at the $Co(CO)_3$ centre, whilst the branched derivative is favoured by increasing acidity of the hydride (see Section 3.1) or the intervention of radical processes (H• addition to the alkene). The ratio of linear to branched products is usually around 3–4:1. The process then evolves classically:

$$R'\text{-}Co(CO)_3 + CO \rightleftharpoons R'\text{-}Co(CO)_4 \rightleftharpoons R'\text{-}CO\text{-}Co(CO)_3$$
$$(16e) \qquad\qquad (18e) \qquad\qquad (16e)$$

$$\begin{array}{c} +\ CO \\ \rightleftharpoons \\ -\ CO \end{array} \quad R'\text{-}CO\text{-}Co(CO)_4 \qquad R' = R\text{-}CH_2\text{-}CH_2\text{-}\ or\ R\text{-}CH(CH_3)\text{-}$$
$$(18e)$$

The acylcobalt tetracarbonyl has been observed *in situ* (150°C, 250 atm) by IR.[15] Its hydrogenolysis regenerates the catalyst and liberates the aldehyde, through a mechanism which is still controversial.

$$R'\text{-}CO\text{-}Co(CO)_n + H_2 \longrightarrow R'CHO + HCo(CO)_n \quad n = 3,4$$

After many cycles, the catalyst loses its activity, through oxidation, poisoning, etc. . . . However, it may easily be regenerated by treatment with pressurized $H_2 + CO$ in the absence of olefin; the $HCo(CO)_4$ product, which is easy to purify because of its volatility, can then be re-injected into the catalytic cycle.

Shell Procedure

The Shell procedure, developed by Slaugh and Mullineaux in 1966,[16] involves modifying the cobalt catalyst by adding a trialkylphosphine such as $(n\text{-}Bu)_3P$. This addition increases the catalytic activity sufficiently to permit operation under $H_2 + CO$ at around 80 bars. It also improves the linear/branched ratio to around 10:1. Furthermore, the formation of aldehydes is suppressed; the corresponding alcohols are formed:

$$R\text{-}CH=CH_2 + CO + 2H_2 \longrightarrow R\text{-}CH_2\text{-}CH_2\text{-}CH_2\text{-}OH$$

Thus, it is no longer a hydroformylation in the strict sense! It is easy to rationalize all these observations. The working catalyst is now $[HCo(CO)_2(PR_3)]$, which means that the steric hindrance of the cobalt is much higher than before and the hydride character of the H–Co bond is enormously increased (see Section 3.1). These two factors favour the formation of linear products. The hydride character is also implicated in the reduction of the first-formed aldehydes into alcohols, through the following formal process:

$$R'CHO + H^\ominus \longrightarrow R'CH_2O^\ominus \overset{H^\oplus}{\longrightarrow} R'CH_2OH$$

Unfortunately, this Shell procedure has not lived up to its intrinsic promise, because problems concerning the recycling of the catalyst have never been fully resolved.

Rhodium Process

A rhodium process was introduced by Union Carbide in 1976.[17] Using a mixture of $H_2 + CO$ under pressure, the stable rhodium carbonyls $Rh_4(CO)_{12}$ and $Rh_6(CO)_{16}$ are converted to $HRh(CO)_4$. The latter is a very active hydroformylation catalyst for olefins but also tends to promote their hydrogenation and isomerization. In addition, its stability is low. To stabilize the mononuclear rhodium species, Union Carbide adds a large excess of triphenylphosphine (Ph_3P). This system works at normal temperatures and pressures and shows an excellent selectivity in favour of the linear aldehyde (up to 30/1) without inducing hydrogenation or isomerization of the alkene. The precatalyst may be $[RhCl(CO)(PPh_3)_2] + Et_3N$, $[Rh(acac)(CO)_2]$, or $[RhH(CO)(PPh_3)_2]$. The industrial conversion of propene into butyraldehyde is carried out in molten triphenylphosphine (m.p. 79°C) at 100°C under $H_2 + CO$ at 50 bars. The selectivity for the linear product is around 92%. The mechanism of the process is the following (where $L = PPh_3$):

1st step, catalyst formation:

$$HRh(CO)_2L_2 \underset{+CO}{\overset{-CO}{\rightleftharpoons}} HRh(CO)L_2$$
$$\text{(18e)} \qquad\qquad \text{(16e)}$$

2nd step, fixing the olefin:

$$H\text{-}Rh(CO)L_2 + RCH=CH_2 \rightleftharpoons \begin{matrix} R\diagdown\!\!= \\ | \\ H\!-\!Rh(CO)L_2 \end{matrix} \longrightarrow R\text{-}CH_2\text{-}CH_2\text{-}Rh(CO)L_2$$

The addition of H—Rh to the olefin is *cis*.

3rd step, carbonylation:

$$R\text{-}CH_2\text{-}CH_2\text{-}Rh(CO)L_2 \overset{CO}{\longrightarrow} R\text{-}CH_2\text{-}CH_2\text{-}CO\text{-}Rh(CO)L_2$$

4th step, hydrogenolysis:

$$R\text{-}CH_2\text{-}CH_2\text{-}CO\text{-}Rh(CO)L_2 \overset{H_2}{\longrightarrow} R\text{-}CH_2\text{-}CH_2\text{-}CHO + HRh(CO)L_2$$

The rate-limiting step is this fourth transformation. Therefore, the process is first order with respect to the H_2 pressure. One of the possible hydrogenolysis mechanisms is the following:

$$R'CO\text{-}Rh(CO)L_2 + H_2 \xrightarrow[\text{addition}]{\text{oxidative}} \begin{matrix} H\diagdown\,\diagup H \\ R'CO\!-\!Rh(CO)L_2 \end{matrix} \overset{\text{fast}}{\longrightarrow} R'CHO + HRh(CO)L_2$$

A vacant site is required on the rhodium, so the excess PPh_3 tends to reduce the rate. The increased CO pressure accelerates the third step and reduces the tendency towards isomerization.

Rhône-Poulenc Process

In spite of its exceptional performance, the rhodium-catalysed 'oxo' process has one crucial disadvantage: the price and rarity of the metal. Rhodium, whose principal producers are Russia, South Africa and Canada, constitutes only 0.0001% of the earth's crust and is the most expensive of all commercially available metals. Consequently, the efficiency of the catalyst recycling procedure becomes an essential economic parameter. A Rhône-Poulenc process, introduced in 1984–85, brought a very elegant solution to this problem. The basic idea consisted of using water-soluble sulfonated derivative, which is prepared from triphenylphosphine as follows:

The protonation of phosphorus in the acidic medium renders it electron-attracting, which directs the sulfonation of the benzene rings to the *meta* position. This *meta* substitution hardly modifies the properties of phosphorus ligand (little or no effect on the donor–acceptor character of phosphorus or the Tolman cone angle) and its rhodium complex $[HRh(CO)_2L_2]$ shows the same activity as the PPh_3 analogue. The only major difference lies in its water solubility. The hydroformylation reaction is carried out in a reaction vessel using a two-phase water/organic medium system under vigorous agitation. The alkene is mainly in the organic phase but has a degree of water solubility; the catalyst is found mainly in the aqueous phase but the lipophilic properties of the Rh–H fragment mean that it tends to concentrate at the phase boundary. Thus catalysis remains possible. When the hydroformylation of the alkene has been achieved, the aqueous phase is decanted to recover and recycle the catalyst. Numerous homogeneous catalytic reactions have been transposed to two-phase media, by application of the same principle; see reference 18.

Recent Developments

The most recent developments in the 'oxo' reaction concern asymmetric hydroformylation and the hydroformylation of internal or functional olefins. If hydroformylation could be carried out selectively on a single face of an olefin, optically active aldehydes would be obtained:

There is currently no catalyst that gives a sufficiently high enantiomeric excess to be commercially viable. One of the best is depicted in the following scheme.[19] Union Carbide has very recently introduced rhodium catalysts containing phosphites of very high steric hindrance for the hydroformylation of internal and functionalized olefins.[20] Finally, another recent trend is the use of bimetallic hydroformylation catalysts, in general binuclear Rh_2 compounds. They permit very high degrees of activity and regioselectivity.[21]

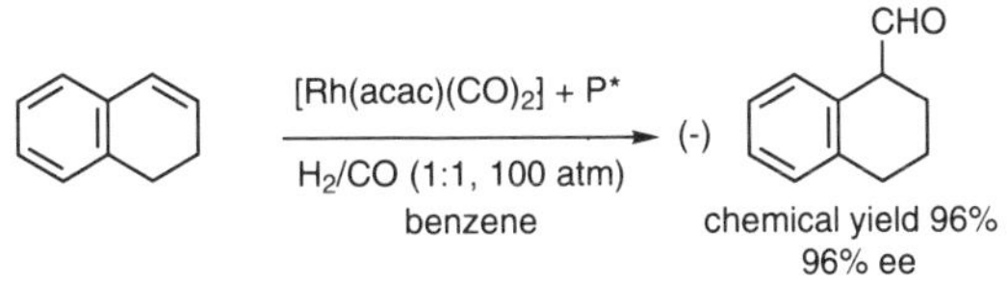

[Rh(acac)(CO)$_2$] + P*
$\xrightarrow{\text{H}_2/\text{CO (1:1, 100 atm)}}$ (-)
benzene

CHO

chemical yield 96%
96% ee

P* = [R,S- BINAPHOS] =

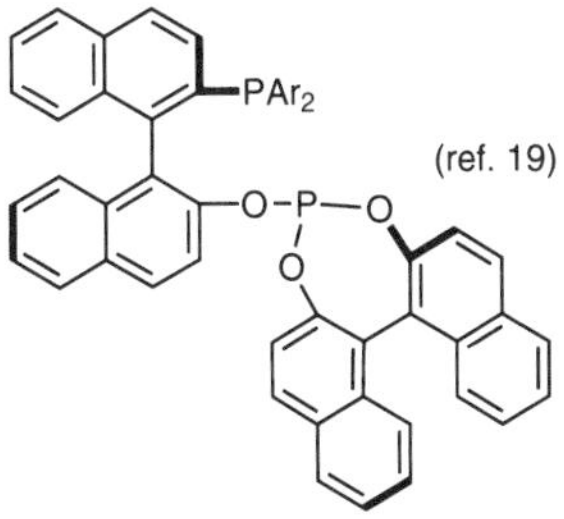

—PAr$_2$

O–P–O

(ref. 19)

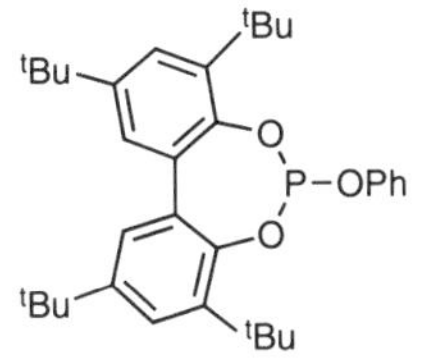

^{t}Bu
^{t}Bu
^{t}Bu
^{t}Bu

O
P–OPh   (ref. 20)
O

5.4 The Syntheses of Acetic Acid and Glycol

The industrial synthesis of acetic acid is now carried out by the carbonylation of methanol:

$$\text{Me-OH} + \text{CO} \longrightarrow \text{Me-COOH}$$

In 1971, Monsanto introduced a catalytic process which promotes this transformation;[22] it currently produces over one million tons of acetic acid per annum. The catalyst is based on rhodium and iodide and operates at a temperature of 180°C under CO pressures of between 30 and 40 bars. The yield of acetic acid is around 99%. The rhodium can be introduced in the form of RhCl$_3$ or as [RhCl(CO)(PPh$_3$)$_2$]; the true catalyst is [Rh(CO)$_2$I$_2$]$^-$ formed by the addition of a little hydroiodic acid (HI) to the system. The catalytic cycle comprises the following steps:

1. oxidative addition:

$$\text{MeOH} + \text{IH} \rightleftharpoons \text{MeI} + \text{H}_2\text{O}$$

[Rh(CO)$_2$I$_2$]$^{\ominus}$ + MeI $\longrightarrow$ $\left[\begin{array}{c} \text{CO} \\ | \\ \text{Me}-\text{Rh}-\text{I} \\ | \\ \text{I} \quad \text{CO} \end{array} \right]^{\ominus}$

(16e)   (18e)

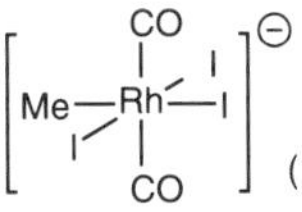

2. carbonylation:

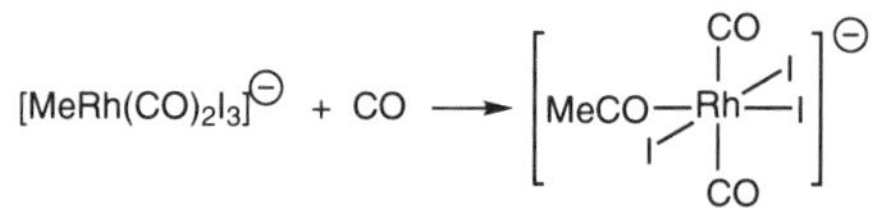

$$[MeRh(CO)_2I_3]^{\ominus} + CO \longrightarrow \left[MeCO{-}\overset{\overset{\displaystyle CO}{|}}{\underset{\underset{\displaystyle CO}{|}}{Rh}}\overset{\displaystyle I}{\underset{\displaystyle I}{\diagup}} \right]^{\ominus}$$

3. reductive elimination:

$$[MeCORh(CO)_2I_3]^{\ominus} \longrightarrow [Rh(CO)_2I_2]^{\ominus} + MeCOI$$

4. hydrolysis:

$$MeCOI + H_2O \longrightarrow MeCOOH + IH$$

The rate-limiting step in the cycle is the oxidative addition. It obeys a SN_2 mechanism:

$$Rh^{\ominus} + \overset{\displaystyle H}{\underset{\underset{\displaystyle H}{H}}{C}}{-}I \longrightarrow Rh{-}\overset{\displaystyle H}{\underset{\displaystyle H}{C}}{\diagup}H + I^{\ominus}$$

The unusual choice of iodide as the anion is justified in two ways. I^- is:

(a) a good leaving group which means that the SN_2 reaction is accelerated;
(b) a good ligand (soft in Pearson terms) which favours the formation of the Rh^- species.

The related synthesis of glycol from carbon monoxide has not been mastered. In 1974, Union Carbide proposed the following scheme:

$$3\,H_2 + 2\,CO \xrightarrow[250°C,\ 3000\ atm]{[Rh_{12}(CO)_{34}]^{2-}} HO{-}CH_2CH_2{-}OH$$

whose mechanism is probably:

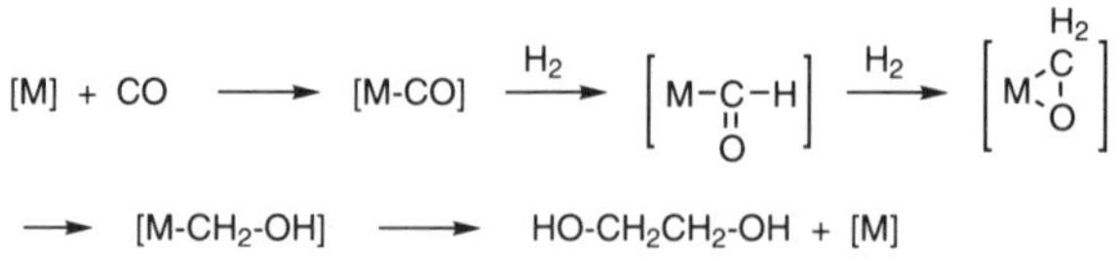

$$[M] + CO \longrightarrow [M{-}CO] \xrightarrow{H_2} \left[M{-}\overset{\overset{\displaystyle}{}}{\underset{\underset{\displaystyle O}{\|}}{C}}{-}H \right] \xrightarrow{H_2} \left[M\overset{\overset{\displaystyle H_2}{}}{\underset{\underset{\displaystyle O}{}}{\overset{\displaystyle C}{\diagdown}}} \right]$$

$$\longrightarrow [M{-}CH_2{-}OH] \longrightarrow HO{-}CH_2CH_2{-}OH + [M]$$

This approach was not followed up because of the severity of the operating conditions. A more viable process, devised by Monsanto (1983), is based on the hydroformylation of formaldehyde. The industrial sequence is as follows:

$$CO + 2\,H_2 \xrightarrow{Fischer\text{ - }Tropsch} CH_3OH$$

$$CH_3OH \xrightarrow[V_2O_5]{O_2} H_2C{=}O$$

$$H_2C{=}O \xrightarrow{hydroformylation} HO{-}CH_2CH_2{-}OH$$

The proposed hydroformylation catalyst is $[RhH(CO)_2(PPh_3)]$, but the conditions of temperature and pressure are still poorly defined. Nonetheless, the carbon monoxide pressure is higher than required for the traditional 'oxo' process. The most likely mechanism is as follows:

1. Formation of the catalyst:

$$[HRh(CO)(PPh_3)_3] \xrightarrow[\text{- 2 PPh}_3]{\text{+ CO}} [HRh(CO)_2(PPh_3)]$$

It should be noted that the precursor is the same as the traditional 'oxo' but the real catalyst is different as a result of the higher CO pressure.

2. Formation of the η^2-formaldehyde complex:

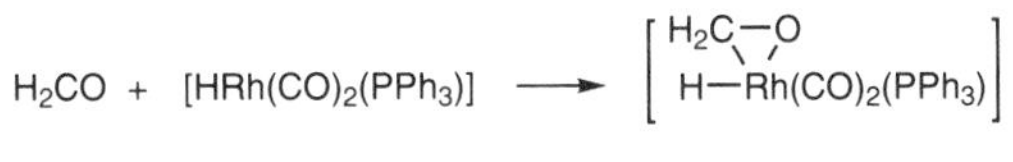

$$H_2CO + [HRh(CO)_2(PPh_3)] \longrightarrow \begin{bmatrix} H_2C\!-\!O \\ \diagdown\diagup \\ H\!-\!Rh(CO)_2(PPh_3) \end{bmatrix}$$

3. Transfer of H from Rh to H_2CO within the metal coordination sphere:

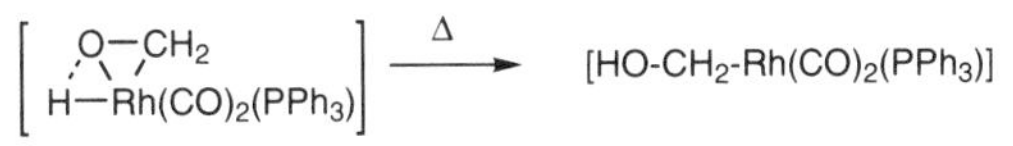

$$\begin{bmatrix} O\!-\!CH_2 \\ \diagdown\diagup \\ H\!-\!Rh(CO)_2(PPh_3) \end{bmatrix} \xrightarrow{\Delta} [HO\text{-}CH_2\text{-}Rh(CO)_2(PPh_3)]$$

4. Carbonylation:

$$[HO\text{-}CH_2\text{-}Rh(CO)_2(PPh_3)] \xrightarrow{CO} [HO\text{-}CH_2\text{-}CO\text{-}Rh(CO)_2(PPh_3)]$$

5. Hydrogenolysis:

$$[HO\text{-}CH_2\text{-}CO\text{-}Rh(CO)_2(PPh_3)] \xrightarrow{H_2} HO\text{-}CH_2\text{-}CHO + [HRh(CO)_2(PPh_3)]$$

$$\downarrow H_2$$

$$HO\text{-}CH_2CH_2\text{-}OH$$

Nonetheless, this procedure has still not been applied on an industrial scale.

5.5 Alkene Polymerization

The most important of all industrial catalytic reactions is the polymerization of alkenes. This explains why the titanium-based polymerization catalysts introduced in 1955 caused a chemical revolution and brought Ziegler and Natta the Nobel prize in 1963. These catalysts, such as $TiCl_3 + Et_2AlCl$, transformed a very difficult process ($200°C$ and 1000 atm are required for the thermal polymerization of ethylene) into a technically straightforward reaction which occurs at $25°C$ under atmospheric pressure. Nowadays 15 million tons of polyethylene and polypropylene are produced annually by such routes. The general scheme for a Ziegler-Natta catalysed polymerization is the following:

1. Initiation; formation of the metal–alkyl or metal–hydrogen bond:

$$[L_nM]^+ \xrightarrow{\;R^{\ominus}\text{(or }H^{\ominus}\text{)}\;} L_nM\text{-}R \quad \text{(or } L_nM\text{-}H\text{)}$$

2. Propagation:

$$L_nM\text{-}R \; + \; \diagdown\hspace{-0.5em}=\hspace{-0.5em}\diagup \longrightarrow \; L_nM\text{-}\underset{|}{\overset{|}{C}}\text{-}\underset{|}{\overset{|}{C}}\text{-}R$$

3. Termination:

$$L_nM\text{-}\Big(\!\underset{|}{\overset{|}{C}}\text{-}\underset{|}{\overset{|}{C}}\!\Big)_{\!m}\!\text{-}R \longrightarrow \Big(\!\underset{|}{\overset{|}{C}}\text{-}\underset{|}{\overset{|}{C}}\!\Big)_{\!m} \; + \; L_nM\text{-}R$$
$$\text{(polymer)}$$

In general, the initiator is formed from the metallic chloride and an alkylaluminium:

$$L_nM\text{-}Cl \; + \; AlR_3 \longrightarrow L_nM\text{-}R \; + \; ClAlR_2$$

The propagation stage comprises an initial π coordination of the olefin followed by the insertion step:

$$L_nM\text{-}R \; + \; H_2C\text{=}CH_2 \longrightarrow \overset{\textstyle H_2C\text{=}CH_2}{\underset{\textstyle |}{L_nM\text{--}R}} \longrightarrow L_nM\text{-}CH_2\text{-}CH_2\text{-}R$$

Therefore it is necessary that the metal is unsaturated ($\leqslant$ 16 electrons) and desirable that the olefin is a good ligand with little hindrance. Thus the ease of polymerization decreases in the order: ethylene ($H_2C\text{=}CH_2$) > propylene ($CH_3-CH\text{=}CH_2$) > 1-butene ($CH_3-CH_2-CH\text{=}CH_2$). For the same reason, ethylene blocks the polymerization of 1-butene. There are a number of termination mechanisms. One can distinguish:

1. β-elimination:

$$L_nM\overset{\textstyle H}{\underset{\textstyle \overset{|}{C}}{\diagdown\;CH}}\text{-}(CH_2\text{-}CH_2)_{m\text{-}1}\text{-}R \longrightarrow L_nM\text{-}H \; + \; H_2C\text{=}CH\text{-}(CH_2\text{-}CH_2)_{m\text{-}1}\text{-}R$$
$$\underset{\textstyle H_2}{}$$

2. Hydrogenation:

$$L_nM\text{—}(CH_2\text{-}CH_2)_m\text{-}R \; + \; H_2 \longrightarrow L_nM\text{-}H \; + \; H\text{—}(CH_2\text{-}CH_2)_m\text{-}R$$

3. Internal hydrogen transfer:

$$H_2C=CH_2 \quad H$$

$$L_nM \quad CH-(CH_2-CH_2)_{m-1}-R \longrightarrow L_nM-C_2H_5 + H_2C=CH-(CH_2-CH_2)_{m-1}-R$$

$$\underset{H_2}{C}$$

In any polymer, the chain length depends on the relative rates of the propagation (k_P) and termination (k_T) steps. Thus, $k_P \gg k_T$ gives high molecular weight polymers, $k_P \approx k_T$ favours oligomers and $k_T \gg k_P$ leads to dimers. Group IV metal systems such as Cp_2TiRX are, therefore, ideal for polymer preparation because their d^0 configurations inhibit the β-eliminations which lead to chain termination (β-elimination is facilitated by electron transfer from the metal d orbitals into the β C–H σ^* combination which, obviously, does not occur at a d^0 centre: see also Section 3.3). Conversely, the use of electron-rich metal systems from the right side of the table ($NiCl_2 + EtAlCl_2$, for example), gives dimers or oligomers because β-elimination is favoured (high k_T).

Polyethylene

Two sorts of polyethylene exist:

(a) High density polyethylene ($d \approx 0.95$, m.p. $\approx 135°C$) which is composed of linear chains $(-CH_2-CH_2-)_n$.

(b) Low density polyethylene ($d \approx 0.92$, m.p. badly defined) whose chains are branched:

$$-(CH_2)_m\text{-}\underset{\underset{CH_3}{|}}{CH}\text{-}(CH_2)_n\text{-}$$

The original Ziegler catalyst was prepared by reacting hydrocarbon solutions of $AlEt_3$ with $TiCl_4$, which gives a suspension capable of promoting the polymerization of C_2H_4 under 1 bar pressure. One of the best catalysts for the synthesis of 'high density' polyethylene is prepared as follows:

$$Cp_2Cr + \begin{matrix} Si-OH \\ O \\ Si-OH \end{matrix} \longrightarrow \begin{matrix} Si-O \\ O \quad Cr \\ Si-O \quad Cp \end{matrix} \begin{matrix} H \end{matrix} + C_5H_6$$

silica   heterogeneous
                        catalyst

The polymer which is produced has a wide molecular weight distribution, a characteristic which reflects the presence of several different catalytic sites. A much 'narrower' distribution is obtained with the $Cp_2ZrCl_2 + [MeAlO]_n$[23] system, which has only one type of active centre. If the methylalumoxane is used in large excess ($Al:Zr > 1000$), impressive activities of up to 2.5×10^7 g of polyethylene per g (Zr) $\times$ h $\times$ atm (C_2H_4) are obtained. There are numerous examples of the exceptional power of this series of catalysts.

Thus, cyclopentene gives a 1,3 polymer rather than its normal polymerization by metathesis:[24]

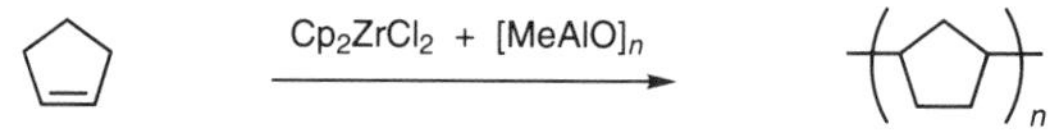

It is possible to introduce various substituents at the cyclopentadienyl ring, in order to exert control over the molecular weight distribution and the properties of the polymeric products. We will return to this point when we look at polypropylene.

Due to the importance of Ziegler-Natta catalysts, numerous studies of their metal centres have been carried out and it is nowadays accepted that the active species is an aluminate salt of the $[Cp_2ZrMe]^+$ cation.[25] Considerable effort has been put into the characterization of such ions. The following system is interesting; it is active but sufficiently stable to be characterized:[26]

$$Cp_2ZrMe_2 + B(C_6F_5)_3 \xrightarrow{\text{pentane}} [Cp_2ZrMe]\,[MeB(C_6F_5)_3]$$

Cp stands for Cp, 1,2-Me$_2$Cp, Cp*

Polypropylene

Polypropylene exists in three forms; these are:

(a) Isotactic polypropylene, where each chiral carbon atom has the same relative configuration:

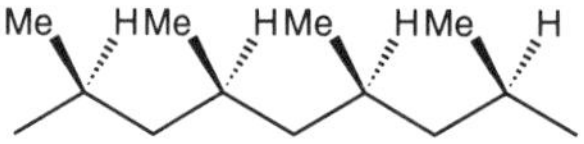

isotactic polypropylene

(b) Syndiotactic polypropylene, where chiral carbons have alternating configurations:

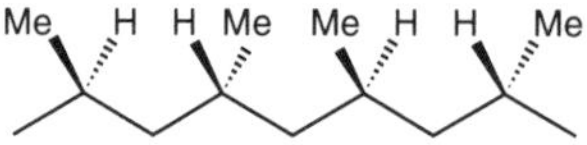

syndiotactic polypropylene

(c) Atactic polypropylene, where the chiral carbons have configurations which are randomly oriented.

The most sought-after polymer is isotactic polypropylene because it is the least soluble, shows the highest mechanical resistance and is high melting. To date, only Ziegler-Natta-type catalysts allow the preparation of highly isotactic polymers. The mechanism which imposes the isotactic geometry involves the preferential attack of a chiral catalyst on one of the polypropylene faces (see the discussion about the principles of asymmetric hydrogenation in Section 5.2).

Generally, the catalyst is composed of equal numbers of (R) and (S) centres, so the polymer is a racemic mixture of (R) and (S) chains. At present, the best control of the tacticity of the polypropylene is obtained with the zirconium catalysts mentioned below.

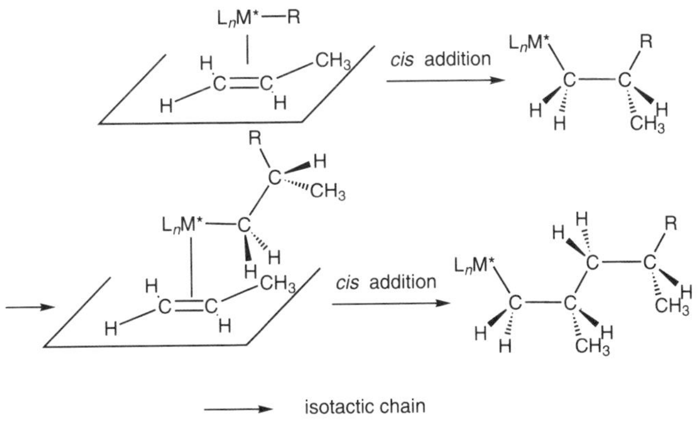
cis addition
cis addition
isotactic chain

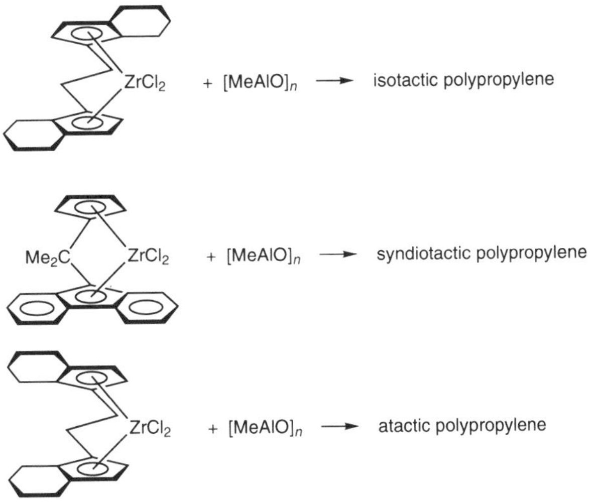
ZrCl2
+ [MeAlO]n
isotactic polypropylene
Me2C
ZrCl2
+ [MeAlO]n
syndiotactic polypropylene
ZrCl2
+ [MeAlO]n
atactic polypropylene

See reference 27 for more details.

Alkene Oligomerization

The majority of alkene oligomerization catalysts involve nickel centres. One of the most thoroughly studied systems is the following:

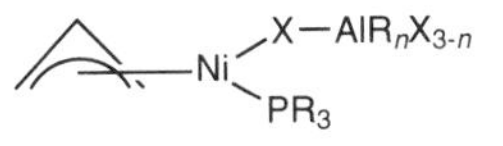
X—AlRnX3-n
Ni
PR3

X = Cl, Br

The degree of alkene polymerization is controlled by the nature of the phosphine. Butene is the main product for phosphines such as Me_3P, which have a small Tolman angle (see Section 3.8). However, a bulky phosphine such as $^tBu_3P$ will give a polymer because the transition state for insertion is less hindered (thus favoured by 'large' phosphines) than the transition state for β-elimination.[28]

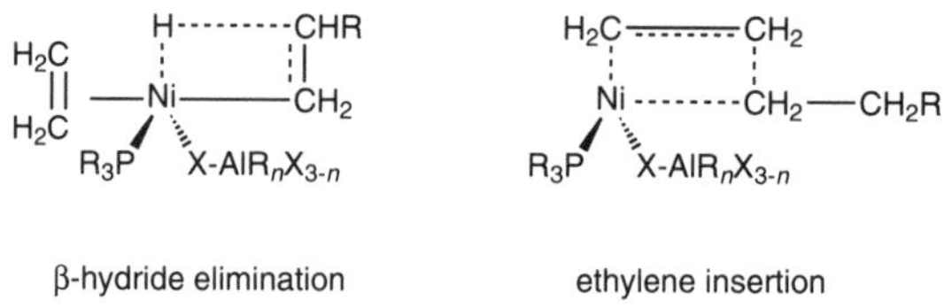

β-hydride elimination   ethylene insertion

The controlled oligomerization of ethylene by a nickel catalyst is one of the two key steps in the SHOP process (see Section 5.6).

Butadiene Cyclooligomerization

It is possible to obtain a whole series of cyclic oligomers from butadiene using various catalysts. The most important products are described below:[29]

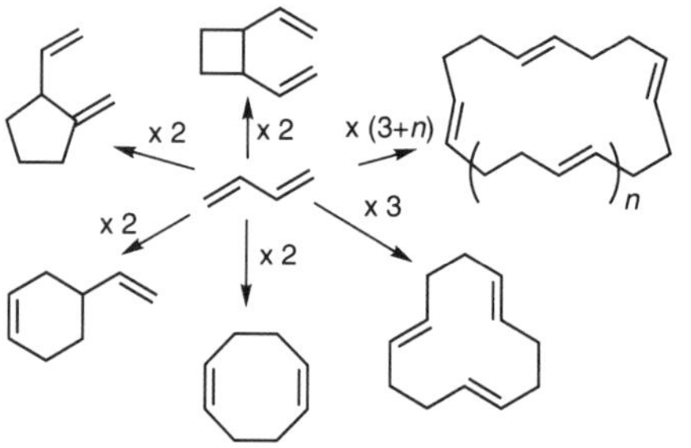

Again, the majority of catalysts are nickel based and operate under very mild conditions (room temperature or below). We will describe a few examples in detail. Cyclotrimerization was developed by Wilke, using bis-(η^3-allyl)-nickel as the catalyst. The mechanism is summarized in the schemes below:

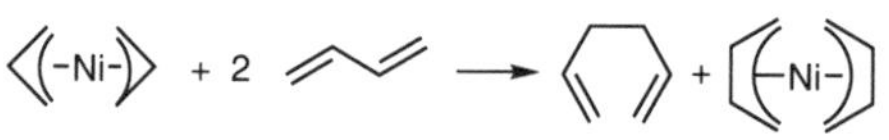

The nickel (II) precursor is reduced to an Ni(0) active species, with oxidation of the allyl ligand to a biallyl.

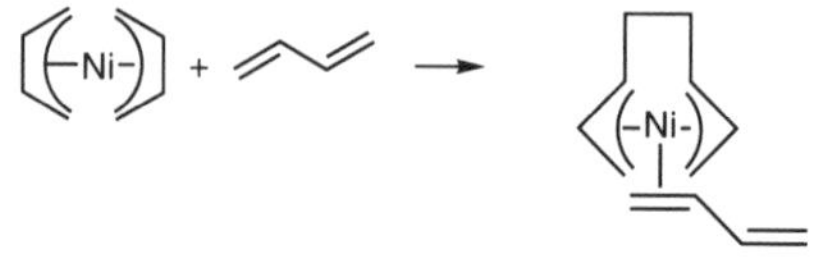

The C−C coupling of two butadiene ligands liberates a coordination site that is filled by a third diene molecule.

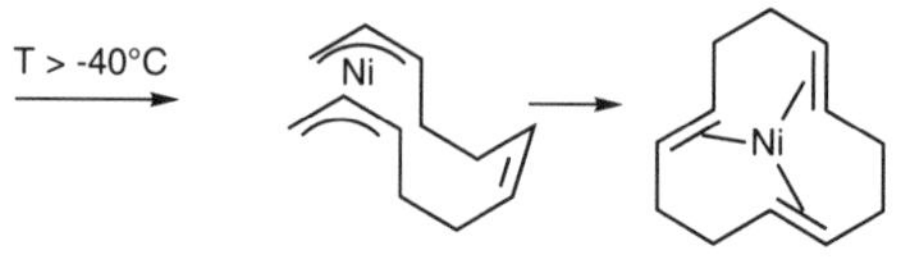

After hydrogenation of the butadiene cyclotrimer to cyclododecane, an oxidative cleavage (with O_2 and HNO_3) produces dodecanedioic acid, which is used as a feedstuff for polyamide production.

From this mechanism, it is easy to see that the addition of a phosphine (L) to the mixture will hinder the coordination of the third diene molecule. Thus, a cyclodimerization is observed:

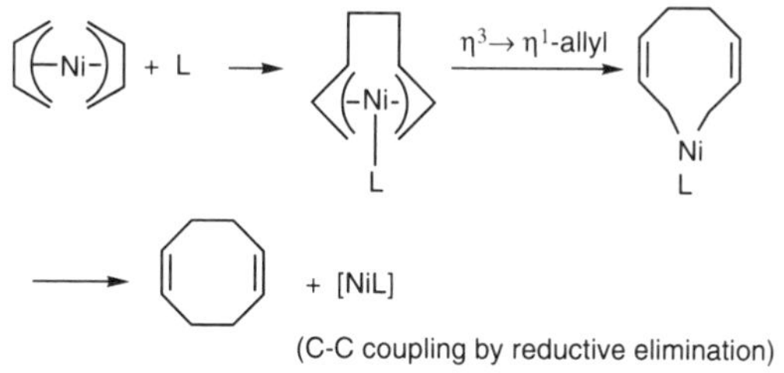

(C-C coupling by reductive elimination)

5.6 Alkene Metathesis

Alkene metathesis involves a statistical redistribution of methylene fragments between the alkene participants. It can be represented as follows:

$$RCH{=}CHR + R'CH{=}CHR' \rightleftharpoons 2\ RCH{=}CHR'$$

It is an interesting process because it has no equivalent in traditional organic chemistry. Furthermore, it has practical as well as academic significance, being involved in industrial processes such as cycloalkene polymerization and the SHOP process (Shell Higher Olefin Process). Metathesis, which was discovered by Banks in 1964,[30] can be promoted by heterogeneous silica- or alumina-supported catalysts such as $Mo(CO)_6$ or $W(CO)_6$, or even Re_2O_7 immobilized on alumina, at temperatures between 150 and 500°C. Alternatively, homogeneous catalysts such as $WCl_6 + [EtAlCl_2]_2$ in EtOH, $WCl_6 + Me_4Sn$, $WOCl_4 + [EtAlCl_2]_2$, $MoCl_2(NO)_2(PPh_3)_2 + Me_3Al_2Cl_3$, etc., may be employed at more moderate temperatures. The currently accepted Chauvin[31] mechanism invokes a carbene complex as the effective catalyst:

and so on...

The majority of precatalysts are obtained by the alkylation of a metallic chloride. A subsequent α-hydrogen elimination then gives the carbene complex.

$$L_nMX_2 \;+\; 2\,R_1CH_2M' \;\longrightarrow\; L_nM\!\!\left\langle\!\!\begin{array}{c}CH_2R_1\\CH_2R_1\end{array}\!\!\right. \;\xrightarrow{\;\Delta\;}\; R_1CH_3 \;+\; L_nM{=}CHR_1$$

in most cases, M = Mo, W
and M' = Al

This Chauvin mechanism requires a $[2+2]$ cycloaddition, which is forbidden in organic chemistry by the Woodward–Hoffmann rules. However, calculations on the $Cl_2Ti{=}CH_2 + H_2C{=}CH_2$ system[32] have shown that the presence of the low energy d orbitals on the transition metal is sufficient to allow this pathway. We have already seen that stable carbene complexes can induce stoichiometric metathesis (see the paragraph on $Ph_2C{=}W(CO)_5$, in Section 3.4) and recent research work has allowed the development of well defined, highly active carbene catalysts. Osborn[33] has developed the following system:

$$\begin{array}{c}X\\ RO{\cdots}\!\!\underset{\underset{X}{|}}{\overset{|}{W}}\!\!{=}C\!\!\left\langle\!\!\begin{array}{c}{\cdots}H\\R\end{array}\right.\quad +\;\; AlX_3\\ RO\end{array}$$

The Lewis acid serves to create a vacant site at the tungsten, thereby allowing the olefin to coordinate to the metal. Schrock[34] has described a similar catalyst which allowed the isolation and structural characterization of the metallacyclobutane intermediate implicated in the metathesis.

$$\begin{array}{c}CH^tBu\\ \|\\ R_fO{\cdots}\!\!\overset{|}{W}\!\!{\equiv}NAr\\ R_fO\end{array}\qquad R_f = -C\!\!\left\langle\!\!\begin{array}{c}Me\\CF_3\\CF_3\end{array}\right.\qquad Ar = \begin{array}{c}\text{(xylyl)}\end{array}$$

$$\begin{array}{c}NAr\\ \|\|\|\\ R_fO{-}W{-}tBu\\ R_fO\;\;\diagdown\;\diagup\;R\\ R'\end{array}$$

Finally we mention the Re(VII) molecular precatalysts developed recently by Herrmann and colleagues.[35] Amongst these $MeReO_3$ is particularly notable; its stability is a surprise in itself. The working catalyst is obtained by heating alumina containing 13% silica to 550°C before doping it with $MeReO_3$. The resulting solid catalyses olefin metathesis in a refluxing solvent.

Numerous applications of metathesis have been developed. These include the transformation of monofunctional olefins to bifunctional analogues.

$$H_2C{=}CH(CH_2)_8\text{-}CO_2Me \;\xrightarrow{\;WOCl_4 + Cp_2TiMe_2\;}\; MeO_2C\text{-}(CH_2)_8CH{=}CH(CH_2)_8\text{-}CO_2Me$$
$$+\;H_2C{=}CH_2\quad(71\%,\;\text{ref. 36})$$

This technique can also be used to lengthen or shorten an unsaturated ester:

$$CH_3(CH_2)_7CH{=}CH(CH_2)_7\text{-}CO_2Me \;+\; H_2C{=}CH_2 \;\xrightleftharpoons{\;WCl_6 + Me_4Sn\;}$$
$$H_2C{=}CH(CH_2)_7\text{-}CO_2Me \;+\; CH_3(CH_2)_7CH{=}CH_2$$
$$(68\%)\qquad(\text{ref. 37})$$

A simple preparation of rings from diolefins has also been described:

This method has been recently extended to include the synthesis of heterocycles containing oxygen[38] and nitrogen.[39] In the same way, polymers can be prepared from unsaturated rings. The metathesis of cyclooctatetrene[40] constitutes an excellent preparation of polyacetylene, which may be doped to give a useful electrical conductor:

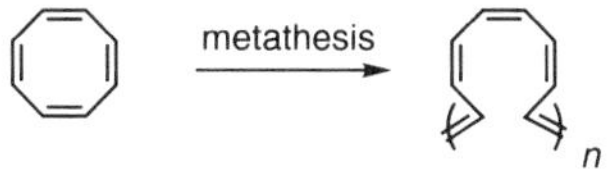

Under the same conditions, benzvalene gives an explosive polymer. The ring strain associated with the cyclopropane units causes this instability.

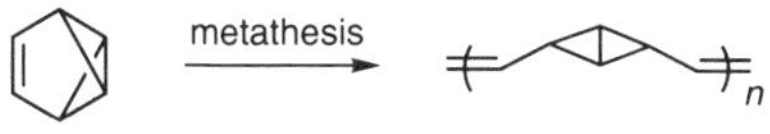

Another recent application concerns the synthesis of PbS microparticles of controlled size (20–100 Å), which show particularly interesting non-linear optical properties. To make them, a plumbocene species is grafted onto a norbornene which is then submitted to metathesis:[41]

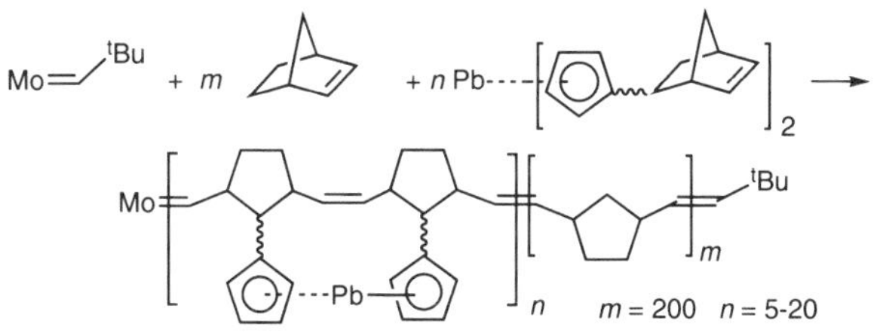

The polymeric product is treated with H_2S; this generates PbS microparticles suspended in a polynorbornene matrix. Fundamentally, the reaction is simply the acid decomposition of plumbocene:

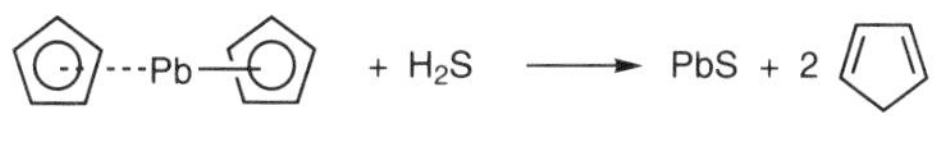

A number of industrial processes involve olefin metathesis. Amongst these, the SHOP[42] procedure for the preparation of long chain linear alcohols from methylene is particularly important (see scheme).

The key to the process is a nickel-based catalyst; it oligomerizes the ethylene to give terminal alkenes having variable lengths ($\approx C_4$–C_{40}) but excellent linearity (99%). Only the C_{10}–C_{14} fraction gives the C_{11}–C_{15} alcohols which are required for detergents, so this

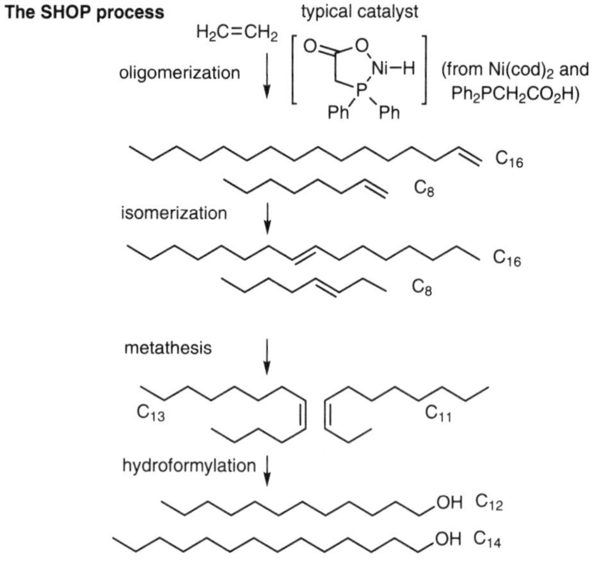

The SHOP process

typical catalyst

$H_2C=CH_2$

oligomerization

(from Ni(cod)$_2$ and Ph$_2$PCH$_2$CO$_2$H)

C_{16}

C_8

isomerization

C_{16}

C_8

metathesis

C_{13} C_{11}

hydroformylation

OH C_{12}

OH C_{14}

useful fraction is separated directly. The C_4–C_9 and C_{15}–C_{40} fractions are then isomerized and resubmitted to metathesis to give a new fraction of C_{10}–C_{14}. Currently, this method produces around a million tons of linear alcohols per annum.

A number of industrial polymers are also obtained by a ring metathesis method (ROMP or ring opening metathesis polymerization), thus:

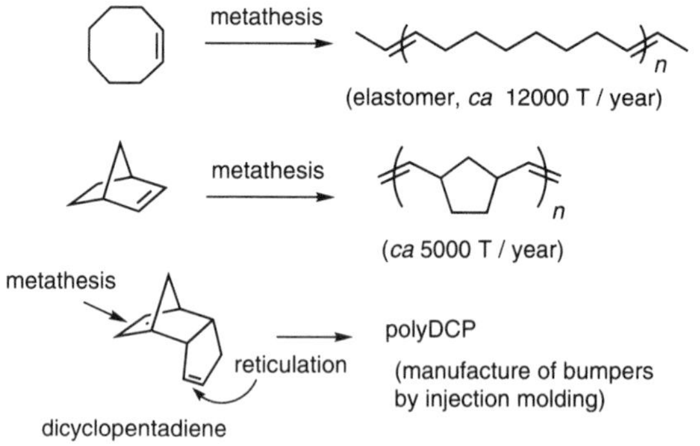

metathesis

(elastomer, *ca* 12000 T / year)

metathesis

(*ca* 5000 T / year)

metathesis

reticulation → polyDCP

(manufacture of bumpers by injection molding)

dicyclopentadiene

A recent review summarizes the state of the art in ROMP polymers.[43] To finish, we point out that it is possible to generalize the metathesis process to alkynes by using carbyne complexes as catalysts. The reaction scheme, which is principally due to Schrock, is the following:

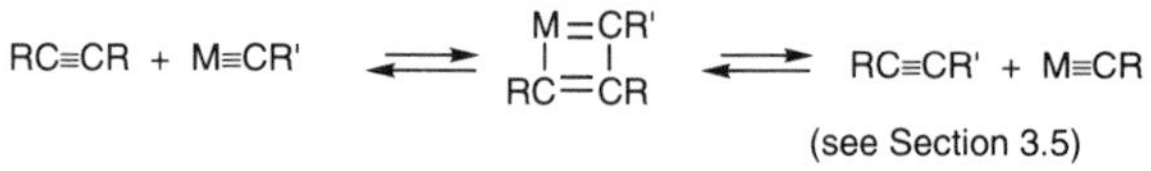

$$RC\equiv CR + M\equiv CR' \rightleftharpoons \begin{matrix} M=CR' \\ | \quad | \\ RC=CR \end{matrix} \rightleftharpoons RC\equiv CR' + M\equiv CR$$

(see Section 3.5)

An example of functional alkyne metathesis has been described in the literature:[44]

$$PhC\equiv C(CH_2)_2Z \xrightleftharpoons{Mo(CO)_6 + C_6H_5OH} PhC\equiv CPh + Z(CH_2)_2C\equiv C(CH_2)_2Z$$

$$Z = OH, OAc, Br, COOH, COOMe, CN$$

5.7 Some Catalytic Applications Involving Palladium

One of the most important industrial catalytic applications of palladium is the 1953 Wacker process for the oxidation of ethylene into acetaldehyde:

$$H_2C=CH_2 \xrightarrow{O_2} H_3C\text{-}CHO$$

Annually, it produces around 4 million tons of acetaldehyde. The background to the discovery of this process is interesting. It was known even in the nineteenth century that palladium chloride oxidizes ethylene to acetaldehyde in the presence of water:

$$PdCl_2 + H_2O + H_2C=CH_2 \longrightarrow H_3C\text{-}CHO + 2\,HCl + [Pd]\downarrow$$

However, the loss of palladium must be avoided to make this process practical. This problem was solved elegantly, by re-oxidizing the palladium with copper (II) chloride before it can precipitate:

$$[Pd] + 2\,CuCl_2 \longrightarrow PdCl_2 + 2\,CuCl$$

Finally the cupric chloride itself is regenerated by atmospheric oxidation:

$$2\,CuCl + 2\,HCl + O_2 \longrightarrow 2\,CuCl_2 + H_2O$$

This gives an overall cycle which may be formulated as follows:

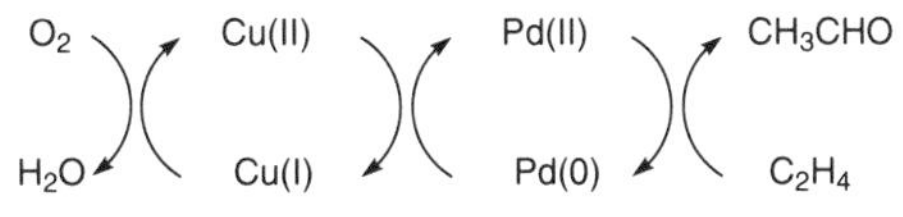

$$O_2 \quad Cu(II) \quad Pd(II) \quad CH_3CHO$$
$$H_2O \quad Cu(I) \quad Pd(0) \quad C_2H_4$$

The rate equation is of the type:

$$v = k\,\frac{[PdCl_4{}^{2-}]\,[C_2H_4]}{[Cl^-]^2\,[H^+]}$$

The $[PdCl_4]^{2-}$ anion is the normal form of palladium chloride in solution. It can be deduced from this rate equation that the transition state controlling the overall turnover involves a C_2H_4 molecule and a palladium centre that has lost two chloride ions and a proton. The mechanism accepted nowadays was proposed by Bäckvall in 1979:[45]

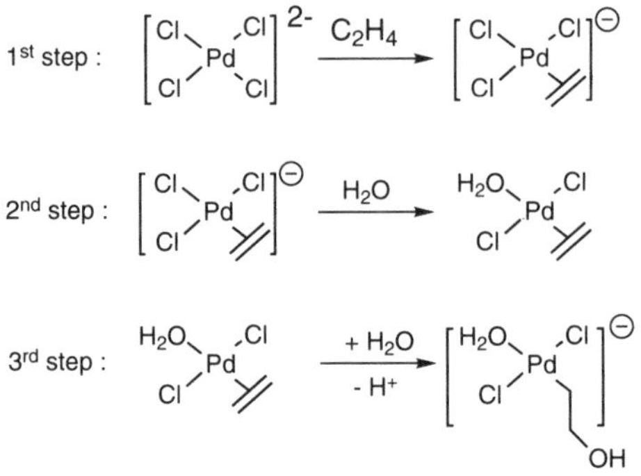

Although this third step involves a formal addition of OH^- to coordinated ethylene, it is actually a non-coordinated water molecule that plays the role of the nucleophile. The neutrality of palladium complex facilitates this nucleophilic attack.

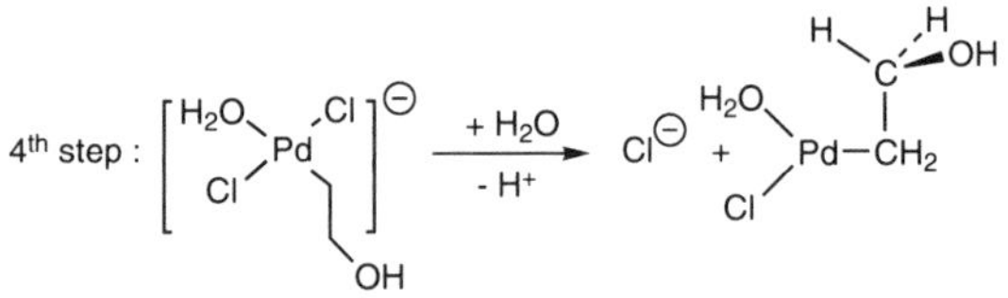

This fourth step is rate-limiting.

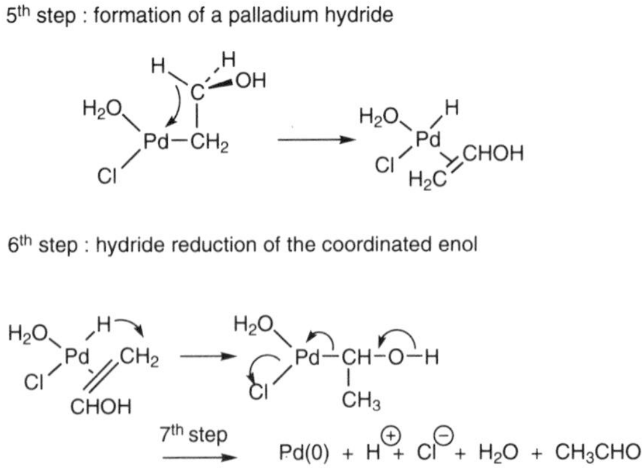

5th step : formation of a palladium hydride

6th step : hydride reduction of the coordinated enol

It may seem logical to draw a shorter cycle, with departure of the enol from the palladium coordination sphere after the 5th step, but experiments in heavy water solution give no incorporation of deuterium into the aldehyde. Thus this shortened mechanism obviously does not operate:

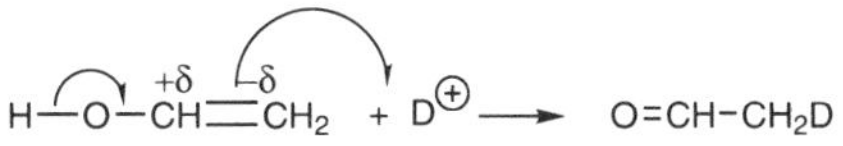

$$H{-}O{-}\overset{+\delta}{CH}{=}\overset{-\delta}{CH_2} + D^{\oplus} \longrightarrow O{=}CH{-}CH_2D$$

Only a mechanism taking place entirely within the palladium coordination sphere can explain this observation.

Although the incoming nucleophile in the Wacker process is water, palladium can be used to catalyse the attack of a whole variety of nucleophiles (alcohols, amines, carboxylates...) upon coordinated olefins. The attack takes place preferentially on the internal position of terminal olefins:

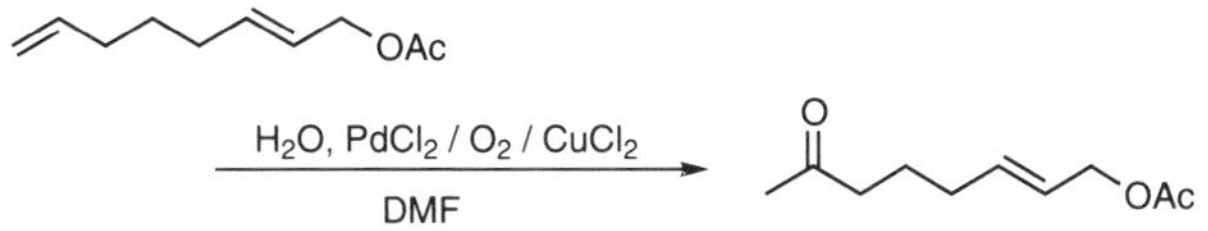

The nucleophile can be included into the unsaturated hydrocarbon. The following example combines this type of intramolecular attack with carbonylation of the palladium–carbon bond:

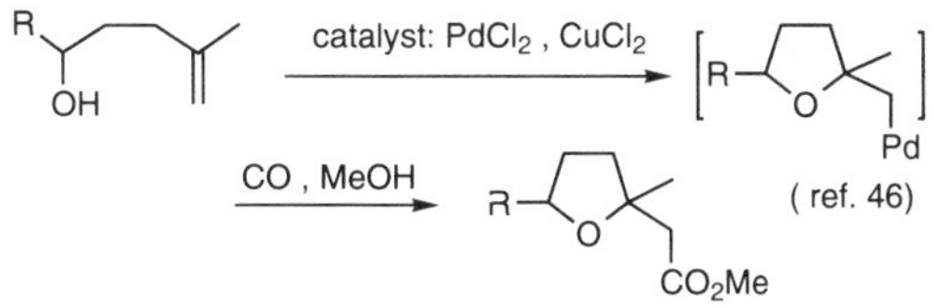

Generally, with an amine as the nucleophile, the reaction functions stoichiometrically because the amine coordinates strongly to the palladium, and blocks the catalytic cycle. However, in the intramolecular version the catalysis is not impeded:

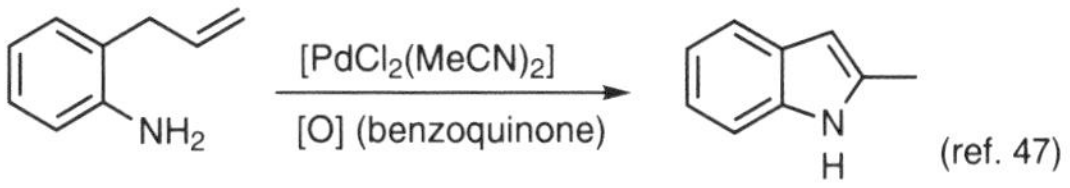

The use of carbanionic nucleophiles is usually difficult because of the oxidizing nature of palladium (II). Only stabilized carbanions that are relatively resistant to coupling can be employed:

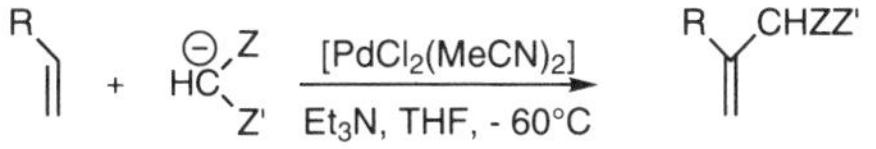

Z, Z' = CO$_2$R, CN, CHO... (ref. 48)

The arylation/vinylation reaction of olefins, often called the Heck reaction, is another important application of palladium chemistry in catalysis.[49,50] Here, a vinylic hydrogen is replaced by an aryl or vinyl group. The following example gives an illustration:

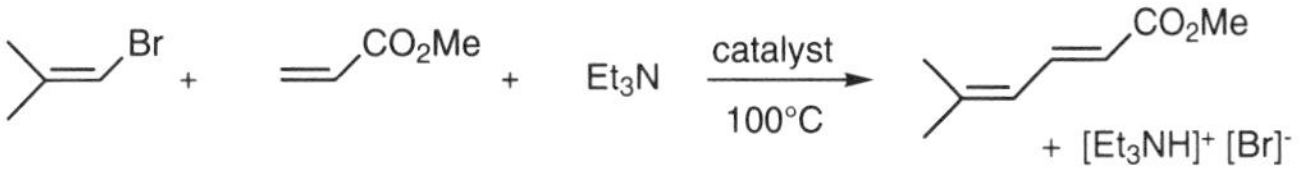

The effective reaction catalyst, PdL_2 ($L = PPh_3$), is formed by the reduction of Pd(II) by the phosphine. The reaction proceeds by the following mechanism:

1st step: oxidative addition of Ar−X to PdL_2:

$$[PdL_2] + Ar\text{-}X \longrightarrow \begin{array}{c} Ar \quad L \\ Pd \\ L \quad X \end{array}$$
$$(14e) \hspace{5cm} (16e)$$

The stereochemistry of this addition is *trans*, as in an SN_2 type reaction. If Ar−X is replaced by R−X, the 16-electron complex may decompose by β-hydride elimination.

2nd step: olefin insertion into the Pd−Ar bond

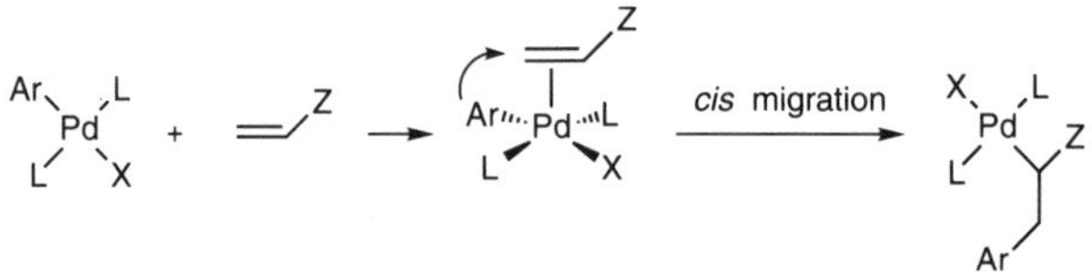

3rd step: β-elimination of H:

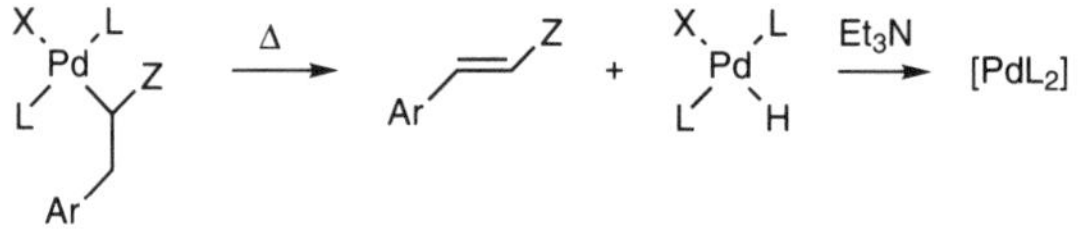

It should be noted that reaction is stoichiometric in Et_3N. A few uses are briefly described below:

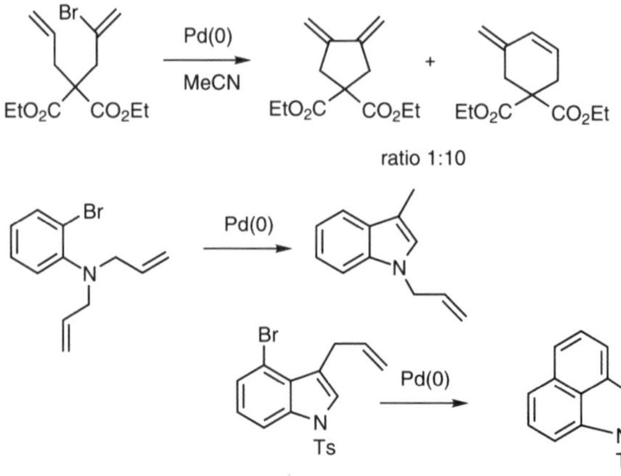

ratio 1:10

The oxidative addition of an organic halide to palladium(0) also forms the basis of a coupling process involving organotin compounds, which has been developed by Stille:[51]

$$RX + Bu_3Sn\text{-}R' \xrightarrow{Pd(0)} R\text{-}R'$$

The key process is the alkylation of an intermediate alkyl, aryl or acyl palladium halide by the tin derivative:

$$\underset{L}{\overset{L}{X\text{-}Pd\text{-}R}} + Bu_3Sn\text{-}R' \longrightarrow \underset{L}{\overset{L}{R'\text{-}Pd\text{-}R}} + Bu_3SnX$$

The ease of transfer from the tin groups to the palladium is:

$$RC{\equiv}C- \; > \; RCH{=}CH- \; > \; Ph \; > \; PhCH_2 \; > \; Me \; > \; Bu$$

Ketones, aldehydes, ... etc. can also be prepared:

$$RC(O)Cl + R'_4Sn \xrightarrow{[PdL_2]} RC(O)R' + R'_3SnCl$$

$$RX + CO + Bu_3SnH \xrightarrow[50°C]{[PdL_2]} RCHO \quad (\text{ref. 52})$$

R = aryl, vinyl, allyl (here, R' = H)

Numerous variations of this basic process are known. Some examples showing the diversity of R and R' groups which are compatible with the reaction are described below:

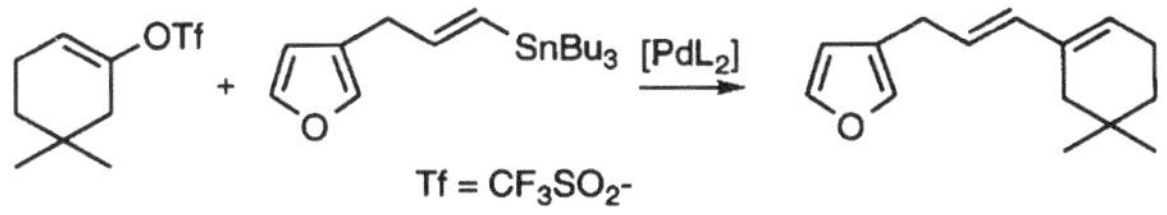

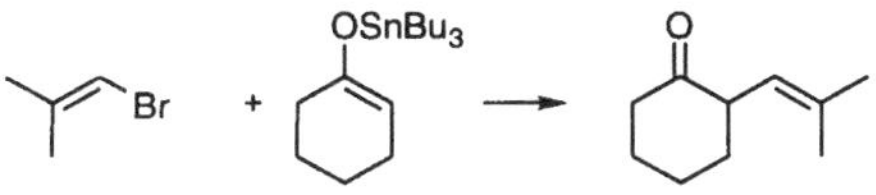

Tf = CF$_3$SO$_2$-

This last example combines the C-alkylation of an enolate with a Stille reaction. Another illustration demonstrates the possibility of alkyne insertion into the intermediate Pd−C bond prior to functionalization:[53]

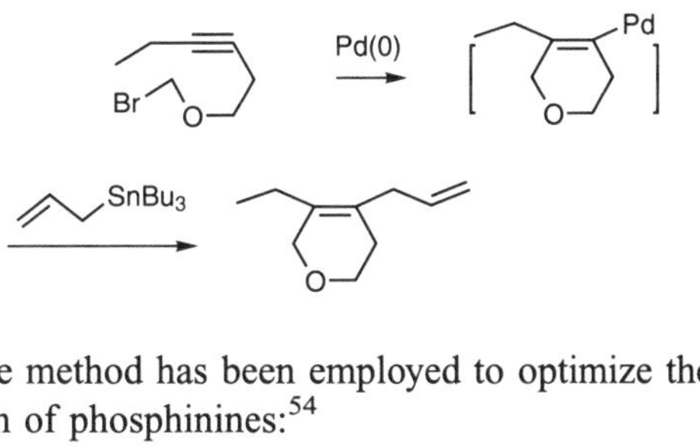

Quite recently, the Stille method has been employed to optimize the first general process for the functionalization of phosphinines:[54]

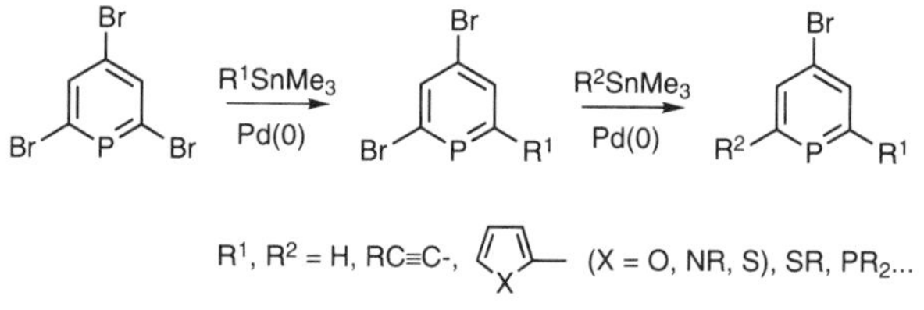

$R^1, R^2 = H, RC\equiv C-,$ $(X = O, NR, S), SR, PR_2...$

The selectivity for the α positions may be explained by assuming that, initially, the palladium gives a σ complex with the ring phosphorus. This is subsequently transformed into an η^2 complex of the P=C(Br) bond (see phosphaalkenes in Section 3.8), which brings the palladium close to the C—Br bond, promoting the insertion. Finally, we should point out that alkylation or arylation agents other than the stannanes can be employed in Stille-type coupling reactions: boranes R'_3B, borates R'_4B^- and alanes R'_3Al are often used.

The palladium(0)-catalysed carbonylation of organic halides is another widely used variation on the same basic process. In the presence of alcohols, the products are esters. An illustration of this reaction is given here:[55]

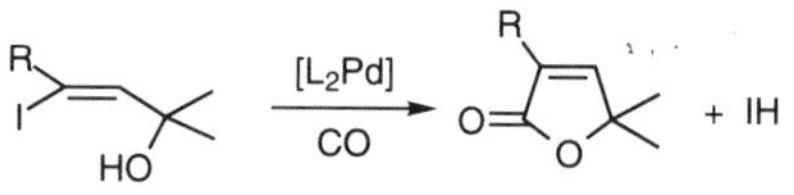

This example is important because it underlines that oxidative addition and carbonylation both occur with retention of stereochemistry at the double bond.

We have now reviewed some important applications of η^2-olefin palladium complexes and the palladium–carbon σ bond. The last major applications of palladium in synthesis involve η^3-allyl-palladium(II)-type intermediates. These are easily prepared from allyl halides:

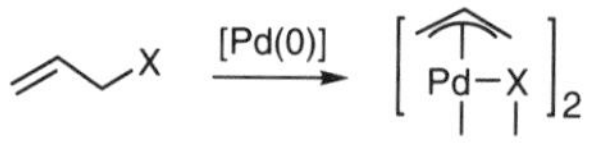

These rather stable complexes may be activated towards nucleophiles by the addition of strongly coordinating ligands such as phosphines, DMSO and HMPT.[56]

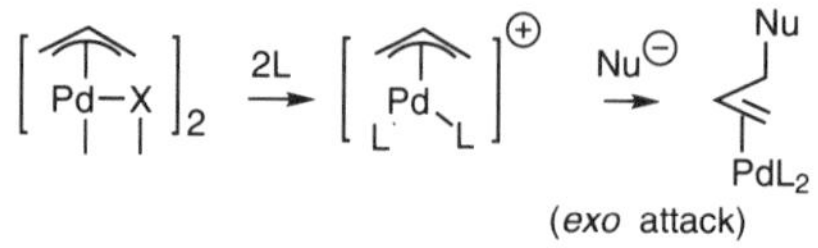

(*exo* attack)

This approach has been extensively studied, particularly by Trost's group. A good example is given hereafter.[57] It can be seen that palladation takes place at the non-conjugated double bond and alkylation occurs at the less hindered terminal (*) position. The understanding of these stoichiometric reactions has led to major developments in the area of catalytic Pd(0) promoted reactions of allyl substrates.

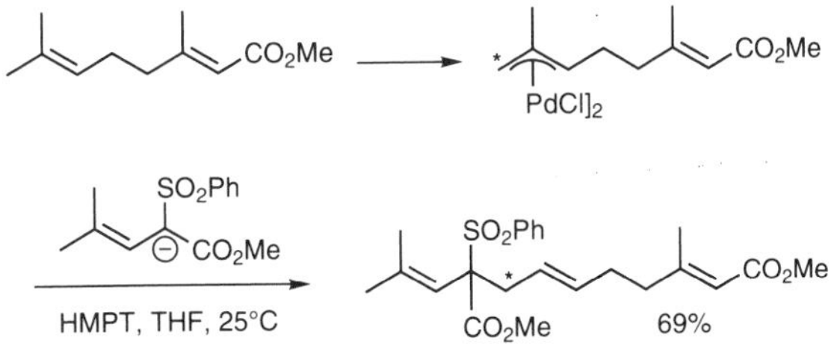

The essence of this chemistry is as follows:

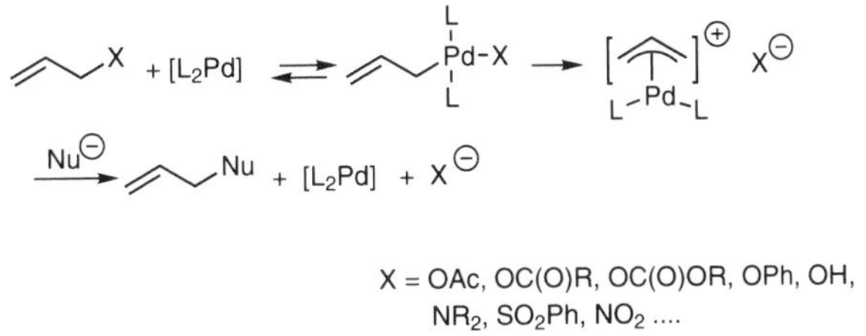

$$X = OAc, OC(O)R, OC(O)OR, OPh, OH,$$
$$NR_2, SO_2Ph, NO_2 ....$$

The applications of this process are innumerable, but lie outside the scope of a book such as this. The interested reader should consult reference 58.

5.8   Asymmetric Epoxidation and Dihydroxylation of Alkenes

Amongst the recent developments in asymmetric catalysis, many of the most useful concern oxidation reactions of alkenes.

The first significant advance was reported in 1980, when Sharpless and coworkers[59] described the first 'practical method of asymmetric epoxidation' of allyl alcohols. The epoxidation agent is *tert*-butyl hydroperoxide and the chirality is induced by an optically active complex of diethyl tartrate (DET) and titanium tetraisopropoxide. The epoxidation can be switched from the upper to the lower side of the allyl alcohol (see scheme) by changing the tartrate from D-(−) to L-(+). Enantiomeric excesses for this process are very high (> 90%).

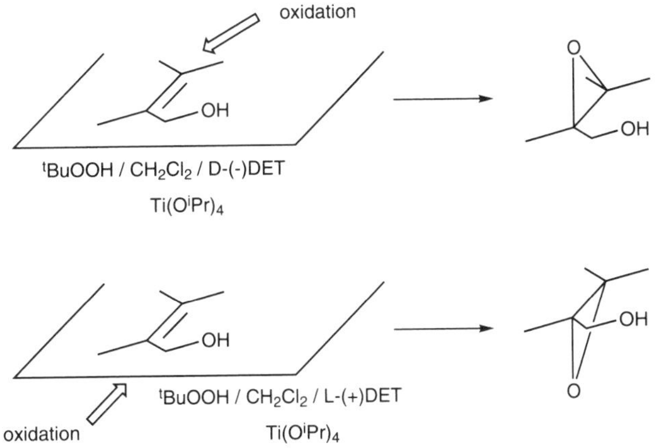

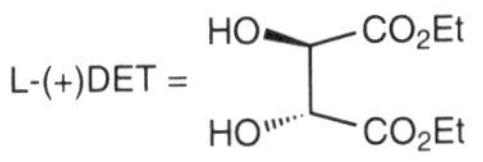

The reaction is performed using 5–10% catalyst in the presence of molecular sieves. The alcohol function is essential for induction, but the precise nature of the complex between the titanium and the tartrate remains unknown. Some typical examples of the procedure are summarized below:[60]

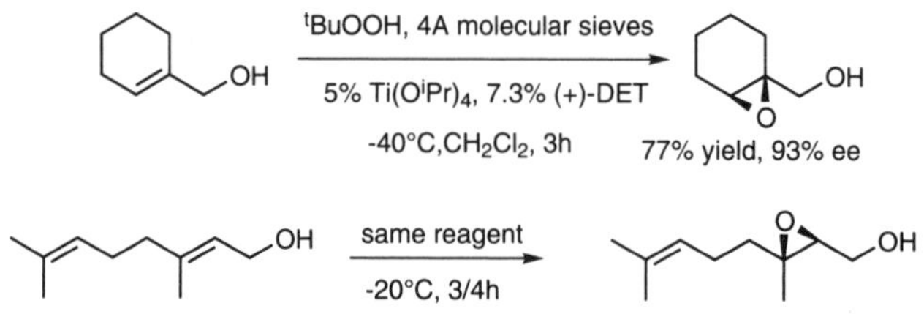

In the second case, the oxidizing agent selectively attacks the double bond adjacent to the alcohol function.

A recent variation on the Sharpless process involves the direct hydroxyepoxidation of olefins according to the scheme below:[61]

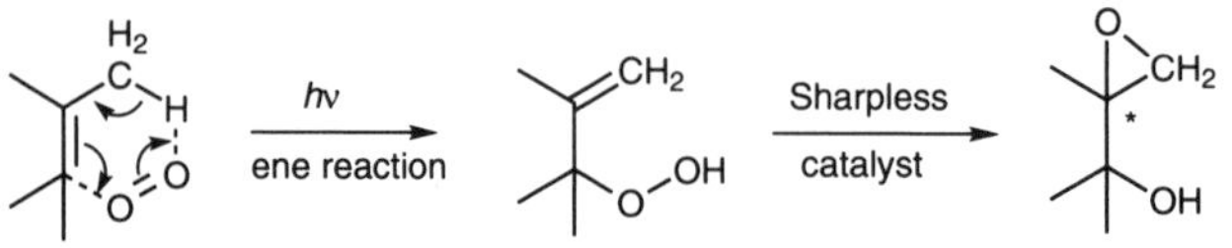

The photooxidation of the alkene by singlet oxygen is coupled to an intramolecular enantioselective oxygen transfer to the double bond which is formed as an intermediate. The reaction can be carried out in a single pot.

Another recent development concerns the epoxidation of non-functional olefins by manganese based catalysts,[62] thus:

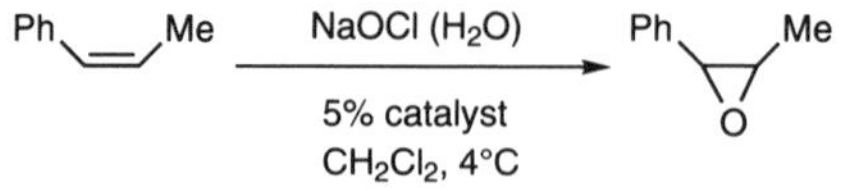

with the following catalyst:

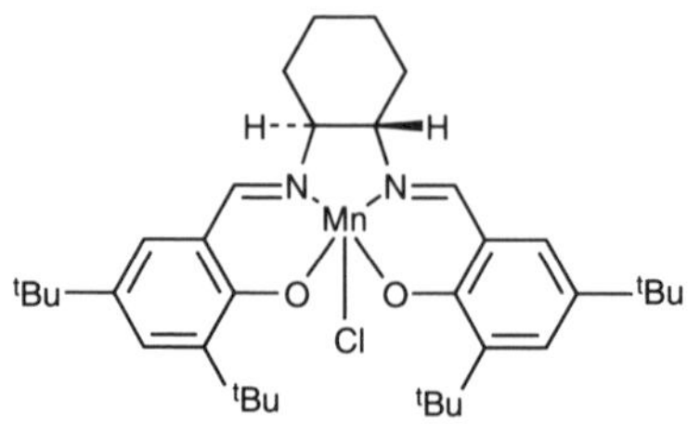

The 1S, 2R-(−) configured epoxide is obtained in 84% chemical yield with an ee of 92%. This powerful process has numerous applications, including the recent preparation of cyclic vinyl epoxides from the corresponding 1,3-dienes.[63]

Since 1988,[64] the Sharpless group has developed an asymmetric variation of the olefin hydroxylation reaction which, again, does not require a functional group adjacent to the double bond. The oxidizing agent is *N*-methylmorpholine oxide in aqueous acetone and the catalyst is osmium tetroxide (generally 0.2–0.4%). The optically active coligands (L) were chosen through optimization of parameters relating to their induction performance, high affinity constants for OsO_4 and geometry.[65] Remember that the mechanism for the *cis* dihydroxylation by OsO_4 follows the following scheme:

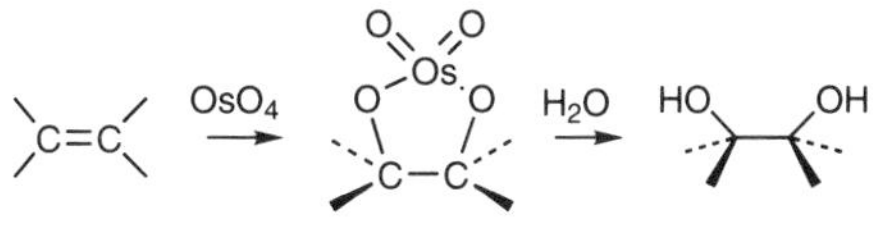

The dihydroquinidine derivative L_1 induces reaction on the upper side of the olefin while the dihydroquinine derivative L_2 promotes attack from the lower side (see scheme):

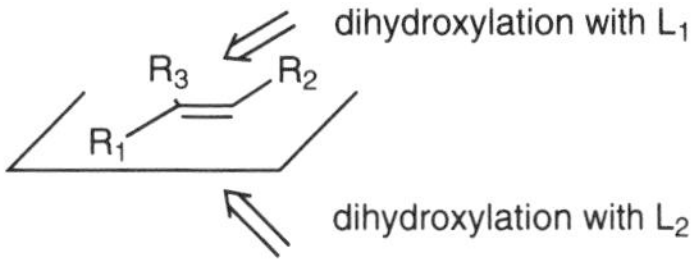

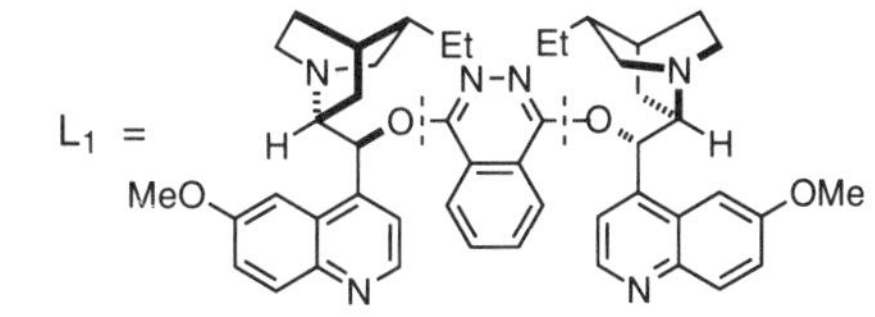

(derivative of dihydroquinidine -side units- and phthalazine central unit)

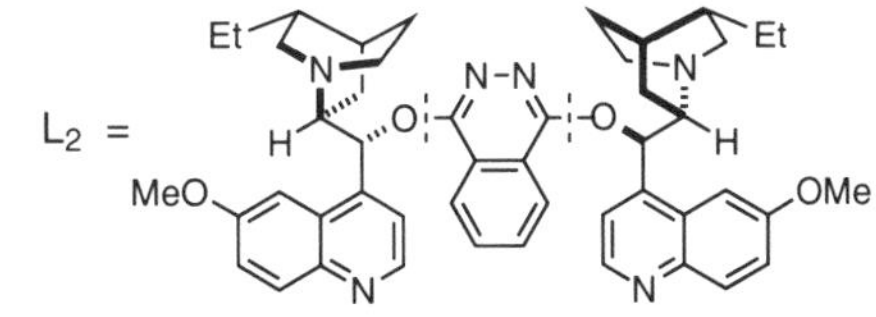

(derivative of dihydroquinine -side units- and phthalazine central unit)

The hydrolysis of the osmate is accelerated by the use of organic sulfonamides.[65] The latest version of the process gives the spectacular results below:

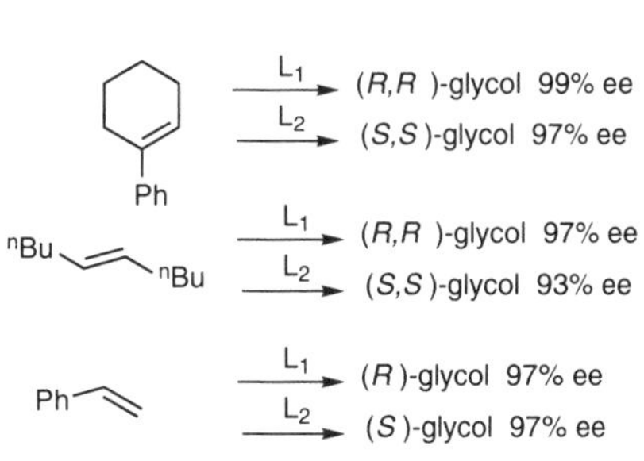

But :

$$^nC_8H_{17}\diagdown\diagup \quad \xrightarrow{\quad L_1 \quad} (R)\text{-glycol } 84\% \text{ ee}$$
$$\xrightarrow{\quad L_2 \quad} (S)\text{-glycol } 80\% \text{ ee}$$

A recent finesse has involved the use of a pyrimidine-type ligand[66] to optimize the process for terminal olefins.

DHQD = dihydroquinidine   (ref 66)

5.9   References

1. J. F. Young, J. A. Osborn, F. H. Jardine and G. Wilkinson, *J. Chem. Soc., Chem. Commun.*, 1965, 131.
2. R. R. Schrock and J. A. Osborn, *J. Am. Chem. Soc.*, 1976, **98**, 2134, 2143, 4450.
3. R. H. Crabtree, *Acc. Chem. Res.*, 1979, **12**, 331.
4. P. G. Jessop, T. Ikariya and R. Noyori, *Nature*, 1994, **368**, 231; see also: M. T. Reetz, W. Konen and T. Strack, *Chimia*, 1993, **47**, 493.
5. J. L. Speier, *Adv. Organomet. Chem.*, 1979, **17**, 407.
6. For a recent review on hydrosilylation catalysis, see: J. F. Harrod in *Encyclopedia of Inorganic Chemistry*, ed. R. B. King, Wiley, New York, 1994, vol. 3, pp. 1486–1496.
7. For a recent review on hydrocyanation catalysis, see: A. L. Casalnuovo, R. J. McKinney and C. A. Tolman in *Encyclopedia of Inorganic Chemistry*, ed. R. B. King, Wiley, New York, 1994, vol. 3, pp. 1428–1432.
8. H. B. Kagan and T. P. Dang, *J. Am. Chem. Soc.*, 1972, **94**, 6429.
9. B. D. Vineyard, W. S. Knowles, M. J. Sabacky, G. L. Bachman and D. J. Weinkauff, *J. Am. Chem. Soc.*, 1977, **99**, 5946.
10. M. D. Fryzuk and B. Bosnich, *J. Am. Chem. Soc.*, 1977, **99**, 6262.
11. R. Noyori and H. Takaya, *Acc. Chem. Res.*, 1990, **23**, 345.
12. M. Kitamura, M. Tokunaga, T. Okhuma and R. Noyori, *Tetrahedron Lett.*, 1991, **32**, 4163.
13. W. S. Knowles, *Acc. Chem. Res.*, 1983, **16**, 106.
14. R. D. Broene and S. L. Buchwald, *J. Am. Chem. Soc.*, 1993, **115**, 12569.
15. R. J. Whyman, *J. Organomet. Chem.* 1974, **81**, 97.
16. L. H. Slaugh and R. H. Mullineaux, *J. Organomet. Chem.* 1968, **13**, 469.
17. R. Fowler, H. Connor and R. A. Bachl, *Chemtech*, 1976, 772.
18. W. A. Herrmann and C. W. Kohlpaintner, *Angew. Chem., Int. Ed. Engl.*, 1993, **32**, 152.
19. N. Sakai, K. Nozaki and H. Takaya, *J. Chem. Soc., Chem. Commun.*, 1994, 395.
20. See for example: G. D. Cuny and S. L. Buchwald, *J. Am. Chem. Soc.*, 1993, **115**, 2066.
21. For a short review, see: G. Süss-Fink, *Angew. Chem., Int. Ed. Engl.*, 1994, **33**, 67.
22. D. Forster, *Adv. Organomet. Chem.*, 1979, **17**, 255.
23. W. Kaminsky and M. Miri, *J. Polym. Sci., Polym. Chem. Ed.*, 1985, **23**, 2151.
24. S. Collins and W. M. Kelly, *Macromolecules*, 1992, **25**, 233.
25. R. F. Jordan, *Adv. Organomet. Chem.*, 1991, **32**, 325; see also: Z. Wu, R. F. Jordan and J. L. Petersen, *J. Am. Chem. Soc.*, 1995, **117**, 5867.
26. X. Yang, C. L. Stern and T. J. Marks, *J. Am. Chem. Soc.*, 1991, **113**, 3623.
27. P. C. Möhring and N. J. Coville, *J. Organomet. Chem.*, 1994, **479**, 1.
28. B. Bogdanovic, *Adv. Organomet. Chem.*, 1979, **17**, 105.

29. W. Keim, A. Behr and M. Röper, in *Comprehensive Organometallic Chemistry*, eds G. Wilkinson, F. G. A. Stone and E. W. Abel, Pergamon, Oxford, 1982, vol. 8, pp. 371–462.
30. R. L. Banks and G. C. Bailey, *Ind. Eng. Chem., Prod. Res. Develop.*, 1964, **3**, 170.
31. J.-L. Hérisson and Y. Chauvin, *Makromol. Chem.*, 1970, **141**, 161; J.-P. Soufflet, D. Commereuc and Y. Chauvin, *C. R. Acad. Sci. Paris*, 1973, **276C**, 169.
32. T. H. Upton and A. K. Rappé, *J. Am. Chem. Soc.*, 1985, **107**, 1206.
33. J. Kress, M. Wesolek and J. A. Osborn, *J. Chem. Soc., Chem. Commun.*, 1982, 514.
34. R. R. Schrock, R. T. DePue, J. Feldman, C. J. Schaverien, J. C. Dewan and A. H. Liu, *J. Am. Chem. Soc.*, 1988, **110**, 1423.
35. W. A. Herrmann, W. Wagner, U. N. Flessner, U. Volkhardt and H. Komber, *Angew. Chem., Int. Ed. Engl.*, 1991, **30**, 1636.
36. J. Tsuji and S. Hashiguchi, *J. Organomet. Chem.*, 1981, **218**, 69.
37. R. H. A. Bosma, F. Van den Aardweg and J .C. Mol, *J. Chem. Soc., Chem. Commun.*, 1981, 1132.
38. G. C. Fu and R. H. Grubbs, *J. Am. Chem. Soc.*, 1992, **114**, 5426.
39. G. C. Fu and R. H. Grubbs, *J. Am. Chem. Soc.*, 1992, **114**, 7324.
40. F. L. Klavetter and R. H. Grubbs, *J. Am. Chem. Soc.*, 1988, **110**, 7807.
41. V. Sankaran, C. C. Cummins, R. R. Schrock, R. E. Cohen and R. J. Silbey, *J. Am. Chem. Soc.*, 1990, **112**, 6858.
42. W. Keim *et al.*, *Angew. Chem., Int. Ed. Engl.*, 1978, **17**, 446; *Organometallics*, 1983, **2**, 594; 1986, **5**, 2356.
43. R. R. Schrock, *Acc. Chem. Res.*, 1990, **23**, 158.
44. D. Villemin and P. Cadiot, *Tetrahedron Lett.*, 1982, **23**, 5139.
45. J. E. Bäckvall, B. Akermark and S. O. Ljunggren, *J. Am. Chem. Soc.*, 1979, **101**, 2411.
46. M. F. Semmelhack and C. Bodurow, *J. Am. Chem. Soc.*, 1984, **106**, 1496.
47. L. S. Hegedus, G. F. Allen, J. J. Bozell and E. L. Waterman, *J. Am. Chem. Soc.*, 1978, **100**, 5800.
48. L. S. Hegedus, R. E. Williams, M. A. McGuire and T. Hayashi, *J. Am. Chem. Soc.*, 1980, **102**, 4973.
49. A. de Meijere and F. E. Meyer, *Angew. Chem., Int. Ed. Engl.*, 1994, **33**, 2379. For recent data on the stereochemistry of this reaction, see ref. 50.
50. A. Madin and L. E. Overman, *Tetrahedron Lett.*, 1992, **33**, 4859.
51. J. W. Labadie and J. K. Stille, *J. Am. Chem. Soc.*, 1983, **105**, 6129.
52. V. P. Baillargeon and J. K. Stille, *J. Am. Chem. Soc.*, 1983, **105**, 7175.
53. R. K. Bhatt, D.-S. Shin, J. R. Falck and C. Mioskowski, *Tetrahedron Lett.*, 1992, **33**, 4885.
54. P. Le Floch, D. Carmichael, L. Ricard and F. Mathey, *J. Am. Chem. Soc.*, 1993, **115**, 10665.
55. A. Cowell and J. K. Stille, *Tetrahedron Lett.*, 1979, 133.
56. J. Tsuji, H. Takahashi and M. Morikawa, *Tetrahedron Lett.*, 1965, 4387.
57. B. M. Trost, L. Weber, P. Strege, T. J. Fullerton and T. J. Dietsche, *J. Am. Chem. Soc.*, 1978, **100**, 3426.
58. R. F. Heck, *Palladium Reagents in Organic Synthesis*, Academic Press, New York, 1985.
59. T. Katsuki and K. B. Sharpless, *J. Am. Chem. Soc.*, 1980, **102**, 5374.
60. Y. Gao, R. M. Hanson, J. M. Klunder, S. Y. Ko, H. Masamune and K. B. Sharpless, *J. Am. Chem. Soc.*, 1987, **109**, 5765.
61. W. Adam and M. J. Richter, *Acc. Chem. Res.*, 1994, **27**, 57.
62. E. N. Jacobsen, W. Zhang, A. R. Muci, J. R. Ecker and L. Deng, *J. Am. Chem. Soc.*, 1991, **113**, 7063.
63. S. Chang, R. M. Heid and E. N. Jacobsen, *Tetrahedron Lett.*, 1994, **35**, 669.
64. E. N. Jacobsen, I. Marko, W. S. Mungall, G. Schröder and K. B. Sharpless, *J. Am. Chem. Soc.*, 1988, **110**, 1968.
65. K. B. Sharpless, W. Amberg, Y. L. Bennani, G. A. Crispino, J. Hartung, K.-S. Jeong, H.-L. Kwong, K. Morikawa, Z.-M. Wang, D. Xu and X.-L. Zhang, *J. Org. Chem.* 1992, **57**, 2768.
66. G. A. Crispino, K.-S. Jeong, H. C. Kolb, Z.-M. Wang, D. Xu and K. B. Sharpless, *J. Org. Chem.*, 1993, **58**, 3785.

Appendix: Group Theory and Molecular Orbitals in Selected Model Complexes

Introduction

If we are to define and understand the symmetry properties of a molecule unequivocally, the concept of group theory is extremely useful, if not indispensable. In this section, we give a very qualitative introduction to group theory and explain how it may be applied to standard organometallic compounds, without exploring its limits or going into great mathematical depth. Our relatively modest objectives will be an understanding of the function of *character tables*, how they may be used to characterize different types of AO and MO, and how to employ them to monitor the evolution of hybrid orbitals by perturbation techniques. We refer the reader to more mathematically oriented works[1] for an understanding of group theory *per se*.

The ligands in an ML_n complex often have shapes which are not perfectly regular and the overall geometry of the molecule may differ slightly from ideality. Where $n = 4$, for example, complexes are often described as tetrahedral (to make a distinction from square planar), in spite of the fact that their angles deviate fractionally from the perfect value of $109°\ 28'$. X-ray studies indicate that real complexes generally have structures which suffer from minor distortions due to steric constraints, electronic repulsion and attraction, etc. However, both experiment and theory show that it is still valid to discuss the ligand fields of such compounds in ideal terms, provided the steric deformations and electronic constraints are not very large.

As we have seen throughout this book, ligands (L) may form many different kinds of bonds to metals (M). We have explained these bonds in traditional terms, as σ interactions (axial electron density, maximum overlap, strong bonds) and π interactions (lateral electron density, poor overlap, weak bonds) and this has allowed us to distinguish between two main classes of complex:

(a) those having ligands L (L = CO, PR_3, NO, X, H, CH_3, ...) which give σ bonds (sometimes reinforced by π backdonation), and

(b) complexes whose bonding interactions are essentially π type, as for ligands such as allyl, cyclopentadienyl, benzene, etc. .... We will treat these two classes of complex separately, in the order above.

In any complex, different ligands having similar bonding properties will often be present. Take PR_3 and CO, for example, which both give σ-type bonds having comparable electronic descriptions. To a first approximation, it is usually valid to assume that the ligand field in a mixed complex such as $ML_nL'_x$ can be treated in the same way as the ideal field in $ML_{(n+x)}$. Thus *the most symmetrical field* gives an acceptable approximation of reality in the vast majority of cases; where precision is vital, the perturbation can be subsequently estimated by classical methods. Whilst many isoelectronic square-planar complexes such as $[Co(CO)_4]^+$, $[RhCl(CO)(PR_3)_2)]$ and $[Ir(CO)(PR_3)_3]^+$ have geometries which are close to ideal, only the first is truly 'pure square planar' (symmetry group D_{4h}) because the others have geometries which are not *exactly* square (as shown by the possibility of isomers for the second). Nevertheless, they can be treated using the same formalisms as for $[Co(CO)_4]^+$, by making the assumption that the orbitals of the square plane are 'quasi-degenerate'.

square planar   pseudo square planar

In this section, we will treat a number of model complexes using the methods and approximations which we have outlined above. The reader should bear in mind that the precise MO levels in a complex will be marginally different from those which we derive if secondary π interactions or weak geometrical distortions are present.

Elements of Group Theory

First of all, we will revise the fundamental properties of finite groups. Let's consider a set $G = \{A, B, \dots \}$ comprising g elements. It is subject to a composition law, called the product, which requires that any two elements A and B possess a unique product $C = AB$. The set G is termed a gth order group if the following conditions are satisfied:

1. The group G must have a closing relationship: each pair of A, B has a corresponding unique element C, such that $AB = C$.
2. The set obeys an associative law: if A, B, C, are elements of G, then $(AB)C = A(BC) = ABC$.
3. A unitary element (often termed the identity) must be present: any element A belonging to the set G is left unchanged when multiplied by the unitary element E, thus the relationship $AE = EA = A$ is obeyed.
4. Inverse symmetry elements are found. If the set G contains an element A, then it must also contain a related element B such that product of A and B is the identity. Thus $AB = BA = E$ and we generally write $B = A^{-1}$.

A group where we have the relationship $AB = BA$ for all pairs of elements is said to be *Abelian*.

We present, rather than derive, the following fundamental concepts and properties:

- *The symmetry operations which can be performed on a molecule define its point group. This point group comprises several irreducible representations* (which we define below).
- *The MOs associated with a molecule are all defined within the irreducible representations of the molecular point group.*
- *Only atomic orbitals (or groups of atomic orbitals) which belong to the same irreducible representation of the molecular point group can overlap to form molecular orbitals.* (This merely translates the constraints of the pictorial orbital overlap model, used throughout the text, into the language of group theory).

To explain the practical applications of these rules, it is easiest to treat a model case. We take a simple MH_2 complex as a prototype.

The Model Complex MH_2: Symmetry Elements

We have a complex whose two M—H bonds are equivalent, in that they cannot be distinguished from each other chemically. The HMH angle is unimportant in the terms of our discussion, but is given as 120° for the sake of argument. Four symmetry operations leave the overall appearance of the system unchanged; as shown below, they are the identity, a C_2 second-order (180°) rotation and the two symmetry planes σ_{xz} and σ_{yz} (noted σ_v and σ'_v respectively).

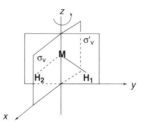

None of these operations has any effect upon M, but it is important to note that C_2 and σ_v exchange the positions of H_1 and H_2, whilst the identity and σ'_v leave them unchanged. In our analysis, the result of a symmetry operation is noted mathematically as $+1$ if the element under consideration undergoes no change and -1 if it is exchanged. We must consider the effects of each symmetry element upon the *$H_1 + H_2$ and $H_1 - H_2$ combinations* rather than the H atoms in isolation.

		$M \rightarrow M$
C_2	$H_1 \rightarrow H_2$	$H_1 + H_2 \rightarrow H_2 + H_1 \ (+1)$
	$H_2 \rightarrow H_1$	$H_1 - H_2 \rightarrow H_2 - H_1 \ (-1)$

We can use this analysis to look at the four symmetry operations through a vectorial analogy. If the four symmetry operations under consideration form the matrix vectors, the results of the above analysis give four different representations within the matrix, which are conventionally given the labels A_1, A_2, B_1 and B_2.

	E	C_2	$\sigma_{v_{xz}}$	$\sigma'_{v_{yz}}$
A_1	1	1	1	1
A_2	1	1	-1	-1
B_1	1	-1	1	-1
B_2	1	-1	-1	1

A_1 and A_2 have $+1$ for the operation C_2, B_1 and B_2 have -1. Multiplying the components column by column, giving the equivalent of a scalar product, shows that the product of any two vectors gives the characteristics of another group vector. Furthermore, if the product of any two vectors is added, it becomes clear that each pair of vectors is orthogonal. For example: $A_2 \times B_1 = (1 \times 1) + (1 \times -1) + (-1 \times 1) + (-1 \times -1)$. The sum of the vector product is 0 and the resulting scheme $\{1, -1, -1, 1\}$ corresponds to B_2. One thus has $A_2 \times B_1 = B_2$. The four preceding representations are termed *irreducible*, which implies that they form the foundations for the representation of all the MOs of any given species and, more generally, its electronic properties. The complete multiplication table of vectors is easy to obtain:

	A_1	A_2	B_1	B_2
A_1	A_1	A_2	B_1	B_2
A_2	A_2	A_1	B_2	B_1
B_1	B_1	B_2	A_1	A_2
B_2	B_2	B_1	A_2	A_1

(representations $A \times A$, $B \times B$ give A; $A \times B$ give B; and also representations 1×1 and $2 \times 2 = 1$; $1 \times 2 = 2$), thus the four irreducible representations clearly form a group with all the criteria that we previously defined.

The MOs of MH_2

After assigning the symmetry group and the reference axes of the molecule, we can employ a very general method which allows a rapid and unequivocal determination of the individual AOs or groups which combine to generate the MOs of any given molecule. Clearly, the physical solution to a real problem is independent of an essentially arbitrary choice of axes (it is always possible to transform a group of MOs into another by a given spatial transformation) but the simplification resulting from the use of a universal language is enormous. An unambiguous method for defining these axes is given later.

We begin by examining the AOs of M and the combinations of H_1 and H_2 which are available before mixing, to define the irreducible representations to which they belong. The following table gives the results of this analysis:

	E	C_2	$\sigma_{v_{xz}}$	$\sigma'_{v_{yz}}$	
s	1	1	1	1	
p_z	1	1	1	1	
$d_{x^2-y^2}$	1	1	1	1	A_1
d_{z^2}	1	1	1	1	
$H_1 + H_2$	1	1	1	1	
p_x	1	-1	1	-1	B_1
d_{xz}	1	-1	1	-1	
p_y	1	-1	-1	1	
d_{yz}	1	-1	-1	1	B_2
$H_1 - H_2$	1	-1	-1	1	
d_{xy}	1	1	-1	-1	A_2

The three independent families, A_1, B_1 and B_2, will eventually lead to specific hybrid MOs. For example, the three B_2 components give three MOs which resemble the classical MOs of H_2O given below:

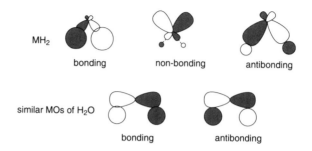

MH₂   bonding   non-bonding   antibonding

similar MOs of H_2O bonding   antibonding

(Note that for H_2O, only two types of AO are combined and only two MOs result.)

As the present problem is to explain the use of the group theory rather than to elaborate the methods employed to obtain MOs, we will return to the assignment of our reference axes.

Rules for Determining Symmetry Operations and Assigning Axes in a Molecule

For a given molecule, the axis of highest symmetry (for example the C_2 in the case of ML_2) must be identified from the outset. It is generally referred to as the 'principal axis' and is always orientated along the z direction in the reference frame.

Any second order-axes which are oriented perpendicular to the principal axis are annotated C'_2.

Symmetry planes are denoted by σ. They fall into three types denoted:

- σ_h for reflections through a plane perpendicular to the principal axis,
- σ_v where reflection is through a plane including the principal axis,
- σ_d where reflection is through a plane which incorporates the principal axis and also bisects two C'_2 axes.

Other symmetry operations are:

- S_n which corresponds to an nth order rotation followed by a reflection through a plane perpendicular to the principal axis.
- i involving an inversion of position relative to a point termed the centre of inversion.

It is clear that the presence or absence of a particular symmetry element in any molecule will be determined by the precise structure of the compound involved. Many molecules and complexes have few symmetry elements. The symbols used to describe the irreducible representations of the point group are explained below:

- A and B are one-dimensional representations: A indicates conservation of form after rotation around the principal axis, B indicates a change.
- The degenerate second-order representations are defined using the symbol E, and third-order cases are noted T.
- Should there be a molecular inversion centre, representations which are symmetrical about i are termed g, those which are not are denoted u.

- For each group there is an irreducible representation which is unchanged by every symmetry operation which can be applied to the group (consequently, it has notation of 1 for every operation). It is termed A_1 (or sometimes Σ^+) for groups having linear symmetry; in cases where an inversion centre is present, the suffix g or u is added, as appropriate.

Since an exhaustive list of point groups would require considerable space, we refer the reader to any of the reference works for a comprehensive compilation. A list of the more common structures, which suffices for the majority of chemically important geometries, is given below.

MOs of Ligand Arrays

It is useful to understand the symmetry profiles of multiligand arrays (L_n) for values of n between 2 and 6, because these are found in the majority of chemically important structures. We conceptually remove the metal from the centre of a complex and treat the resulting ligand array in isolation. The overall MO scheme for the complex can subsequently be obtained by combining the MOs of the ligand group with metal orbitals having the corresponding symmetry.

From the analysis above, we have already seen above that two H ligands (or more generally L_2) give rise to two combinations $L1 + L2$ and $L1 - L2$. The case where three ligands occupy the corners of an equilateral triangle is also common. There are two possible configurations in this case: the symmetry and orientation of the ligand orbitals depend upon whether the metal lies in the ligand plane (giving a planar D_{3h} complex) or above it (giving a pyramidal C_{3v} structure).

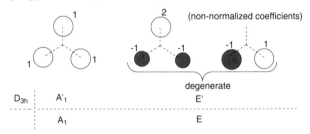

For four ligands located at the corners of a square, D_{4h} (square planar) or C_{4v} (square pyramidal) arrays are possible. The following MOs are obtained:

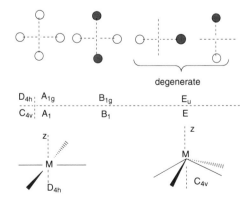

The situation generated by an array of five ligands may be understood as a combination of the cases above; this is most clearly evident in the trigonal bipyramidal structure, which may be treated as a simple superimposition of two independent sets: one formed by the three equatorial ligands and the other by the two axial ligands.

For six ligands occupying the apices of an octahedron, there are the following combinations:

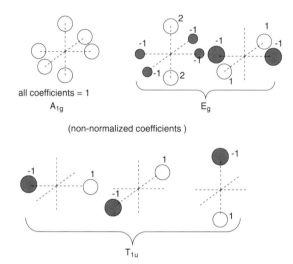

The majority of complexes which are formed by σ-type ligands can be readily understood in terms of these arrays of ligand MOs. A series of examples will be given at the end of the appendix.

π Complexes

We will cover only a few typical cases. Furthermore, we will not deal with ligands which may be described as a non-interacting combination of a classical L_n component and an isolated π ligand, because the two fragments can usually be treated independently in such cases.

Often, it is not necessary to construct the entire set of MO levels to understand the metal–ligand interactions in a complex. As we have already given detailed treatments of many common π complexes in the text, we will consider only the cases of an isolated metal in the presence of allyl, cyclopentadienyl or benzene groups here. We present the MOs of these ligands and their normalized coefficients without further discussion.

Allyl Ligand

The MOs of the allyl group are reiterated below:

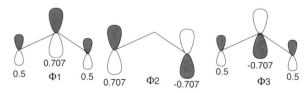

The presence of only a single symmetry plane (and thus a point group C_s) means that the interaction with the metal is particularly simple. Orbitals $\Phi1$ and $\Phi3$ lie symmetrically with respect to the molecular symmetry plane and interact with the six metal MOs having the correct symmetry. Orbital $\Phi2$ interacts with the three remaining metal orbitals.

one symmetry plane
C_s

In this first case, the overall interaction can only be understood through calculation. The case of a metal bound to two allyl groups is easier to treat. Two limiting geometrical forms, C_{2h} or C_{2v} (pseudo-chair and pseudo-boat, respectively, in organic terminology) are possible.

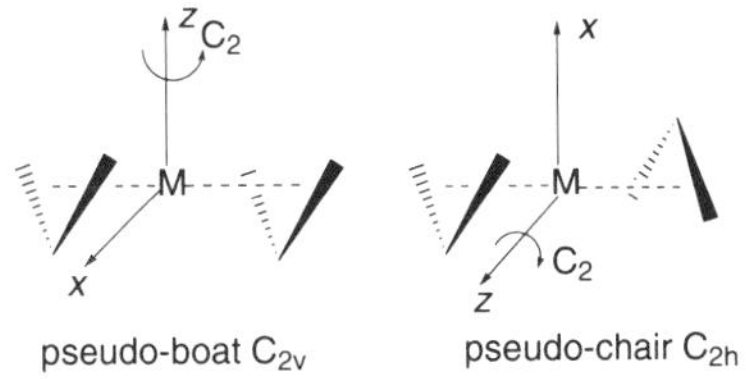

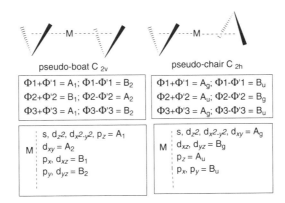

pseudo-boat C_{2v} pseudo-chair C_{2h}

We will consider each of these configurations in the absence of the metal and, since the distance between the two allyl ligands is large, build the ligand array by first adding and then subtracting MOs $\Phi1$ to $\Phi3$ on one allyl and $\Phi'1$ to $\Phi'3$ on the other. The addition gives the in-phase overlap, whilst the subtraction yields the out-of-phase combination. The result is the following:

pseudo-boat C_{2v}

$\Phi1+\Phi'1 = A_1$; $\Phi1-\Phi'1 = B_2$
$\Phi2+\Phi'2 = B_1$; $\Phi2-\Phi'2 = A_2$
$\Phi3+\Phi'3 = A_1$; $\Phi3-\Phi'3 = B_2$

pseudo-chair C_{2h}

$\Phi1+\Phi'1 = A_g$; $\Phi1-\Phi'1 = B_u$
$\Phi2+\Phi'2 = A_u$; $\Phi2-\Phi'2 = B_g$
$\Phi3+\Phi'3 = A_g$; $\Phi3-\Phi'3 = B_u$

M	
	s, d_{z^2}, $d_{x^2-y^2}$, $p_z = A_1$
	$d_{xy} = A_2$
	p_x, $d_{xz} = B_1$
	p_y, $d_{yz} = B_2$

M	
	s, d_{z^2}, $d_{x^2-y^2}$, $d_{xy} = A_g$
	d_{xz}, $d_{yz} = B_g$
	$p_z = A_u$
	p_x, $p_y = B_u$

The in-phase $\Phi2 + \Phi'2$ and out-of-phase $\Phi2 - \Phi'2$ combinations (having A_u and B_g symmetry respectively for the C_{2h} point group) are depicted in the following figure, along with the metal MOs having the same symmetry. We show only the bonding combinations: the two corresponding antibonding combinations may be obtained simply by changing the sign of the metal MOs in each orbital lobe.

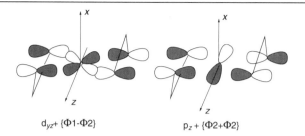

$d_{yz} + \{\Phi1-\Phi2\}$ $p_z + \{\Phi2+\Phi2\}$

Cyclopentadienyl

The isolated cyclopentadienyl radical is planar and delocalized, which means that it belongs to the D_{5h} point group. Its MOs, characterized by A''_1, E''_1 and E''_2 are shown in the following figure:

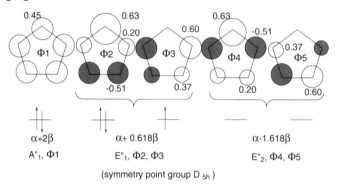

(symmetry point group D_{5h})

Inspection of an M (Cp) complex having the metal centre situated upon the C_5 axis reveals the following orbital relationships which are imposed by the C_{5v} point group. In turn, they govern the molecular orbital interactions between the metal and the Cp ligand:

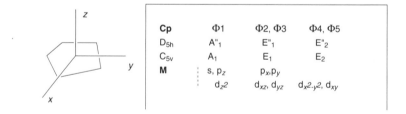

Cp	$\Phi1$	$\Phi2, \Phi3$	$\Phi4, \Phi5$
D_{5h}	A''_1	E''_1	E''_2
C_{5v}	A_1	E_1	E_2
M	s, p_z	p_x, p_y	
	d_{z^2}	d_{xz}, d_{yz}	$d_{x^2-y^2}, d_{xy}$

The MOs which were originally of the form E''_1 (D_{5h}) become E_1 in the group C_{5v} and their interactions with the metal AOs having complementary symmetry are shown in the figure. We illustrate the mixtures of $p_x + d_{xz}$ and $p_y + d_{yz}$ (see next page).

Metallocenes

We will consider only the simplest cases, wherein the two Cp ligands lie perpendicular to the C_5 axis. In these structures, we have two limiting geometries D_{5h} (where the ring carbon atoms are eclipsed) and D_{5d} (where they are staggered). In the eclipsed geometry the interactions are as illustrated on the next page.

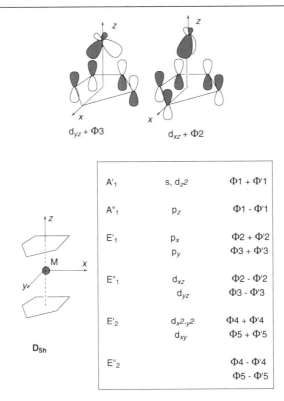

$d_{yz} + \Phi3$ $d_{xz} + \Phi2$

A'_1	s, d_{z^2}	$\Phi1 + \Phi'1$
A''_1	p_z	$\Phi1 - \Phi'1$
E'_1	p_x	$\Phi2 + \Phi'2$
	p_y	$\Phi3 + \Phi'3$
E''_1	d_{xz}	$\Phi2 - \Phi'2$
	d_{yz}	$\Phi3 - \Phi'3$
E'_2	$d_{x^2-y^2}$	$\Phi4 + \Phi'4$
	d_{xy}	$\Phi5 + \Phi'5$
E''_2		$\Phi4 - \Phi'4$
		$\Phi5 - \Phi'5$

D_{5h}

(note, again, that the + sign means in-phase addition of the isolated MOs of the fragment). We will not show the resulting MOs. However, both eclipsed and staggered geometries have occupied ligand MOs which are suitable for interaction with corresponding orbitals on the metal to allow the formation of three bonding combinations occupied by six electrons. As we saw in Chapter 3, this formal 6π electron configuration satisfies the electronic requirements of an aromatic Cp^- anion.

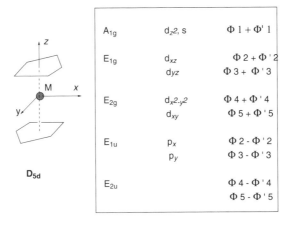

A_{1g}	d_{z^2}, s	$\Phi 1 + \Phi' 1$
E_{1g}	d_{xz}	$\Phi 2 + \Phi' 2$
	d_{yz}	$\Phi 3 + \Phi' 3$
E_{2g}	$d_{x^2-y^2}$	$\Phi 4 + \Phi' 4$
	d_{xy}	$\Phi 5 + \Phi' 5$
E_{1u}	p_x	$\Phi 2 - \Phi' 2$
	p_y	$\Phi 3 - \Phi' 3$
E_{2u}		$\Phi 4 - \Phi' 4$
		$\Phi 5 - \Phi' 5$

D_{5d}

Benzene

The MOs of benzene ($\Phi 1$ to $\Phi 6$ in order of decreasing stability) have the structures given in the following figure. We show their relationship towards a metal orientated along the C_6 axis within the C_{6v} framework.

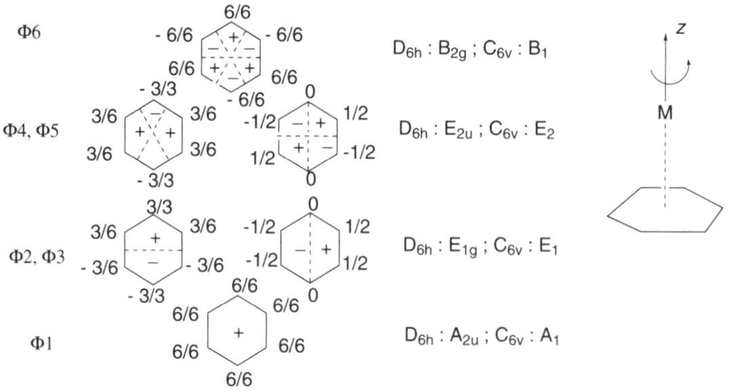

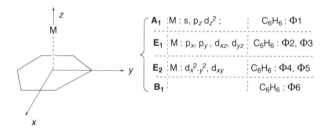

The matches between the MOs with the metal are indicated below:

	M	C_6H_6
A$_1$	M : s, p_z d_z^2 ;	C_6H_6 : $\Phi 1$
E$_1$	M : p_x, p_y, d_{xz}, d_{yz}	C_6H_6 : $\Phi 2$, $\Phi 3$
E$_2$	M : $d_{x^2-y^2}$, d_{xy}	C_6H_6 : $\Phi 4$, $\Phi 5$
B$_1$		C_6H_6 : $\Phi 6$

It is straightforward to use the table to derive the MOs for any given complex, through the processes of addition and subtraction of interacting orbitals which were outlined above. Sandwich complexes of the type M(benzene)$_2$ can also be analysed using formalisms similar to those which were employed for the metallocenes. For brevity, we will leave the verification of this (perfectly reasonable) statement as an exercise for the reader.

Structures and Nomenclature in Complexes

The use of group theory is essential in this area. We have employed the standard notation found in traditional character tables for the description of MOs. The ordering of orbital energy levels is necessarily approximate, because for a given geometry the relative spacing and absolute positions of any orbitals within the set will depend upon bond lengths, valence angles and dihedral angles. Full calculations, at least at the Hückel + overlap level, are necessary for a more precise insight. Such calculations can be easily carried out on all modern minicomputers.

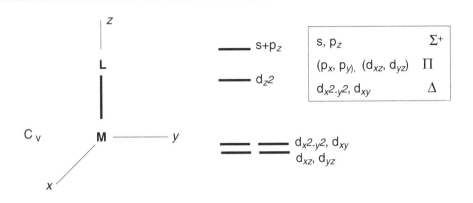

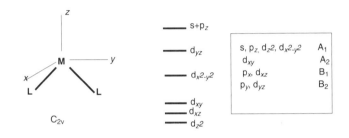

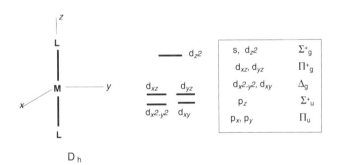

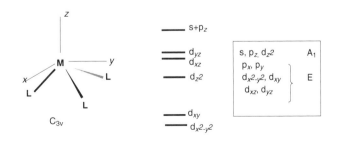

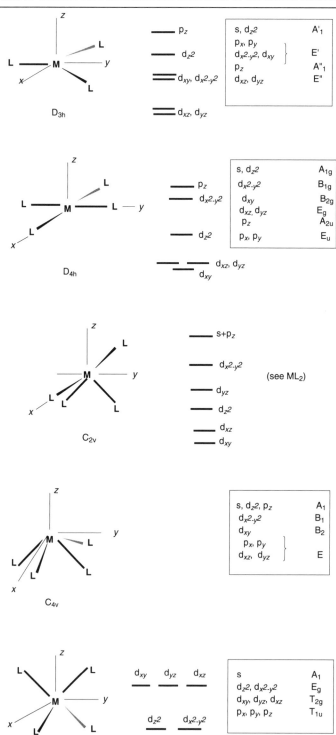

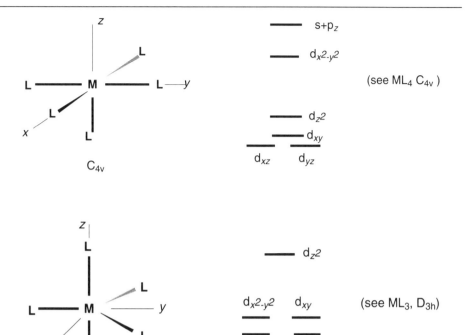

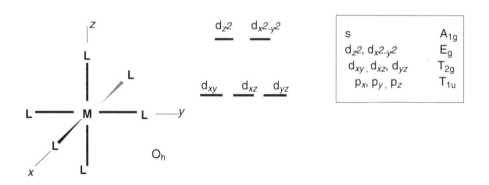

References

1. Various textbooks dealing with these topics are available. The following list is only indicative:

    (a) F. A. Cotton, *Group Theory and its Applications*, 2nd edn, Wiley, New York, 1971.

    (b) J. M. Hollas, *Symmetry in Molecules*, Chapman & Hall, London, 1972.

    (c) P. R. Bunker, *Molecular Symmetry and Spectroscopy*, Academic Press, London, 1979.

    (d) Applications are found in T. A. Albright, J. K. Burdett and M. H. Whangbo, *Orbital Interactions in Chemistry*, Wiley, New York, 1985.

Index